VOLUME ONE HUNDRED AND THIRTY FIVE

Advances in CANCER RESEARCH

miRNA and Cancer

VOLUME ONE HUNDRED AND THIRTY FIVE

ADVANCES IN CANCER RESEARCH

miRNA and Cancer

Edited by

CARLO M. CROCE
Wexner Medical Center, The Ohio State University, Columbus, OH, United States

PAUL B. FISHER
VCU Institute of Molecular Medicine; VCU Massey Cancer Center, Virginia Commonwealth University, School of Medicine, Richmond, VA, United States

ELSEVIER

ACADEMIC PRESS
An imprint of Elsevier

Academic Press is an imprint of Elsevier
50 Hampshire Street, 5th Floor, Cambridge, MA 02139, United States
525 B Street, Suite 1800, San Diego, CA 92101-4495, United States
The Boulevard, Langford Lane, Kidlington, Oxford OX5 1GB, United Kingdom
125 London Wall, London, EC2Y 5AS, United Kingdom

First edition 2017

ISBN: 978-0-12-811922-8
ISSN: 0065-230X

Publisher: Zoe Kruze
Acquisition Editor: Zoe Kruze
Editorial Project Manager: Fenton Coulthurst
Production Project Manager: James Selvam
Senior Cover Designer: Mathew Limbert

Typeset by SPi Global, India

CONTENTS

Contributors vii
Preface ix

1. MicroRNAs and Cancer: A Long Story for Short RNAs 1
Alessandra Drusco and Carlo M. Croce

2. The Enigma of miRNA Regulation in Cancer 25
Anjan K. Pradhan, Luni Emdad, Swadesh K. Das, Devanand Sarkar, and Paul B. Fisher

1. Introduction 26
2. Tumor Suppressor miRs and OncomiRs 27
3. Factors Regulating miRNAs 28
4. Regulation of the miRNA Biogenesis Pathway 28
5. Direct (Transcriptional) Regulation 32
6. Indirect Regulation 35
7. Conclusions and Future Perspectives 43

Acknowledgments 43
References 44

3. Animal Models to Study MicroRNA Function 53
Arpita S. Pal and Andrea L. Kasinski

1. Introduction 54
2. Generation of Model Organisms and Their Use in MiRNA Functional Studies 68
3. Conclusion 97

References 105

4. Cancer Hallmarks and MicroRNAs: The Therapeutic Connection 119
Katrien Van Roosbroeck and George A. Calin

1. Introduction 119
2. Regulation of Cell Proliferation and Cell Growth 120
3. Regulation of Apoptosis and Induction of Genomic Instability and Mutations 129

4. Regulation of Angiogenesis 129
5. Regulation of Invasion and Metastasis 133
6. Regulation of Immune Response 136
7. Conclusion 137
References 138

5. microRNAs in Cancer Susceptibility 151

Bríd M. Ryan

1. Introduction 152
2. Functional Consequences of miRNA SNPs 152
3. Emerging and Unresolved Concepts in miR-SNP Biology 158
4. Conclusions 163
References 163

6. Role of the tRNA-Derived Small RNAs in Cancer: New Potential Biomarkers and Target for Therapy 173

Veronica Balatti, Yuri Pekarsky, and Carlo M. Croce

1. Regulatory Noncoding RNAs: Long ncRNAs, Small ncRNAs, and tRNA-Derived ncRNAs 174
2. tRNA-Derived ncRNAs: tRFs, tsRNAs, and tiRNA and the Differences in Their Biogenesis, Structures, and Nomenclature 175
3. tsRNA in Cancer: Signatures, Functions, and Mutations 179
4. Concluding Remarks 184
References 185

7. MicroRNAs and Epigenetics 189

Catia Moutinho and Manel Esteller

1. Introduction 190
2. Epigenetics and Its Mechanisms 191
3. Epigenetic Regulation of miRNAs in Cancer 193
4. miRNAs Control of the Epigenetic Machinery 204
5. Exploiting miRNAs Epigenetics to Fight Cancer 208
6. Conclusion 210
References 210

CONTRIBUTORS

Veronica Balatti
The Ohio State University, Columbus, OH, United States

George A. Calin
Center for RNA Interference and Non-Coding RNAs, The University of Texas MD Anderson Cancer Center, Houston, TX, United States

Carlo M. Croce
Wexner Medical Center, The Ohio State University, Columbus, OH, United States

Swadesh K. Das
VCU Institute of Molecular Medicine; VCU Massey Cancer Center, Virginia Commonwealth University, School of Medicine, Richmond, VA, United States

Alessandra Drusco
Wexner Medical Center, The Ohio State University, Columbus, OH, United States

Luni Emdad
VCU Institute of Molecular Medicine; VCU Massey Cancer Center, Virginia Commonwealth University, School of Medicine, Richmond, VA, United States

Manel Esteller
Cancer Epigenetics and Biology Program (PEBC), Bellvitge Biomedical Research Institute (IDIBELL); Institució Catalana de Recerca i Estudis Avançats (ICREA), Barcelona; School of Medicine and Health Sciences, University of Barcelona (UB), Catalonia, Spain

Paul B. Fisher
VCU Institute of Molecular Medicine; VCU Massey Cancer Center, Virginia Commonwealth University, School of Medicine, Richmond, VA, United States

Andrea L. Kasinski
Purdue Center for Cancer Research, Purdue University, West Lafayette, IN, United States

Catia Moutinho
Cancer Epigenetics and Biology Program (PEBC), Bellvitge Biomedical Research Institute (IDIBELL), Barcelona, Catalonia, Spain

Arpita S. Pal
PULSe Graduate Program, Purdue University, West Lafayette, IN, United States

Yuri Pekarsky
The Ohio State University, Columbus, OH, United States

Anjan K. Pradhan
Virginia Commonwealth University, School of Medicine, Richmond, VA, United States

Bríd M. Ryan
Laboratory of Human Carcinogenesis, Center for Cancer Research, National Cancer Institute, Bethesda, MD, United States

Devanand Sarkar
VCU Institute of Molecular Medicine; VCU Massey Cancer Center, Virginia Commonwealth University, School of Medicine, Richmond, VA, United States

Katrien Van Roosbroeck
The University of Texas MD Anderson Cancer Center, Houston, TX, United States

PREFACE

It is well established that less than 2% of the human genome codes for proteins, while the vast majority of the genome is noncoding, although most of this noncoding genetic information is transcribed. For multiple reasons, it was thought that this noncoding part of the genome was just useless genetic information. Although it seemed quite unreasonable to think that nature designed a genome with most of its content unimportant, the relevance of the noncoding genome in all cellular functions, including normal physiology and in disease, was not evident until the last 15 years. The first suggestion that noncoding RNA could play a role in normal physiology was the discovery that mutations of a gene, lin 4, of the worm (*C. elegans*) affected its development. It was found that this gene encoded a very short noncoding RNA, a microRNA. After this discovery and that of another microRNA, let7, not much interest in microRNAs and noncoding RNA was expressed by the scientific community. Such short RNAs were found to bind, for the most part, to sites at the 3′ untranslated region of mRNA, causing a block of translation and/or degradation of the targeted mRNA. The situation dramatically changed with the discovery of siRNA in 1998, since microRNAs seemed to function in a way similar to that of siRNA. Accordingly, a renewed interest in microRNAs emerged and, as a result, it was found that Drosophila carried microRNA genes (2000), and in 2001 it was found that mice, rats, and human beings carried microRNA genes, which were seen as negative regulators of gene expression.

In 2002 investigation of the deletions at 13q14 in chronic lymphocytic leukemia (CLL), the most common human leukemia, led to a focus on two microRNA genes, miR-15a and miR-16-1, that map on the same polycistronic RNA. These studies indicated that in most CLL a loss of miR-15/16 occurs leading to the development of a malignancy. Since microRNAs have multiple targets, experiments focused on identifying a target of miR-15/16 that could explain the development of the malignant disease. In 2005 a specific target of miR-15/16, BCL2, a gene discovered in 1984 in the Croce laboratory that is involved in translocation with human immunoglobulin loci and responsible for follicular lymphoma was identified. Thus, loss of miR-15/16, two negative regulators of BCL2, results in constitutive overexpression of BCL2 and malignancy. The circle of discovery was closed when Abbott developed a small-molecule inhibitor of BCL2, ABT199 or venetoclax. This drug is a phenomenal selective killer of leukemic cells with little side effects and has been approved by the FDA for the treatment of CLL in April 2016. A historical perspective on the discovery and development of the miRNA field can be found in Chapter 1 (Drusco and Croce: MicroRNAs and Cancer: A Long Story for Short RNAs) and a review of current therapeutics focused on miRNA-protein targets can be found in Chapter 4 (Van Roosbroeck and Calin: Cancer Hallmarks and MicroRNAs: The Therapeutic Connection).

During the past 10 years, it has become clear that microRNAs contribute to the function of each biochemical pathway and to all cellular functions. It has also become clear that dysregulation of microRNAs occurs in and contributes to every malignancy. Thus, we can exploit microRNA dysregulation for the diagnosis, prognosis, and monitoring of human cancer. MicroRNAs and antimicroRNAs could also be seen as drugs. MicroRNAs could be therapeutic in malignancies that depend on their loss for transformation and antimicroRNAs could be therapeutic for malignancies caused by overexpression of microRNAs, such as overexpression of miR-155 in aggressive lymphomas. In Chapter 5 (Ryan: microRNAs in Cancer Susceptibility) the role of miRNAs in cancer etiology and susceptibility is reviewed. This chapter also discusses the genotype–phenotype relationship between miRNA-SNPs (single nucleotide polymorphisms).

Animal models are essential to assess the role of genes in cancer, and animal models carrying dysregulated microRNA genes have been developed and have provided important validations for their role in cancer pathogenesis and progression. Chapter 3 (Pal and Kasinski: Animal Models to Study MicroRNA Function) discusses various strategies that have been used to develop transgenic organisms with a review of the application of transgenic mice for investigating preclinical efficacy of microRNA-based cancer therapeutics.

The major problem with the development of microRNAs and antimicroRNAs in cancer therapeutics is that at present it is extremely difficult to target enough microRNA/antimicroRNA into 100% of the cancer cells. Targeting technologies are, however, improving at a fast rate; thus, it is not difficult to predict that in the near future microRNAs and antimicroRNAs will be drugs. Insights into this area will benefit from a better understanding of the upstream regulatory molecules that regulate miRNAs and the role of epigenetic changes in regulating miRNA expression. These topics are discussed, respectively, in Chapter 2 (Pradhan et al.: The Enigma of miRNA

Regulation in Cancer) and in Chapter 7 (Moutinho and Esteller: MicroRNAs and Epigenetics).

Because of these extraordinary developments in studies of microRNA biology, function, and as biomarkers, it is extremely appropriate and timely to attempt to put together a volume that summarizes some of the most exciting developments in many of the aspects of microRNA biology and of the role of microRNA in disease, and of the use of microRNA dysregulation for the diagnosis, prognosis, and monitoring of cancer. Chapter 6 (Balatti et al.: Role of the tRNA-Derived Small RNAs in Cancer: New Potential Biomarkers and Target for Therapy) discusses the role of tRNA-derived small RNAs (tsRNAs) in cancer by regulating gene expression. We are optimistic that future developments in this exciting area of investigation will permit the efficient and successful entry of miRNA-targeted therapeutics and diagnostics into the clinic for the therapy of cancer and other pathological diseases.

Carlo M. Croce, MD
Paul B. Fisher, MPh, PhD

CHAPTER ONE

MicroRNAs and Cancer: A Long Story for Short RNAs

Alessandra Drusco, Carlo M. Croce[1]
Wexner Medical Center, The Ohio State University, Columbus, OH, United States
[1]Corresponding author: e-mail address: carlo.croce@osumc.edu

Abstract

More than six decades ago Watson and Crick published the chemical structure of DNA. This discovery revolutionized our approach to medical science and opened new perspectives for the diagnosis and treatment of many diseases including cancer.

Since then, progress in molecular biology, together with the rapid advance of technologies, allowed to clone hundreds of protein-coding genes that were found mutated in all types of cancer. Normal and aberrant gene functions, interactions, and mechanisms of mutations were studied to identify the intricate network of pathways leading to cancer. With the acknowledgment of the genetic nature of cancer, new diagnostic, prognostic, and therapeutic strategies have been attempted and developed, but very few have found their way in the clinical field. In an effort to identify new translational targets, another great discovery has changed our way to look at genes and their functions. MicroRNAs have been the first noncoding genes involved in cancer.

This review is a brief chronological history of microRNAs and cancer. Through the work of few of the greatest scientists of our times, this chapter describes the discovery of microRNAs from *C. elegans* to their debut in cancer and in the medical field, the concurrent development of technologies, and their future translational applications.

The purpose was to share the exciting path that lead to one of the most important discoveries in cancer genetics in the past 20 years.

Until the early 1990s, molecular genetic studies were based on hunting protein-coding DNA regions to understand their function, regulation, and role in physiologic and pathologic processes. The remaining DNA was considered by many as "junk." Complex protein networks were built to find targets for future applications, ignoring the silent presence of other players that, in the last 15 years, have become leading actors of the biomedical world.

It is in 1993 that Victor Ambros (Lee, Feinbaum, & Ambros, 1993) (Fig. 1) discovered the existence of a short noncoding region (lin-4) negatively regulating the expression of LIN-14 during the development of *C. elegans*.

Advances in Cancer Research, Volume 135
ISSN 0065-230X
http://dx.doi.org/10.1016/bs.acr.2017.06.005

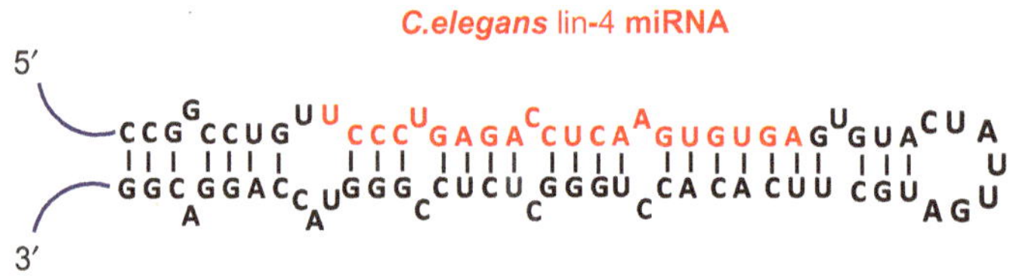

Fig. 1 *C. elegans* lin-4 miRNA is discovered by Ambros in 1993 (Lee et al., 1993).

In the same year and *Cell* volume, Gary Ruvkun (Wightman, Ha, & Ruvkun, 1993) showed that lin-4 binds to the lin-14 3′untranslated region (UTR) "that bears multiple conserved elements complementary to the lin-4 RNAs."

For several years, the scientific community regarded the lin-4 phenomenon with skepticism, but in 2000 Pasquinelli (Pasquinelli et al., 2000) and Reinhart (Reinhart et al., 2000) found another 21-nucleotide-long RNA, let-7, that regulates *C. elegans* development. It is in 2001 (Lagos-Quintana, Rauhut, Lendeckel, & Tuschl, 2001; Lau, Lim, Weinstein, & Bartel, 2001; Lee & Ambros, 2001) that these tiny RNAs were classified as a new class of genes and were named microRNAs, miRNAs, or miRs.

Hundreds of miRs were identified in worms, flies, mice, human cells, and plants (Lagos-Quintana et al., 2001, 2002; Lau et al., 2001; Lee & Ambros, 2001; Llave, Xie, Kasschau, & Carrington, 2002; Park, Li, Song, Messing, & Cheng, 2002; Reinhart, Weinstein, Rhoades, Bartel, & Bartel, 2002; Válóczi et al., 2004).

MicroRNA regulation is still debated because several exceptions to the general rules have been observed in the past few years. MicroRNAs are located by themselves or reside within intronic or exonic regions of coding and noncoding genes, can be single, bi- or polycistronic, are transcribed dependently and independently of other genes (Monteys et al., 2010; Veronese et al., 2015), but all of them derive from a larger double-stranded precursor of about 70 nucleotides (pre-miRNA), that originated from primary transcripts transcribed by RNA Polymerase II, named pri-miRNA. Once translocated to the cytoplasm by Exportin 5, their hairpin region is cleaved by the Dicer complex into 21–25-nucleotide RNA molecules that form the mature miRNA (Krol, Loedige, & Filipowicz, 2010; Li & Rana, 2012). Mature microRNAs function as guide RNAs to posttranscriptionally regulate mRNAs fate. The seed region is a sequence of 6–8 nucleotides at the 5′ end of microRNAs with partial or total complementarity to the

3′UTRs of target mRNAs. Partial or total microRNA::mRNA pairing results in repression of translation and/or in mRNA degradation (Oliveto, Mancino, Manfrini, & Biffo, 2017).

Each microRNA has dozens or hundreds of target mRNAs, while microRNAs with a common seed region regulate the same targets (Bartel, 2009; Di Leva, Garofalo, & Croce, 2014).

Recent reports have suggested that microRNAs also bind to the 5′UTR of mRNAs and instead of repressing, activate the translation of the target mRNA (Fabian, Sonenberg, & Filipowicz, 2010; Vasudevan, Tong, & Steitz, 2007, 2008).

It was only in 2002 that microRNAs made a debut in the medical research field. For years, several scientists looked for a coding gene on a region of the long arm of chromosome 13, at 13q14, that is frequently deleted in chronic lymphocytic leukemia (CLL) (Bullrich et al., 2001; Liu et al., 1997; Mabuchi et al., 2001; Migliazza et al., 2001; Rondeau et al., 2001; Rowntree et al., 2002; Wolf et al., 2001). In Croce's laboratory the deleted chromosome gap was narrowed down over and over until 2002, when George Calin et al. isolated the smallest deleted region, containing the precursor of a microRNA cluster, miR-15a/miR-16-1 (Calin et al., 2002) (Fig. 2). A chromosomal translocation t(2:13) within a breakpoint at 13q14 in a case of CLL involved the same precise region (Calin et al., 2002). Deletion or downregulation of miR-15 and miR-16 is found in approximately 70% of CLLs. Moreover, the latter is often associated with the indolent form of CLLs that frequently progresses into the aggressive form. Thus, the "junk" DNA is involved in cancer. This finding opened new perspectives in cancer research and became the milestone of a new era of biomedical science.

Other findings confirmed Calin et al. studies, showing that mutations affecting microRNAs protein processing complex can not only impair development and cell homeostasis, but can cause diseases and cancer (Carmell, Xuan, Zhang, & Hannon, 2002; McManus, 2003).

The following year Calin et al. found that 19% of 186 microRNAs are located at chromosomal fragile sites and 52.5% of 186 microRNAs are found in minimal regions of loss of heterozygosity or minimal regions of amplifications (Calin, Sevignani, et al., 2004). Interestingly, some microRNAs are organized in clusters of at least two microRNAs. These findings suggested that microRNAs might be located at chromosomal sites in which previous studies have failed to find deregulated coding genes, thus, behaving as tumor suppressor genes or oncogenes.

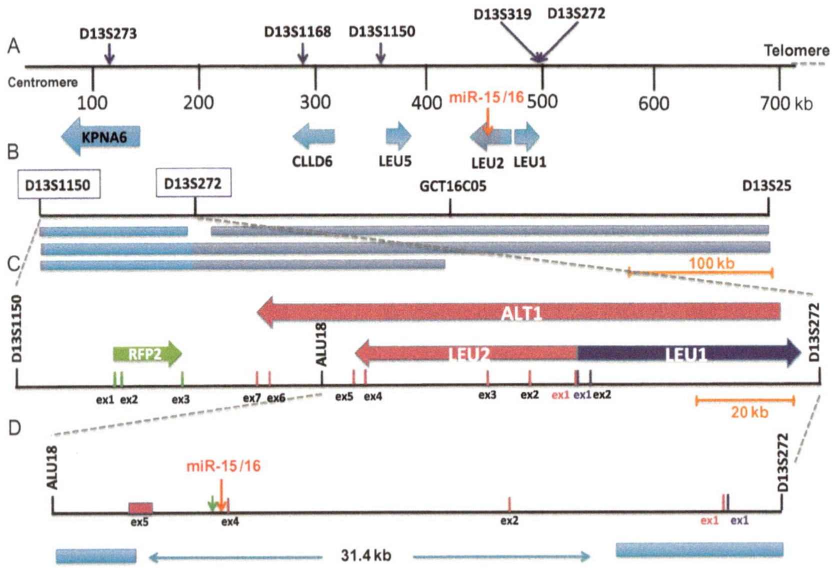

Fig. 2 (A) Map of the chromosomal region at 13q14 often deleted in CLL. Genes and genetic markers mapping locations are shown. (B) 13q14 locus between D13S1150 and D13S25 genetic markers. *Horizontal blue boxes* under the 13q14 locus represent reported deletions. (C) Area of the 13q14 locus delimited by D13S1150 and D13S272. Genes are represented by *arrows* that point the gene orientation. (D) Area of the 13q14 locus delimited by ALU18 and D13S272. The *green arrow* points at the breakpoint in CLL-B carrying a t(2;13)(q32;14) translocation. The *filled box* represents the portions of chromosome 13 found in the hybrid clone derived by a patient with retinoblastoma and CLL with a deletion of 31.4 kb. MiR-15/16 are highlighted in *red* in all maps. *Modified from Calin, G. A., Dumitru, C. D., Shimizu, M., Bichi, R., Zupo, S., Noch, E., et al. (2002). Frequent deletions and down-regulation of micro-RNA genes miR-15 and miR-16 at 13q14 in chronic lymphocytic leukemia.* Proceedings of the National Academy of Sciences of the United States of America, 99*(24), 15524–15529.*

Given these promising premises Chan-Gong Liu, in Croce's laboratory, developed the first oligonucleotide microchip array for the detection of microRNA expression patterns across human and mouse tissue samples (Liu et al., 2004; Liu, Calin, Volinia, & Croce, 2008), allowing to quantify the expression of 281 precursor microRNAs.

Thomson and Babak designed oligonucleotide microarrays to identify differentially expressed microRNAs in mouse tissues (Babak, Zhang, Morris, Blencowe, & Hughes, 2004; Thomson, Parker, Perou, & Hammond, 2004).

Meanwhile, Schmittgen set up the first experiments for microRNA detection and quantification by RT-PCRs (Schmittgen, Jiang, Liu, &

Yang, 2004) and Exiqon laboratories synthesized locked nucleic acid (LNA)-modified oligonucleotide probes to improve sensitivity and specificity of microRNA northern blot hybridization (Thomsen, Nielsen, & Jensen, 2005). The same probes will be used for in situ hybridization of paraffin-embedded tissues, assessing microRNAs localization and their targets (Kloosterman, Wienholds, de Bruijn, Kauppinen, & Plasterk, 2006; Nuovo et al., 2009).

The following years of cancer research were characterized by intense profiling of all kinds of tumor tissues (Fig. 3). The foremost articles remain the most exciting, each of them revealing a piece of the microRNA world that led to gradually foresee and understand the immense potentials of these small RNAs.

Chang-Gong microarray platform was first used by Calin (Calin, Liu, et al., 2004) in 2004. A total of 47 samples comprehensive of 42 samples of blood mononuclear cells derived by 38 CLL patients and 6 normal specimens (1 lymph node, 2 tonsillar CD5+ B cells, and 3 blood mononuclear cells) were processed and analyzed, defining clustering signatures that could differentiate between CLL and normal controls and, within the CLL cluster, between ZAP-70 and mutated IgV_H prognostic groups. Increased ZAP-70 levels in lymphoid cells were an early predictor of progression, while mutated IgV_H is an indicator of favorable outcome (Hamblin, Davis, Gardiner, Oscier, & Stevenson, 1999; Orchard et al., 2004; Oscier et al., 2002). Deletion at 13q14 strongly associated with IgV_H mutation and, given the new findings, with downregulation of miR-15 and miR-16.

Hence, microRNAs are involved in cancer pathogenesis as much as coding genes and, more importantly, seem to relate to patients' prognosis.

One year later, the same group published in the *New England Journal of Medicine* a profiling study of 94 CLL cases, confirming that loss of expression of miR-15 and miR-16 together with mutation of IgV_H and deletion at 13q14 is associated with a more favorable prognosis (Calin et al., 2005). They also found germ line or somatic mutations in 15% of CLL samples and in 11 miRs sequenced in 42 cases of CLL. In 8 of the 11 patients with microRNA abnormalities, 8 had a family history of cancer and, of the latter, 2 patients harbored a single nucleotide difference in the miR-15/16 pre-miR precursor. *In vitro* and *in vivo* experiments further showed that the abnormal base difference decreased the mature miR-15/16 expression levels and that, in both cases, the mutation was associated with deletion of the other allele at 13q14.

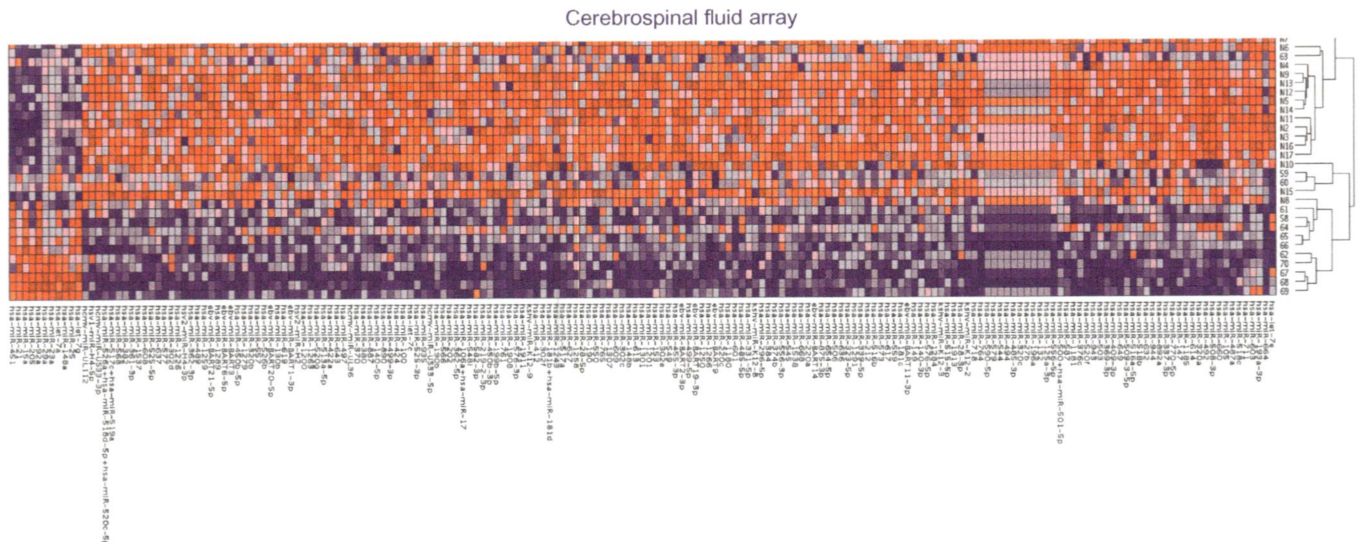

Fig. 3 Array profile of samples of cerebrospinal fluid coming from patients with brain metastasis from breast cancer vs controls.

Therefore, microRNAs with tumor suppressor functions followed the Knudson's two hits hypothesis and, as for coding genes, their expression could be affected by other type of mutations.

In 2008, Shen demonstrated that microRNAs may also be associated with breast/ovarian familial cancers: in a group of 42 patients, the polymorphic miR-146a bound BRCA1 with more avidity than its normal sequence (Shen et al., 2008). In 2010, Calin conducted a case–control population study in which he observed that single nucleotide polymorphism (SNP) distribution is significantly different between breast cancer and normal population, proving that microRNA SNPs could affect breast cancer susceptibility (Nicoloso et al., 2010).

The last finding of the 2004 article by Calin represented the very first insight of the most intriguing microRNA potential: microRNA clusters distinguish between hematopoietic and nonhematopoietic tissue/cells, which means that microRNA signatures might be able to differentiate among different types of tissues/cells. The following year Lu, using a bead-based flow cytometry profiling method, analyzed 334 samples including human cancer tissues and found that, differently from mRNA, microRNA profiles were able to classify undifferentiated tumors according to the tissue of origin and differentiation stage (Lu, Getz, et al., 2005). These findings were confirmed and expanded in a 2008 study that analyzing 400 samples of primary tumors and their recurrences demonstrated that through microRNA signatures, it was possible to determine the tissue of origin of metastatic lesions (Rosenfeld et al., 2008). The resulting clinical application would benefit a 3%–5% of patients with a newly diagnosed metastatic lesion of unknown origin, for whom the inability to determine the primary tumor would deny any chance of receiving the appropriate therapy and care, accelerating the unfavorable outcome (Ferracin et al., 2011; Pimiento, Teso, Malkan, Dudrick, & Palesty, 2007). In future studies on metastatic tumors, microRNA signatures were addressed as "metastamiR signatures" and will include prometastatic as well as antimetastatic genes (White et al., 2011).

The first microarray profiling article on solid tumors came again from Croce's laboratory (Iorio et al., 2005). In 2005, Iorio defined a breast cancer signature of 11 microRNAs of which 5 are the most significantly deregulated: miR-10b, miR-125b, and miR-145 are downregulated, while miR-21 and miR-155 are upregulated. These microRNAs were extensively mentioned in subsequent literature and will not only be linked to breast cancer but also to other tumors. Each of them would definitely require an entire chapter, but, for the historical purpose, we will mention the most renown and relevant findings.

In 2007, Weinberg's laboratory showed that miR-10b promotes breast cancer metastasis *in vitro* and *in vivo*. MiR-10b was upregulated by Twist and inhibited HOXD10 protein translation, which, in turn, increased the expression of the prometastatic RHOC protein (Ma, Teruya-Feldstein, & Weinberg, 2007). MiR-125b was extensively studied because it is the human orthologue of lin-4 in *C. elegans*. MiR-125b plays a dual role in cancer, behaving as a tumor suppressor or an oncogene in a tissue-dependent manner. It is upregulated in some leukemia, in gastric, pancreatic, and in colorectal cancers associated with a poor prognosis, and it is downregulated in breast, colorectal, gastric, glioblastoma, nonsmall-cell lung tumors and ovarian cancer (Banzhaf-Strathmann & Edbauer, 2014; Iorio et al., 2007). MiR-125b maps at 11q24, a chromosomal region often mutated in hematologic malignancies. Translocation t(2,11)(p21;q23) is observed in myelodysplastic syndromes (MDSs) and acute myelogenous leukemia and results in overexpression of miR-125b (Shaham, Binder, Gefen, Borkhardt, & Izraeli, 2012). A reduced expression of miR-125b is characteristic of indolent and aggressive forms of CLL harboring deletion at 11q23 (Tili et al., 2012). While the molecular mechanism of miR-125b upregulation remains unclear, its downregulation is attributed to methylation of its promoter.

Saito was the first to show that microRNAs can be regulated epigenetically (Saito et al., 2006). Treatment of the bladder carcinoma cell line T24 with demethylating drugs rescued miR-127 downregulation (Saito et al., 2006), while treatment of the ovarian cell line OVCAR3 with 5-AZA resulted in upregulation of miR-21 (Iorio et al., 2007). MiR-21 is one of the most expressed microRNAs in solid tumors (Volinia, Calin, et al., 2006) where it strongly predicts a poor prognosis. In a profiling study of 2008, in which an American and a Chinese cohort of patients with colon adenocarcinomas were followed up for about 9 years, Schetter et al. found that miR-21 expression increases progressively from the adenoma stage to invasive carcinoma. Moreover, it was associated with poor therapy response, worse survival rates (Schetter et al., 2008), and with lymph node metastasis (Drusco et al., 2014). MiR-21 was also upregulated in hematologic malignancies such as pre-B-cell lymphoma (Gu et al., 2013; Medina, Nolde, & Slack, 2010), CLL (Fulci et al., 2007), and natural killer cell leukemia (Go et al., 2015). In solid tumors, miR-21 targets PTEN and PDCD4, promoting proliferation, tumor initiation, and invasion (Selcuklu,

Donoghue, & Spillane, 2009; Zhu et al., 2008), represses RAS antagonists activating RAS/MAP kinase extracellular signal-regulated kinase pathway, and, in hematologic malignancies through PTEN and forkhead box protein O1 downregulation, activates phosphoinositide-3 kinase (PI3K/AKT) (Feng & Tsao, 2016). In 2010, Medina (Medina et al., 2010) provided *in vivo* evidence that miR-21 is sufficient for tumor development: in a conditional mouse model, miR-21 overexpression induces a pre-B malignant lymphoid phenotype. MiR-21 promotes proliferation, tumor initiation, and invasion (Fang et al., 2017; Feng & Tsao, 2016; Go et al., 2015; Selcuklu et al., 2009; Zhu et al., 2008) and is officially classified as an oncogene.

The immune system is the first barrier to cancer development, and several oncomicroRNAs, including miR-21, miR-125b, and miR-155, were found to be important mediators of inflammation. MiR-21 controls inflammation through induction of pro- and antiinflammatory cytokines; miR-125b regulates macrophage response, inhibiting the inflammatory response to LPS; and miR-155, instead, promotes cytokine production, amplifying the cellular inflammatory response that often induces the initial steps of transformation (Tili, Michaille, & Croce, 2013). MiR-155 expression increases in activated immune cells and seems to orchestrate several responses of both innate and adaptive immunity (Vigorito, Kohlhaas, Lu, & Leyland, 2013). MiR-155 knockout showed defective hematopoiesis, suggesting that it may modulate immune cell differentiation (Thai et al., 2007), while its overexpression in mice B cells led to the development of high-grade lymphoma/leukemia (Costinean et al., 2006; Sandhu et al., 2012). Upregulation of miR-155 in Eμ-mmu-miR-155 transgenic mice led to aggressive B-cell lymphoma and leukemia, suggesting that miR-155 is an oncogene in B cells and providing the first evidence for an oncogenic microRNA (Costinean et al., 2006). MiR-155 is upregulated in most of solid and hematological malignancies (Higgs & Slack, 2013) and is associated with development of breast, lung, gastric cancer and leukemia (Tili et al., 2010).

MiR-155 is another interesting example of epigenetic regulation of microRNA. In breast cancer, deacetylation of the promoter region leads to repression of its expression (Chang et al., 2011).

Another challenge of the early years of microRNA profiling has been identifying microRNA targets.

In a 2 year range, three computer-based approaches develop different algorithms to identify microRNA targets: Targetscan (2003), Miranda

(2004), and Pictar (2005) (John et al., 2004; Krek et al., 2005; Lewis, Shih, Jones-Rhoades, Bartel, & Burge, 2003).

In the following years, computational tools have expanded proportionally to the progressive growth of microRNA field (Veneziano et al., 2016).

Slack's and Croce's laboratories were the first to identify microRNA targets. In 2005, Johnson (Slack's lab) screened the database for encoding genes with complementary sites with the 3′UTR of *C. elegans let-7* family of microRNAs. The matching gene was *let-60*, the nematode homologue to the human RAS family. Loss of *let-7* causes overexpression of RAS, one of the most studied contributors to the oncogenesis of several tumors (Johnson et al., 2005).

On the other hand, Cimmino (Croce's lab) described the mechanism underlying the development of indolent CLLs. MiR-15 and miR-16 negatively regulate BCL-2, the antiapoptotic gene that the same laboratory cloned 21 years earlier (Pegoraro et al., 1984). Loss of miR-15 and miR-16 resulted in uncontrolled expression of BCL-2 which, as for follicular lymphoma, inhibits apoptosis promoting a relentless expansion of malignant cells (Calin et al., 2008). Five years later, Klein et al. confirmed these findings *in vivo*: as in humans, knockout mice for miR-15/16 developed indolent CLL (Klein et al., 2010).

2005 is also the year of the "antagomiRs." Stoffel's group, at The Rockefeller University, designed modified synthetic cholesterol-conjugated oligonucleotides complementary to microRNAs. MiR-122 is liver specific and modulates cholesterol biosynthesis. *In vivo* injection of miR-122 antagomiR showed, as expected, a 40% decrease in plasma cholesterol levels in treated mice. This was the first *in vivo* suggestion of microRNA therapeutic potentials (Krützfeldt et al., 2005).

Finally, in September 2005, Lu published the first paper that uses massively parallel signature sequencing (MPSS) (also known as next-generation sequencing), to sequence the entire small RNA genome of *Arabidopsis* (Lu, Tej, et al., 2005), while, in October, Griffiths-Jones of The Wellcome Trust Sanger Institute announced the creation of miRBase, providing the first database that integrates microRNA nomenclature, sequences, and targets (Griffiths-Jones, Grocock, van Dongen, Bateman, & Enright, 2006).

Cancer initiation, progression, and invasion are the result of genetic alterations of coding and noncoding sequences, bound by an intricated network of complex feedback interactions that are shared by more than one type of tumor.

With this perspective, in 2006, Volinia screened 540 samples of 6 types of solid tumors (breast, colon, lung, prostate, pancreas, and stomach) and identified a common solid tumor microRNA signature that included miR-17, miR-20a, miR-92a, miR-106a, miR-21, and miR-155 (Volinia, Galasso, et al., 2006). The signature contained the best-characterized polycistronic cluster of microRNA. Located on chromosome 3, the miR-17~92 cluster includes miR-17, miR-18, miR-19a, miR-20a, miR-19b, and miR-92a; it is highly conserved in vertebrates and has two mammalian paralogs: miR-106a~363 on chromosome 7 and miR-106b~363 on chromosome X (Concepcion, Bonetti, & Ventura, 2012). The clusters got the attention of scientists in 2005, when O'Donnell discovered that MYC regulates the expression of miR-17, miR-18a, miR-19a, miR-20a, miR-19b, miR-92a, and miR-106 by binding directly to the locus (O'Donnell, Wentzel, Zeller, Dang, & Mendell, 2005). Identified by Croce's laboratory in 1982 as involved in the t(8;14) translocation in Burkitt lymphoma (Dalla-Favera et al., 1982), MYC encodes a transcription factor known for its oncogenic role in several other tumors (Dang, 2012; Stine, Walton, Altman, Hsieh, & Dang, 2015; Zack et al., 2013). MiR-17~92 cluster is overexpressed in neuroblastoma, medulloblastoma, and colon, gastric, and small-cell lung cancer (Hayashita et al., 2005). The three clusters exhibit widespread roles in oncogenesis and their targets seem to be very context specific (Concepcion et al., 2012).

Of the solid tumor signature, miR-155 deserves one more comment. In 2006, Croce's laboratory published the first microRNA mouse model: 6-month-old Eμ-mmu-miR-155 transgenic mice presented with a B-cell polyclonal proliferation typical of preleukemic pre-B stages that later evolved in high-grade B-cell malignancy (Costinean et al., 2006) (Fig. 4). This was the first *in vivo* model demonstrating that dysregulation of a single microRNA is sufficient for the development and progression of tumors (Zanesi, Pekarsky, Trapasso, Calin, & Croce, 2010). In 2007, Sevignani characterized mouse cancer susceptibility loci harboring microRNAs (Sevignani et al., 2007).

Starting in 2002, the microRNA field has expanded exponentially to define cancer tissue signatures, their targets, and their genetic and epigenetic regulation and mutations and, in order to match the clinical expectations, has been extended to biological fluids and synthetic molecules.

Protein tumor markers cannot be used for diagnosis and early detection, nor to predict recurrences. Because of their ability to distinguish among tumors and tissues, microRNAs may serve as diagnostic markers and, since

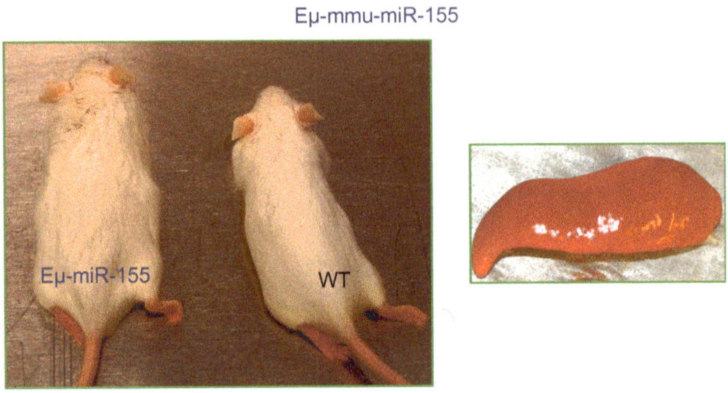

Fig. 4 Eμ-mmu-miR-155 transgenic mice were characterized by an enlarged abdomen when compared to wild-type control mice. The increased abdominal size due to splenomegaly was caused to polyclonal B-cell expansion.

their differential expression can often predict the patient outcome, they may also be suitable for prognosis. They are also very stable in formalin-fixed paraffin-embedded tissues (Nelson et al., 2006; Xi et al., 2007) as well as in plasma/serum (Mitchell et al., 2008), in urine (Mall, Rocke, Durbin-Johnson, & Weiss, 2013), and in cerebrospinal fluid (CSF) (Baraniskin et al., 2011; Drusco et al., 2015). MicroRNAs travel either bound to Argonaute 2 protein (Arroyo et al., 2011) or to high-density lipoproteins (Vickers, Palmisano, Shoucri, Shamburek, & Remaley, 2011), or within extracellular vesicles (Valadi et al., 2007) which account for their stability (Mitchell et al., 2008).

After the detection of miRs into exosomes of human and mouse mast cells in 2007 (Valadi et al., 2007), microRNAs were then isolated from serum of diffuse large B-cell lymphoma patients in 2008 (Lawrie et al., 2008). The same year, Chen found that serum microRNAs show specific expression patterns across different group of disease including cancer (Chen et al., 2008).

A year later, differentially expressed microRNAs were identified in forensic relevant biological fluids (blood, semen, saliva, vaginal secretions, and menstrual blood) (Hanson, Lubenow, & Ballantyne, 2009). It was found that miR-21, miR-92, miR-93, miR-126, and miR-29a were upregulated in serum of patients with ovarian cancer compared to healthy donors (Resnick et al., 2009), and miR-126, miR-182, and miR-199a were differentially expressed in urine of patients with carcinoma of the bladder (Hanke et al., 2010), while miR-21 was upregulated in the sputum of patients with nonsmall lung carcinoma with respect to normal healthy donors (Xie et al., 2010). MicroRNA signatures for colon cancer and ulcerative colitis were tested in feces, where they seemed to show better sensitivity and specificity than traditional tests (Ahmed et al., 2009).

In 2010, microRNAs were found in a series of 12 normal body fluids (amniotic fluid, breast milk, bronchial lavage, CSF, colostrum, peritoneal fluid, plasma, pleural fluid, saliva, seminal fluid, tears, and urine) (Weber et al., 2010).

Currently, the most frequently listed circulating microRNAs are miR-21, miR-221, miR-155, miR-223, miR-92a, miR-16, miR-20a, miR-141, miR-145, and miR-210 (Matsuzaki & Ochiya, 2017).

Despite the large number of studies on biological fluids demonstrating that microRNAs can be powerful biomarkers for diagnosis, prognosis, and follow-up of patients with malignancies, the application into the clinical practice found some technical obstacles.

A univocal scientific protocol to process samples and analyze data from biological fluids has not yet been achieved. Extraction, measurements, technological platforms, normalizations, and statistical analyses are all sources of variability that, in conjunction with the genetic variability of the individual patient, has shown the difficulty of comparing different studies (He et al., 2015; Larrea et al., 2016). Nonetheless, scientists are still trying to find standardized methods to fulfill microRNA diagnostic and prognostic promises. Of help was the development in 2011 of the nCounter platform by Nanostring Technologies. The platform is able to count the effective number of copies of each microRNA and requires only 100 ng RNA per sample (Wyman et al., 2011). Furthermore, in 2012, a research data bank, miRandola, was developed to classify circulating microRNAs according to the source, the biological fluid, and functions of each miR (Russo et al., 2012).

The most fascinating potential of microRNA is the delivery for therapeutic purposes.

In an article of 2010, Volinia et al. built a network of connected clusters comparing microRNA profiles of 4419 samples of 50 normal tissues and 51 type of tumors. The tumor network is composed of independent subnetworks, whereas normal samples organize in clusters linked by analogous functions (Volinia et al., 2010). This finding is essential to understand that mutation of different genes uses common pathways to cancer, and the cancer treatment should be based on the type of mutations rather than on localization.

MicroRNA therapeutic strategies are based on antagonizing the expression levels of crucial oncogenic miRs or replacing tumor suppressor miRs. Antisense naked oligonucleotides showed effectiveness in *Drosophila* (Boutla, Delidakis, & Tabler, 2003), but failed in *C. elegans* (Hutvágner, Simard, Mello, & Zamore, 2004). Different strategies aimed to improve stability, binding affinity, and specificity of these molecules have led to the development of several classes of synthetic therapeutic miRs. Stability is essential to overcome nuclease activity and organ clearance, while a significant binding affinity and specificity are vital to avoid undesirable side effects due to nonspecific delivery of the miR drug.

On the other hand, the logic of the different approaches has defined four major groups of anti-miR drugs: miRNA mimics, anti-miR oligonucleotides (AMOs), microsponges, and small-molecule inhibitors.

MiRNA mimics restore the expression of deleted or downregulated microRNAs by introducing synthetic oligonucleotides. For better delivery, oligonucleotides are combined with synthetic polymers, nanoparticles, liposome, minicells, hydrogels, or viral carriers (Chen, Zhu, Zhang, Liu, & Huang, 2010; Chiarantini et al., 2005; Chirila, Rakoczy, Garrett, Lou, & Constable, 2002; Guo, Zhang, Chen, Garver, & Trottier, 1998; Michelfelder & Trepel, 2009; Rosi et al., 2006; Zhao, Pan, Holt, Lewis, & Lu, 2009; Zuhorn, Engberts, & Hoekstra, 2007). The first microRNA clinical trial, MRX34, was conducted in 2003 and delivered liposomal injections of miR-34 in multiple solid tumors and hematological malignancies (https://clinicaltrials.gov/show/NCT01829971). Unfortunately, the clinical trial was stopped because MRX34 causes severe immune-related adverse effects in some patients.

Liposome may be toxic and has shown low sensitivity and specificity (Bozzuto & Molinari, 2015), but minicells, generated by asymmetric bacterial binary fission, are less toxic and have high stability and specificity allowing *in vivo* delivery of high drug concentration (MacDiarmid et al., 2007, 2009; Reid et al., 2013). TargomiR are EGFR-targeted minicells containing miRNA mimic for the treatment of malignant mesothelioma that has reached the clinical trial in 2004 (https://clinicaltrials.gov/show/NCT02369198) (Reid et al., 2016).

Among viral carriers, adeno-associated viral vectors do not integrate into the genome, are effectively eliminated, and have shown minimal toxicity in Phase I and Phase II clinical trials (Michelfelder & Trepel, 2009).

Positively charged cationic polymers are the most popular nanoparticles but can be cytotoxic. Resistant to endosomal and lysosomal degradation, cationic polymers release efficiently the nucleic acid into the cell nucleus or cytoplasm. The only limit is represented by the poor biodegradability: the accumulation of the polymer can be highly cytotoxic (Catela Ivkovic, Voss, Cornella, & Ceder, 2017; Ganju et al., 2017).

On the other hand, the addition of 2′-O-methyl group to the oligonucleotide has improved delivery and protects from degradation miRNA mimics as well as AMOs.

AMOs are antisense nucleotides that by competitively binding the mature microRNA guide strand ensure their degradation. The cholesterol-modified miR-122 antagomiR is the first example of AMO (Krützfeldt et al., 2005). Second-generation AMO includes LNA anti-miR constructs. MiR-122 is essential for Hepatitis C virus (HCV) RNA stability. LNA anti-miR-122 was developed in 2013 as a potential therapy for HCV-infected patients and was called Miravirsen (Janssen et al., 2013). Miravirsen binds miR-122 and inhibits viral replication. At the time, the drug did not show short- or long-term side effects and was an effective therapy (van der Ree et al., 2014), but its successor, RG-10, although well tolerated by the majority of patients, induced jaundice in two patients, and the clinical trial was temporarily suspended.

Another AMO's subgroup is miR-Mask. MiR-Mask are antisense oligonucleotides that bind the 3′UTR for the target mRNA with 100% complementarity. The high affinity favors miR-Mask binding to the microRNA, masking the seed region and allowing the expression of the target mRNA. This strategy has been applied to inhibit the repression of the TGF-β signaling pathway by miR-430 (Choi, Giraldez, & Schier, 2007).

MicroRNA sponges are molecules with tandem repeats of oligonucleotide antisense sequences to decoy microRNAs from their targets. Although sponges can repress the expression of an entire family of microRNAs, it is difficult to modulate the levels of microRNA inhibition, and overexpression of target genes may be deleterious (Ebert, Neilson, & Sharp, 2007).

As microRNAs regulate target gene expression, there are pathways or genes that control microRNA expression, maturation, and degradation. This is another strategy that was exploited in 2008 with the observation that diazobenzene 1 modulated miR-21 expression (Gumireddy et al., 2008).

Beside all these options, LNA oligonucleotides are the only miRNA therapies still on clinical trials. MLG-106 is an anti-miR-155 LNA oligonucleotide launched in 2015 for Phase I clinical trial (https://clinicaltrials.gov/show/NCT02580552). Intratumoral injection of MLG-106 reduces the cutaneous lesion in patients with mycosis fungoides with few tolerable side effects (Querfeld et al., 2016).

One of the most attractive advantages of microRNAs is the possibility to target multiple pathways with a single oligonucleotide. Cancer is indeed the result of several mutations affecting multiple pathways. In 2014, Laganà created miR-Synth, a computational resource to design synthetic miRNAs to target multiple genes and multiple pathways (Laganà et al., 2014). Recently, Acunzo et al. have developed a new approach designing a miRNA-like artificial molecule to target a point mutation on KRAS mRNA seed region. The molecule effectively inhibits its expression without redundancy (Acunzo et al., 2017).

Despite the numerous obstacles encountered in the development of these new drugs, microRNAs remain one of the most promising translational medical fields, and technologies and computational approaches are continuously growing to help fulfill the microRNA potential.

REFERENCES

Acunzo, M., Romano, G., Nigita, G., Veneziano, D., Fattore, L., Laganá, A., et al. (2017). Selective targeting of point-mutated KRAS through artificial microRNAs. *Proceedings of the National Academy of Sciences of the United States of America, 114*(21), E4203–E4212.

Ahmed, F. E., Jeffries, C. D., Vos, P. W., Flake, G., Nuovo, G. J., Sinar, D. R., et al. (2009). Diagnostic microRNA markers for screening sporadic human colon cancer and active ulcerative colitis in stool and tissue. *Cancer Genomics & Proteomics, 6*(5), 281–295.

Arroyo, J. D., Chevillet, J. R., Kroh, E. M., Ruf, I. K., Pritchard, C. C., Gibson, D. F., et al. (2011). Argonaute 2 complexes carry a population of circulating microRNAs independent of vesicles in human plasma. *Proceedings of the National Academy of Sciences of the United States of America, 108*(12), 5003–5008.

Babak, T., Zhang, W., Morris, Q., Blencowe, B. J., & Hughes, T. R. (2004). Probing microRNAs with microarrays: Tissue specificity and functional inference. *RNA, 10*(11), 1813–1819.

Banzhaf-Strathmann, J., & Edbauer, D. (2014). Good guy or bad guy: The opposing roles of microRNA 125b in cancer. *Cell Communication and Signaling: CCS, 12*, 30.

Baraniskin, A., Kuhnhenn, J., Schlegel, U., Chan, A., Deckert, M., Gold, R., et al. (2011). Identification of microRNAs in the cerebrospinal fluid as marker for primary diffuse large B-cell lymphoma of the central nervous system. *Blood, 117*(11), 3140–3146.

Bartel, D. P. (2009). MicroRNAs: Target recognition and regulatory functions. *Cell, 136*(2), 215–233.

Boutla, A., Delidakis, C., & Tabler, M. (2003). Developmental defects by antisense-mediated inactivation of microRNAs 2 and 13 in Drosophila and the identification of putative target genes. *Nucleic Acids Research, 31*(17), 4973–4980.

Bozzuto, G., & Molinari, A. (2015). Liposomes as nanomedical devices. *International Journal of Nanomedicine, 10*, 975–999.

Bullrich, F., Fujii, H., Calin, G., Mabuchi, H., Negrini, M., Pekarsky, Y., et al. (2001). Characterization of the 13q14 tumor suppressor locus in CLL: Identification of ALT1, an alternative splice variant of LEU2 gene. *Cancer Research, 61*(18), 6640–6648.

Calin, G. A., Cimmino, A., Fabbri, M., Ferracin, M., Wojcik, S. E., Shimizu, M., et al. (2008). MiR-15a and miR-16-1 cluster functions in human leukemia. *Proceedings of the National Academy of Sciences of the United States of America, 105*(13), 5166–5171.

Calin, G. A., Dumitru, C. D., Shimizu, M., Bichi, R., Zupo, S., Noch, E., et al. (2002). Frequent deletions and down-regulation of micro-RNA genes miR-15 and miR-16 at 13q14 in chronic lymphocytic leukemia. *Proceedings of the National Academy of Sciences of the United States of America, 99*(24), 15524–15529.

Calin, G. A., Ferracin, M., Cimmino, A., Di Leva, G., Shimizu, M., Wojcik, S. E., et al. (2005). microRNA signature associated with prognosis and progression in chronic lymphocytic leukemia. *The New England Journal of Medicine, 353*(17), 1793–1801.

Calin, G. A., Liu, C. G., Sevignani, C., Ferracin, M., Felli, N., Dumitru, C. D., et al. (2004). MicroRNA profiling reveals distinct signatures in B cell chronic lymphocytic leukemias. *Proceedings of the National Academy of Sciences of the United States of America, 101*(32), 11755–11760.

Calin, G. A., Sevignani, C., Dumitru, C. D., Hyslop, T., Noch, E., Yendamuri, S., et al. (2004). Human microRNA genes are frequently located at fragile sites and genomic regions involved in cancer. *Proceedings of the National Academy of Sciences of the United States of America, 101*(9), 2999–3004.

Carmell, M. A., Xuan, Z., Zhang, M. Q., & Hannon, G. J. (2002). The Argonaute family: Tentacles that reach into RNAi, developmental control, stem cell maintenance, and tumorigenesis. *Genes & Development, 16*(21), 2733–2742.

Catela Ivkovic, T., Voss, G., Cornella, H., & Ceder, Y. (2017). microRNAs as cancer therapeutics: A step closer to clinical application. *Cancer Letters*, pii: S0304-3835(17)30237-9. http://dx.doi.org/10.1016/j.canlet.2017.04.007. [Epub ahead of print].

Chang, S., Wang, R. H., Akagi, K., Kim, K. A., Martin, B. K., Cavallone, L., et al. (2011). Tumor suppressor BRCA1 epigenetically controls oncogenic microRNA-155. *Nature Medicine, 17*(10), 1275–1282.

Chen, X., Ba, Y., Ma, L., Cai, X., Yin, Y., Wang, K., et al. (2008). Characterization of microRNAs in serum: A novel class of biomarkers for diagnosis of cancer and other diseases. *Cell Research, 18*(10), 997–1006.

Chen, Y., Zhu, X., Zhang, X., Liu, B., & Huang, L. (2010). Nanoparticles modified with tumor-targeting scFv deliver siRNA and miRNA for cancer therapy. *Molecular Therapy, 18*(9), 1650–1656.

Chiarantini, L., Cerasi, A., Fraternale, A., Millo, E., Benatti, U., Sparnacci, K., et al. (2005). Comparison of novel delivery systems for antisense peptide nucleic acids. *Journal of Controlled Release, 109*(1–3), 24–36.

Chirila, T. V., Rakoczy, P. E., Garrett, K. L., Lou, X., & Constable, I. J. (2002). The use of synthetic polymers for delivery of therapeutic antisense oligodeoxynucleotides. *Biomaterials, 23*(2), 321–342.

Choi, W. Y., Giraldez, A. J., & Schier, A. F. (2007). Target protectors reveal dampening and balancing of nodal agonist and antagonist by miR-430. *Science, 318*(5848), 271–274.

Concepcion, C. P., Bonetti, C., & Ventura, A. (2012). The microRNA-17-92 family of microRNA clusters in development and disease. *Cancer Journal, 18*(3), 262–267.

Costinean, S., Zanesi, N., Pekarsky, Y., Tili, E., Volinia, S., Heerema, N., et al. (2006). Pre-B cell proliferation and lymphoblastic leukemia/high-grade lymphoma in E(mu)-miR155 transgenic mice. *Proceedings of the National Academy of Sciences of the United States of America, 103*(18), 7024–7029.

Dalla-Favera, R., Bregni, M., Erikson, J., Patterson, D., Gallo, R. C., & Croce, C. M. (1982). Human c-myc oncogene is regulated on the region of chromosome 8 that is translocated in Burkitt lymphoma cells. *Proceedings of the National Academy of Sciences of the United States of America, 79*(24), 7824–7827.

Dang, C. V. (2012). MYC on the path to cancer. *Cell*, *149*(1), 22–35.

Di Leva, G., Garofalo, M., & Croce, C. M. (2014). MicroRNAs in cancer. *Annual Review of Pathology*, *9*, 287–314.

Drusco, A., Bottoni, A., Laganà, A., Acunzo, M., Fassan, M., Cascione, L., et al. (2015). Differentially expressed set of microRNAs in cerebrospinal fluid (CSF) can diagnose CNS malignancies. *Oncotarget*, *6*(25), 20829–20839.

Drusco, A., Nuovo, G. J., Zanesi, N., Di Leva, G., Pichiorri, F., Volinia, S., et al. (2014). MicroRNA profiles discriminate among colon cancer metastasis. *PLoS One*, *9*(6), e96670.

Ebert, M. S., Neilson, J. R., & Sharp, P. A. (2007). MicroRNA sponges: Competitive inhibitors of small RNAs in mammalian cells. *Nature Methods*, *4*(9), 721–726.

Fabian, M. R., Sonenberg, N., & Filipowicz, W. (2010). Regulation of mRNA translation and stability by microRNAs. *Annual Review of Biochemistry*, *79*, 351–379.

Fang, H., Xie, J., Zhang, M., Zhao, Z., Wan, Y., & Yao, Y. (2017). miRNA-21 promotes proliferation and invasion of triple-negative breast cancer cells through targeting PTEN. *American Journal of Translational Research*, *9*(3), 953–961.

Feng, Y. H., & Tsao, C. J. (2016). Emerging role of microRNA-21 in cancer. *Biomedical Reports*, *5*(4), 395–402.

Ferracin, M., Pedriali, M., Veronese, A., Zagatti, B., Gafà, R., Magri, E., et al. (2011). MicroRNA profiling for the identification of cancers with unknown primary tissue-of-origin. *The Journal of Pathology*, *225*(1), 43–53.

Fulci, V., Chiaretti, S., Goldoni, M., Azzalin, G., Carucci, N., Tavolaro, S., et al. (2007). Quantitative technologies establish a novel microRNA profile of chronic lymphocytic leukemia. *Blood*, *109*(11), 4944–4951.

Ganju, A., Khan, S., Hafeez, B. B., Behrman, S. W., Yallapu, M. M., Chauhan, S. C., et al. (2017). MiRNA nanotherapeutics for cancer. *Drug Discovery Today*, *22*(2), 424–432.

Go, H., Jang, J. Y., Kim, P. J., Kim, Y. G., Nam, S. J., Paik, J. H., et al. (2015). MicroRNA-21 plays an oncogenic role by targeting FOXO1 and activating the PI3K/AKT pathway in diffuse large B-cell lymphoma. *Oncotarget*, *6*(17), 15035–15049.

Griffiths-Jones, S., Grocock, R. J., van Dongen, S., Bateman, A., & Enright, A. J. (2006). miRBase: MicroRNA sequences, targets and gene nomenclature. *Nucleic Acids Research*, *34*(Database issue), D140–D144.

Gu, L., Song, G., Chen, L., Nie, Z., He, B., Pan, Y., et al. (2013). Inhibition of miR-21 induces biological and behavioral alterations in diffuse large B-cell lymphoma. *Acta Haematologica*, *130*(2), 87–94.

Gumireddy, K., Young, D. D., Xiong, X., Hogenesch, J. B., Huang, Q., & Deiters, A. (2008). Small-molecule inhibitors of microRNA miR-21 function. *Angewandte Chemie (International ed. in English)*, *47*(39), 7482–7484.

Guo, P., Zhang, C., Chen, C., Garver, K., & Trottier, M. (1998). Inter-RNA interaction of phage phi129 pRNA to form a hexameric complex for viral DNA transportation. *Molecular Cell*, *2*(1), 149–155.

Hamblin, T. J., Davis, Z., Gardiner, A., Oscier, D. G., & Stevenson, F. K. (1999). Unmutated Ig V(H) genes are associated with a more aggressive form of chronic lymphocytic leukemia. *Blood*, *94*(6), 1848–1854.

Hanke, M., Hoefig, K., Merz, H., Feller, A. C., Kausch, I., Jocham, D., et al. (2010). A robust methodology to study urine microRNA as tumor marker: MicroRNA-126 and microRNA-182 are related to urinary bladder cancer. *Urologic Oncology*, *28*(6), 655–661.

Hanson, E. K., Lubenow, H., & Ballantyne, J. (2009). Identification of forensically relevant body fluids using a panel of differentially expressed microRNAs. *Analytical Biochemistry*, *387*(2), 303–314.

Hayashita, Y., Osada, H., Tatematsu, Y., Yamada, H., Yanagisawa, K., Tomida, S., et al. (2005). A polycistronic microRNA cluster, miR-17-92, is overexpressed in human lung cancer and enhances cell proliferation. *Cancer Research*, *65*(21), 9628–9632.

He, Y., Lin, J., Kong, D., Huang, M., Xu, C., Kim, T. K., et al. (2015). Current state of circulating microRNAs as cancer biomarkers. *Clinical Chemistry*, *61*(9), 1138–1155.

Higgs, G., & Slack, F. (2013). The multiple roles of microRNA-155 in oncogenesis. *Journal of Clinical Bioinformatics*, *3*(1), 17.

Hutvágner, G., Simard, M. J., Mello, C. C., & Zamore, P. D. (2004). Sequence-specific inhibition of small RNA function. *PLoS Biology*, *2*(4), E98.

Iorio, M. V., Ferracin, M., Liu, C. G., Veronese, A., Spizzo, R., Sabbioni, S., et al. (2005). MicroRNA gene expression deregulation in human breast cancer. *Cancer Research*, *65*(16), 7065–7070.

Iorio, M. V., Visone, R., Di Leva, G., Donati, V., Petrocca, F., Casalini, P., et al. (2007). MicroRNA signatures in human ovarian cancer. *Cancer Research*, *67*(18), 8699–8707.

Janssen, H. L., Reesink, H. W., Lawitz, E. J., Zeuzem, S., Rodriguez-Torres, M., Patel, K., et al. (2013). Treatment of HCV infection by targeting microRNA. *The New England Journal of Medicine*, *368*(18), 1685–1694.

John, B., Enright, A. J., Aravin, A., Tuschl, T., Sander, C., & Marks, D. S. (2004). Human microRNA targets. *PLoS Biology*, *2*(11), e363.

Johnson, S. M., Grosshans, H., Shingara, J., Byrom, M., Jarvis, R., Cheng, A., et al. (2005). RAS is regulated by the let-7 microRNA family. *Cell*, *120*(5), 635–647.

Klein, U., Lia, M., Crespo, M., Siegel, R., Shen, Q., Mo, T., et al. (2010). The DLEU2/miR-15a/16-1 cluster controls B cell proliferation and its deletion leads to chronic lymphocytic leukemia. *Cancer Cell*, *17*(1), 28–40.

Kloosterman, W. P., Wienholds, E., de Bruijn, E., Kauppinen, S., & Plasterk, R. H. (2006). In situ detection of miRNAs in animal embryos using LNA-modified oligonucleotide probes. *Nature Methods*, *3*(1), 27–29.

Krek, A., Grün, D., Poy, M. N., Wolf, R., Rosenberg, L., Epstein, E. J., et al. (2005). Combinatorial microRNA target predictions. *Nature Genetics*, *37*(5), 495–500.

Krol, J., Loedige, I., & Filipowicz, W. (2010). The widespread regulation of microRNA biogenesis, function and decay. *Nature Reviews. Genetics*, *11*(9), 597–610.

Krützfeldt, J., Rajewsky, N., Braich, R., Rajeev, K. G., Tuschl, T., Manoharan, M., et al. (2005). Silencing of microRNA in vivo with 'antagomirs'. *Nature*, *438*(7068), 685–689.

Laganà, A., Acunzo, M., Romano, G., Pulvirenti, A., Veneziano, D., Cascione, L., et al. (2014). miR-Synth: A computational resource for the design of multi-site multi-target synthetic miRNAs. *Nucleic Acids Research*, *42*(9), 5416–5425.

Lagos-Quintana, M., Rauhut, R., Lendeckel, W., & Tuschl, T. (2001). Identification of novel genes coding for small expressed RNAs. *Science*, *294*(5543), 853–858.

Lagos-Quintana, M., Rauhut, R., Yalcin, A., Meyer, J., Lendeckel, W., & Tuschl, T. (2002). Identification of tissue-specific microRNAs from mouse. *Current Biology*, *12*(9), 735–739.

Larrea, E., Sole, C., Manterola, L., Goicoechea, I., Armesto, M., Arestin, M., et al. (2016). New concepts in cancer biomarkers: Circulating miRNAs in liquid biopsies. *International Journal of Molecular Sciences*, *17*(5), pii: E627. http://dx.doi.org/10.3390/ijms17050627. Review.

Lau, N. C., Lim, L. P., Weinstein, E. G., & Bartel, D. P. (2001). An abundant class of tiny RNAs with probable regulatory roles in Caenorhabditis elegans. *Science*, *294*(5543), 858–862.

Lawrie, C. H., Gal, S., Dunlop, H. M., Pushkaran, B., Liggins, A. P., Pulford, K., et al. (2008). Detection of elevated levels of tumour-associated microRNAs in serum of patients with diffuse large B-cell lymphoma. *British Journal of Haematology*, *141*(5), 672–675.

Lee, R. C., & Ambros, V. (2001). An extensive class of small RNAs in Caenorhabditis elegans. *Science, 294*, 862–864.

Lee, R. C., Feinbaum, R. L., & Ambros, V. (1993). The C. elegans heterochronic gene lin-4 encodes small RNAs with antisense complementarity to lin-14. *Cell, 75*(5), 843–854.

Lewis, B. P., Shih, I. H., Jones-Rhoades, M. W., Bartel, D. P., & Burge, C. B. (2003). Prediction of mammalian microRNA targets. *Cell, 115*(7), 787–798.

Li, Z., & Rana, T. M. (2012). Molecular mechanisms of RNA-triggered gene silencing machineries. *Accounts of Chemical Research, 45*(7), 1222–1231.

Liu, C. G., Calin, G. A., Meloon, B., Gamliel, N., Sevignani, C., Ferracin, M., et al. (2004). An oligonucleotide microchip for genome-wide microRNA profiling in human and mouse tisuues. *Proceedings of the National Academy of Sciences of the United States of America, 101*(26), 9740–9744.

Liu, C. G., Calin, G. A., Volinia, S., & Croce, C. M. (2008). MicroRNA expression profiling using microarrays. *Nature Protocols, 3*(4), 563–578.

Liu, Y., Corcoran, M., Rasool, O., Ivanova, G., Ibbotson, R., Grandér, D., et al. (1997). Cloning of two candidate tumor suppressor genes within a 10 kb region on chromosome 13q14, frequently deleted in chronic lymphocytic leukemia. *Oncogene, 15*(20), 2463–2473.

Llave, C., Xie, Z., Kasschau, K. D., & Carrington, J. C. (2002). Cleavage of Scarecrow-like mRNA targets directed by class of Arabidopsis miRNA. *Science, 297*(5589), 2053–2056.

Lu, J., Getz, G., Miska, E. A., Alvarez-Saavedra, E., Lamb, J., Peck, D., et al. (2005). MicroRNA expression profiles classify human cancer. *Nature, 435*(7043), 834–838.

Lu, C., Tej, S. S., Luo, S., Haudenschild, C. D., Meyers, B. C., & Green, P. J. (2005). Elucidation of the small RNA component of the transcriptome. *Science, 309*(5740), 1567–1569.

Ma, L., Teruya-Feldstein, J., & Weinberg, R. A. (2007). Tumour invasion and metastasis initiated by microRNA-10b in breast cancer. *Nature, 449*(7163), 682–688.

Mabuchi, H., Fujii, H., Calin, G., Alder, H., Negrini, M., Rassenti, L., et al. (2001). Cloning and characterization of CLLD6, CLLD7, and CLLD8, novel candidate genes for leukemogenesis at chromosome 13q14, a region commonly deleted in B-cell chronic lymphocytic leukemia. *Cancer Research, 61*(7), 2870–2877.

MacDiarmid, J. A., Amaro-Mugridge, N. B., Madrid-Weiss, J., Sedliarou, I., Wetzel, S., Kochar, K., et al. (2009). Sequential treatment of drug-resistant tumors with targeted minicells containing siRNA or a cytotoxic drug. *Nature Biotechnology, 27*, 643–651.

MacDiarmid, J. A., Mugridge, N. B., Weiss, J. C., Phillips, L., Burn, A. L., Paulin, R. P., et al. (2007). Bacterially derived 400 nm particles for encapsulation and cancer cells targeting of chemotherapeutics. *Cancer Cell, 11*(5), 431–445.

Mall, C., Rocke, D. M., Durbin-Johnson, B., & Weiss, R. H. (2013). Stability of miRNA in human urine supports its biomarker potential. *Biomarkers in Medicine*, 7(4), 623–631.

Matsuzaki, J., & Ochiya, T. (2017). Circulating microRNAs and extracellular vesicles as potential cancer biomarkers: A systematic review. *International Journal of Clinical Oncology, 22*(3), 413–420. http://dx.doi.org/10.1007/s10147-017-1104-3.

McManus, M. T. (2003). MicroRNAs and cancer. *Seminars in Cancer Biology, 13*(4), 253–258.

Medina, P. P., Nolde, M., & Slack, F. J. (2010). OncomiR addiction in an in vivo model of microRNA-21-induced pre-B-cell lymphoma. *Nature, 467*(7311), 86–90.

Michelfelder, S., & Trepel, M. (2009). Adeno-associated viral vectors and their redirection to cell-type specific receptors. *Advances in Genetics, 67*, 29–60.

Migliazza, A., Bosch, F., Komatsu, H., Cayanis, E., Martinotti, S., Toniato, E., et al. (2001). Nucleotide sequence, transcription map, and mutation analysis of the 13q14 chromosomal region deleted in B-cell chronic lymphocytic leukemia. *Blood, 97*(7), 2098–2104.

Mitchell, P. S., Parkin, R. K., Kroh, E. M., Fritz, B. R., Wyman, S. K., Pogosova-Agadjanyan, E. L., et al. (2008). Circulating microRNAs as stable blood-based markers for cancer detection. *Proceedings of the National Academy of Sciences of the United States of America*, *105*(30), 10513–10518.

Monteys, A. M., Spengler, R. M., Wan, J., Tecedor, L., Lennox, K. A., Xing, Y., et al. (2010). Structure and activity of putative intronic miRNA promoters. *RNA*, *16*(3), 495–505.

Nelson, P. T., Baldwin, D. A., Kloosterman, W. P., Kauppinen, S., Plasterk, R. H., & Mourelatos, Z. (2006). RAKE and LNA-ISH reveal microRNA expression and localization in archival human brain. *RNA*, *12*(2), 187–191.

Nicoloso, M. S., Sun, H., Spizzo, R., Kim, H., Wickramasinghe, P., Shimizu, M., et al. (2010). Single-nucleotide polymorphism inside microRNA target sites influence tumor susceptibility. *Cancer Research*, *70*(7), 2789–2798.

Nuovo, G. J., Elton, T. S., Nana-Sinkam, P., Volinia, S., Croce, C. M., & Schmittgen, T. D. (2009). A methodology for the combined in situ analyses of the precursor and mature forms of microRNAs and correlation with their putative targets. *Nature Protocols*, *4*(1), 107–115.

O'Donnell, K. A., Wentzel, E. A., Zeller, K. I., Dang, C. V., & Mendell, J. T. (2005). c-Myc-regulated microRNAs modulate E2F1 expression. *Nature*, *435*(7043), 839–843.

Oliveto, S., Mancino, M., Manfrini, N., & Biffo, S. (2017). Role of microRNAs in translation regulation and cancer. *World Journal of Biological Chemistry*, *8*(1), 45–56.

Orchard, J. A., Ibbotson, R. E., Davis, Z., Wiestner, A., Rosenwald, A., Thomas, P. W., et al. (2004). ZAP-70 expression and prognosis in chronic lymphocytic leukemia. *Lancet*, *363*(9403), 105–111.

Oscier, D. G., Gardiner, A. C., Mould, S. J., Glide, S., Davis, Z. A., Ibbotson, R. E., et al. (2002). Multivariate analysis of prognostic factors in CLL: Clinical stage, IGVH gene mutational status, and loss or mutation of the p53 gene are independent prognostic factors. *Blood*, *100*(4), 1177–1184.

Park, W., Li, J., Song, R., Messing, J., & Cheng, X. (2002). CARPEL FACTORY, a dicer homolog, and HEN1, a novel protein, act in microRNA metabolism in Arabidopsis thaliana. *Current Biology*, *12*(17), 1484–1495.

Pasquinelli, A. E., Reinhart, B. J., Slack, F., Martindale, M. Q., Kuroda, M. I., Maller, B., et al. (2000). Conservation of the sequence and temporal expression of let-7 heterochromic regulatory RNA. *Nature*, *408*(6808), 86–89.

Pegoraro, L., Palumbo, A., Erikson, J., Falda, M., Giovanazzo, B., Emanuel, B. S., et al. (1984). A 14;18 and 8;14 chromosome in a cell line derived from an acute B-cell leukemia. *Proceedings of the National Academy of Sciences of the United States of America*, *81*(22), 7166–7170.

Pimiento, J. M., Teso, D., Malkan, A., Dudrick, S. J., & Palesty, J. A. (2007). Cancer of unknown primary origin: A decade of experience in a community-based hospital. *American Journal of Surgery*, *194*(6), 833–837[discussion 837–838].

Querfeld, C., Pacheco, T., Foss, F. M., Pacheco, T., Foss, F. M., HAlwani, A., et al. (2016). Preliminary results of phase 1 trial evaluating MRG-106, a synthetic microRNA antagonist (LNA antimiR) of microRNA-155, in patients with CTCL. *Blood*, *128*, 1829.

Reid, G., Kao, S. C., Pavlakis, N., Brahmbhatt, H., MacDiarmid, J., Clarke, S., et al. (2016). Clinical development of TargomiRs, a miRNA mimic-based treatment for patients with recurrent thoracic cancer. *Epigenomics*, *8*(8), 1079–1085.

Reid, G., Pel, M. E., Kirschner, M. B., Cheng, Y. Y., Mugridge, N., Weiss, J., et al. (2013). Restoring expression of miR-16: A novel approach to therapy for malignant pleural mesothelioma. *Annals of Oncology*, *24*(12), 3128–3135.

Reinhart, B. J., Slack, F. J., Basson, M., Pasquinelli, A. E., Bettinger, J. C., Rougvie, A. E., et al. (2000). The 21-nucleotide let-7 RNA regulates developmental timing in Caenorhabditis elegans. *Nature*, *403*(6772), 901–906.

Reinhart, B. J., Weinstein, E. G., Rhoades, M. W., Bartel, B., & Bartel, D. P. (2002). MicroRNAs in plants. *Genes & Development*, *16*(13), 1616–1626.

Resnick, K. E., Alder, H., Hagan, J. P., Richardson, D. L., Croce, C. M., & Cohn, D. E. (2009). The detection of differentially expressed microRNAs from the serum of ovarian cancer patients using a novel real-time PCR platform. *Gynecologic Oncology*, *112*(1), 55–59.

Rondeau, G., Moreau, I., Bézieau, S., Petit, J. L., Heilig, R., Fernandez, S., et al. (2001). Comprehensive analysis of a large genomic sequence at the putative B-cell chronic lymphocytic leukaemia (B-CLL) tumour suppresser gene locus. *Mutation Research*, *458*(3–4), 55–70.

Rosenfeld, N., Aharonov, R., Meiri, E., Rosenwald, S., Spector, Y., Zepeniuk, M., et al. (2008 Apr). MicroRNA accurately identify cancer tissue origin. *Nature Biotechnology*, *26*(4), 462–469.

Rosi, N. L., Giljohann, D. A., Thaxton, C. S., Lytton-Jean, A. K., Han, M. S., & Mirkin, C. A. (2006). Oligonucleotide-modified gold nanoparticles for intracellular gene regulation. *Science*, *312*(5776), 1027–1030.

Rowntree, C., Duke, V., Panayiotidis, P., Kotsi, P., Palmisano, G. L., Hoffbrand, A. V., et al. (2002). Deletion analysis of chromosome 13q14.3 and characterization of an alternative splice form of LEU1 in B cell chronic lymphocytic leukemia. *Leukemia*, *16*(7), 1267–1275.

Russo, F., Di Bella, S., Nigita, G., Macca, V., Laganà, A., Giugno, R., et al. (2012). miRandola: Extracellular circulating microRNAs database. *PLoS One*, 7(10), e47786.

Saito, Y., Liang, G., Egger, G., Friedman, J. M., Chuang, J. C., Coetzee, G. A., et al. (2006). Specific activation of microRNA-127 with downregulation of the proto-oncogene BCL6 by chromatin-modifying drugs in human cancer cells. *Cancer Cell*, *9*(6), 435–443.

Sandhu, S. K., Volinia, S., Costinean, S., Galasso, M., Neinast, R., Santhanam, R., et al. (2012). miR-155 targets histone deacetylase 4 (HDAC4) and impairs transcriptional activity of B-cell lymphoma 6 (BCL6) in the Eμ-miR-155 transgenic mouse model. *Proceedings of the National Academy of Sciences of the United States of America*, *109*(49), 20047–20052.

Schetter, A. J., Leung, S. Y., Sohn, J. J., Zanetti, K. A., Bowman, E. D., Yanaihara, N., et al. (2008). MicroRNA expression profiles associated with prognosis and therapeutic outcome in colon adenocarcinoma. *Journal of the American Medical Association*, *299*(4), 425–436.

Schmittgen, T. D., Jiang, J., Liu, Q., & Yang, L. (2004). A high throughput method to monitor the expression of microRNA precursors. *Nucleic Acids Research*, *32*(4), e43.

Selcuklu, S. D., Donoghue, M. T., & Spillane, C. (2009). miR-21 as a key regulator of oncogenic processes. *Biochemical Society Transactions*, *37*(Pt. 4), 918–925.

Sevignani, C., Calin, G. A., Nnadi, S. C., Shimizu, M., Davuluri, R. V., Hyslop, T., et al. (2007). MicroRNA genes are frequently located near mouse cancer susceptibility loci. *Proceedings of the National Academy of Sciences of the United States of America*, *104*(19), 8017–8022.

Shaham, L., Binder, V., Gefen, N., Borkhardt, A., & Izraeli, S. (2012). MiR-125 in normal and malignant hematopoiesis. *Leukemia*, *26*(9), 2011–2018.

Shen, J., Ambrosone, C. B., DiCioccio, R. A., Odunsi, K., Lele, S. B., & Zhao, H. A. (2008). Functional polymorphism in the miR-146a gene and age of familial breast/ovarian cancer diagnosis. *Carcinogenesis*, *29*(10), 1963–1966.

Stine, Z. E., Walton, Z. E., Altman, B. J., Hsieh, A. L., & Dang, C. V. (2015). MYC, metabolism, and cancer. *Cancer Discovery*, *5*(10), 1024–1039.

Thai, T. H., Calado, D. P., Casola, S., Ansel, K. M., Xiao, C., Xue, Y., et al. (2007). Regulation of the germinal center response by microRNA-155. *Science*, *316*(5824), 604–608.

Thomsen, R., Nielsen, P. S., & Jensen, T. H. (2005). Dramatically improved RNA in situ hybridization signals using LNA-modified probes. *RNA*, *11*(11), 1745–1748.

Thomson, J. M., Parker, J., Perou, C. M., & Hammond, S. M. (2004). A custom microarray platform for analysis of microRNA gene expression. *Nature Methods*, *1*(1), 47–53.

Tili, E., Michaille, J. J., Adair, B., Alder, H., Limagne, E., Taccioli, C., et al. (2010). Resveratrol decreases the levels of miR-155 by upregulating miR-663, a microRNA targeting JunB and JunD. *Carcinogenesis*, *31*(9), 1561–1566.

Tili, E., Michaille, J. J., & Croce, C. M. (2013). MicroRNAs play a central role in molecular dysfunction linking inflammation with cancer. *Immunological Reviews*, *253*(1), 167–184.

Tili, E., Michaille, J. J., Luo, Z., Volinia, S., Rassenti, L. Z., Kipps, T. J., et al. (2012). The down-regulation of miR-125b in chronic lymphocytic leukemias leads to metabolic adaptation of cells to a transformed state. *Blood*, *120*(13), 2631–2638.

Valadi, H., Ekström, K., Bossios, A., Sjöstrand, M., Lee, J. J., & Lötvall, J. O. (2007). Exosome-mediated transfer of mRNAs and microRNAs is a novel mechanism of genetic exchange between cells. *Nature Cell Biology*, *9*(6), 654–659.

Válóczi, A., Hornyik, C., Varga, N., Burgyán, J., Kauppinen, S., & Havelda, Z. (2004). Sensitive and specific detection of microRNAs by northern blot analysis using LNA-modified oligonucleotide probes. *Nucleic Acids Research*, *32*(22), e175.

van der Ree, M. H., van der Meer, A. J., de Bruijne, J., Maan, R., van Vliet, A., Welzel, T. M., et al. (2014). Long-term safety and efficacy of microRNA-targeted therapy in chronic hepatitis C patients. *Antiviral Research*, *11*, 53–59.

Vasudevan, S., Tong, Y., & Steitz, J. A. (2007). Switching from repression to activation: MicroRNAs can up-regulate translation. *Science*, *318*(5858), 1931–1934.

Vasudevan, S., Tong, Y., & Steitz, J. A. (2008). Cell-cycle control of microRNA-mediated translation regulation. *Cell Cycle*, *7*(11), 1545–1549.

Veneziano, D., Di Bella, S., Nigita, G., Laganà, A., Ferro, A., & Croce, C. M. (2016). Non-coding RNA: Current deep sequencing data analysis approaches and challenges. *Human Mutation*, *37*(12), 1283–1298.

Veronese, A., Pepe, F., Chiacchia, J., Pagotto, S., Lanuti, P., Veschi, S., et al. (2015). Allele-specific loss and transcription of the miR-15a/16-1 cluster in chronic lymphocytic leukemia. *Leukemia*, *29*(1), 86–95.

Vickers, K. C., Palmisano, B. T., Shoucri, B. M., Shamburek, R. D., & Remaley, A. T. (2011). MicroRNAs are transported in plasma and delivered to recipient cells by high-density lipoproteins. *Nature Cell Biology*, *13*(4), 423–433.

Vigorito, E., Kohlhaas, S., Lu, D., & Leyland, R. (2013). miR-155: An ancient regulator of the immune system. *Immunological Reviews*, *253*(1), 146–157.

Volinia, S., Calin, G. A., Liu, C. G., Ambs, S., Cimmino, A., Petrocca, F., et al. (2006). A microRNA expression signature of human solid tumors defines cancer gene targets. *Proceedings of the National Academy of Sciences of the United States of America*, *103*(7), 2257–2261.

Volinia, S., Galasso, M., Costinean, S., Tagliavini, L., Gamberoni, G., Drusco, A., et al. (2010). Reprogramming of miRNA networks in cancer and leukemia. *Genome Research*, *20*(5), 589–599.

Weber, J. A., Baxter, D. H., Zhang, S., Huang, D. Y., Huang, K. H., Lee, M. J., et al. (2010). The microRNA spectrum in 12 body fluids. *Clinical Chemistry*, *56*(11), 1733–1741.

White, N. M., Fatoohi, E., Metias, M., Jung, K., Stephan, C., & Yousef, G. M. (2011). MetastamiRs: A stepping stone towards improved cancer management. *Nature Reviews. Clinical Oncology*, *8*(2), 75–84.

Wightman, B., Ha, I., & Ruvkun, G. (1993). Posttranscriptional regulation of the heterochronic gene lin-14 by lin-4 mediates temporal pattern formation in C. elegans. *Cell*, *75*(5), 855–862.

Wolf, S., Mertens, D., Schaffner, C., Korz, C., Döhner, H., Stilgenbauer, S., et al. (2001). B-cell neoplasia associated gene with multiple splicing (BCMS): The candidate B-CLL gene on 13q14 comprises more than 560 kb covering all critical regions. *Human Molecular Genetics*, *10*(12), 1275–1285.

Wyman, S. K., Knouf, E. C., Parkin, R. K., Fritz, B. R., Lin, D. W., Dennis, L. M., et al. (2011). Post-transcriptional generation of miRNA variants by multiple nucleotidyl transferases contributes to miRNA transcriptome complexity. *Genome Research*, *21*(9), 1450–1461.

Xi, Y., Nakajima, G., Gavin, E., Morris, C. G., Kudo, K., Hayashi, K., et al. (2007). Systematic analysis of microRNA expression of RNA extracted from fresh frozen and formalin-fixed paraffin-embedded samples. *RNA*, *13*(10), 1668–1674.

Xie, Y., Todd, N. W., Liu, Z., Zhan, M., Fang, H., Peng, H., et al. (2010). Altered miRNA expression in sputum for diagnosis of non-small cell lung cancer. *Lung Cancer*, *67*(2), 170–176.

Zack, T. I., Schumacher, S. E., Carter, S. L., Cherniack, A. D., Saksena, G., Tabak, B., et al. (2013). Pan-cancer patterns of somatic copy number alteration. *Nature Genetics*, *45*(10), 1134–1140.

Zanesi, N., Pekarsky, Y., Trapasso, F., Calin, G., & Croce, C. M. (2010). MicroRNAs in mouse models of lymphoid malignancies. *Journal of Nucleic Acids Investigation*, *1*(1), 36–40.

Zhao, X., Pan, F., Holt, C. M., Lewis, A. L., & Lu, J. R. (2009). Controlled delivery of antisense oligonucleotides: A brief review of current strategies. *Expert Opinion on Drug Delivery*, *6*(7), 673–686.

Zhu, S., Wu, H., Wu, F., Nie, D., Sheng, S., & Mo, Y. Y. (2008). MicroRNA-21 targets tumor suppressor genes in invasion and metastasis. *Cell Research*, *18*(3), 350–359.

Zuhorn, I. S., Engberts, J. B., & Hoekstra, D. (2007). Gene delivery by cationic lipid vectors: Overcoming cellular barriers. *European Biophysics Journal*, *36*(4–5), 349–362.

CHAPTER TWO

The Enigma of miRNA Regulation in Cancer

Anjan K. Pradhan*, Luni Emdad*,†,‡, Swadesh K. Das*,†,‡, Devanand Sarkar*,†,‡, Paul B. Fisher*,†,‡,1

*Virginia Commonwealth University, School of Medicine, Richmond, VA, United States
†VCU Institute of Molecular Medicine, Virginia Commonwealth University, School of Medicine, Richmond, VA, United States
‡VCU Massey Cancer Center, Virginia Commonwealth University, School of Medicine, Richmond, VA, United States
[1]Corresponding author: e-mail address: paul.fisher@vcuhealth.org

Contents

1. Introduction 26
2. Tumor Suppressor miRs and OncomiRs 27
3. Factors Regulating miRNAs 28
4. Regulation of the miRNA Biogenesis Pathway 28
   4.1 AGO Proteins 30
   4.2 Drosha Proteins 31
   4.3 DGCR8 Proteins 31
   4.4 DICER Proteins 32
5. Direct (Transcriptional) Regulation 32
   5.1 c-MYC 32
   5.2 E2F1 33
   5.3 p53 34
6. Indirect Regulation 35
   6.1 Nuclear Receptors 36
   6.2 Epigenetic Control 37
   6.3 TGF-β Signaling 38
   6.4 MDA-7/IL-24 38
   6.5 hPNPase 40
   6.6 hnRNPA1 41
   6.7 ADARs 42
   6.8 Stress 42
   6.9 Diet and Natural Products 42
7. Conclusions and Future Perspectives 43
Acknowledgments 43
References 44

Advances in Cancer Research, Volume 135
ISSN 0065-230X
http://dx.doi.org/10.1016/bs.acr.2017.06.001

Abstract

MicroRNAs (miRNAs or miRs) are small 19–22 nucleotide long, noncoding, single-stranded, and multifunctional RNAs that regulate a diverse assortment of gene and protein functions that impact on a vast network of pathways. Lin-4, a noncoding transcript discovered in 1993 and named miRNA, initiated the exploration of research into these intriguing molecules identified in almost all organisms. miRNAs interfere with translation or posttranscriptional regulation of their target gene and regulate multiple biological actions exerted by these target genes. In cancer, they function as both oncogenes and tumor suppressor genes displaying differential activity in various cellular contexts. Although the role of miRNAs on target gene functions has been extensively investigated, less is currently known about the upstream regulatory molecules that regulate miRNAs. This chapter focuses on the factors and processes involved in miRNA regulation.

1. INTRODUCTION

MicroRNAs (miRNAs or miRs) are small noncoding RNAs, approximately19–22 nucleotides long that play a critical role in the posttranscriptional gene silencing by degrading messenger RNAs or by inhibiting translation (He & Hannon, 2004). They constitute approximately 1% of all the predicted genes in different organisms including mammals (Lai, Tomancak, Williams, & Rubin, 2003). Lin-4 and let-7 were initially discovered miRNAs (Bagga et al., 2005), which control the timing of development and fate of adult cell determination in *Caenorhabditis elegans* by binding to the 3′ UTR of target developmental regulators, and causing either degradation or blocking translation of target mRNA. Subsequent studies demonstrated the presence of miRNAs in *Arabidopsis* (Zhang, Wang, & Pan, 2007), and shortly thereafter several investigations uncovered the roles of additional miRNAs in plants, animals, and viruses. Accumulated evidence garnered over the past 10 years confirm an involvement of miRNAs in almost all essential biological processes, i.e., cellular development, cellular proliferation, differentiation, cell metabolism, cell death, and cancer development and progression (Ma, 2016; Reddy, 2015; Sayed & Abdellatif, 2011). Expression levels of specific miRNAs differ during developmental stages, in different organs, and in specific disease states (Lewis & Steel, 2010). Research has focused on the roles of differentially regulated miRNAs during distinct stages of diseases and regulatory pathways (Holohan, Lahiri, Schneider, Foroud, & Saykin, 2012).

Several similarities and differences exist between miRNAs and mRNAs. The chromosomal location of miRNAs coincides with normal functions or abnormal functions in different diseases (Ardekani & Naeini, 2010). miRNAs are dispersed over the genome, with approximately half of them positioned over intronic regions (Lin, Miller, & Ying, 2006). They coexpress with coding genes sharing their promoter regions or they can have their own promoter (Hsu et al., 2006). Additionally, some miRNAs are located at varying distances from their coding genes (Rinn & Chang, 2012). Several features distinguish miRNAs from mRNAs, among these the biogenesis pathway is considered to be most significant (Macfarlane & Murphy, 2010). A large pool of miRNAs are transcribed by polycistronic variants, suggesting that miRNA family members develop in a similar manner (Cammaerts, Strazisar, De Rijk, & Del Favero, 2015). Also, miRNAs are sequentially conserved throughout evolution, regulating expression patterns of mRNAs (Hill, Jabbari, Matyunina, & McDonald, 2014). Studies have predicted that the regulatory mechanism of miRNAs evolved before the evolution of multicellular organisms (Shabalina & Koonin, 2008). It is worth noting that miRNAs also exist in viruses, with viral miRNAs differing from those of vertebrates (Skalsky & Cullen, 2010).

The interplay between miRNA and mRNA is intriguing, since one miRNA can regulate multiple mRNAs and the physiological role of miRNAs depend on the target mRNA. In addition, miRNAs do not require 100% sequence similarity to the binding region of mRNA; the seed sequence (5′ region of the miRNA) is the major contributor to miRNA function, which is predominantly conserved (Huang, Lin, et al., 2012). Target prediction software, algorithms, and bioinformatics programs take into account the sequence information and free energy associated with structural alterations to predict target interacting mRNAs (Witkos, Koscianska, & Krzyzosiak, 2011). These bioinformatics analyses predict thousands of targets that can result in false positives that need further experimentation and validation before being considered as true miRNA targets.

In this chapter, we summarize the past and recent discoveries of factors regulating miRNA biogenesis, expression, and function, particularly as they relate to cancer.

2. TUMOR SUPPRESSOR miRs AND OncomiRs

miRNAs that typically target a tumor suppressor are classified as oncomiRs, which are generally upregulated in different types of cancer

(Svoronos, Engelman, & Slack, 2016). Conversely, miRNAs, which downregulate oncogenes, are defined as tumor suppressor miRs (Svoronos et al., 2016). Tumor suppressor miRs are often lost in cancer and this usually happens through multiple mechanisms including promoter methylation, mutation or deletion, or defective miRNA processing (Mott, 2009; Zhang, Pan, Cobb, & Anderson, 2007). The let-7 family of miRNAs are well-known tumor suppressor miRs, while miR-221/miR-222 are well-characterized oncomiRs (Pradhan et al., 2017).

3. FACTORS REGULATING miRNAs

Since miRNAs play a critical and central role in gene regulation in almost all organisms, defining the mechanisms underlying regulation and expression patterns of miRNAs is important and relevant in comprehending the complexities of both normal and abnormal cellular physiology. This chapter describes the regulation patterns of miRNAs. The expression of miRNA can be regulated in different steps, specifically at the stage of miRNA biogenesis/processing, transcriptional regulation, or indirect regulation by stress and other conditions.

4. REGULATION OF THE miRNA BIOGENESIS PATHWAY

Most eukaryotes have miRNAs, which were initially discovered in *C. elegans* (Miska et al., 2007). The gene coding miRNA is transcribed to primary miRNAs, which undergo cleavage to form the precursor molecule. Precursor primary miRNAs undergo further cleavage ultimately generating mature miRNAs (Fig. 1).

The regulatory loop of miRNAs and their targets have become areas of intense research in the past decades. More than 60% of protein coding genes are reported to be targets of miRNAs (Friedman, Farh, Burge, & Bartel, 2009). The approaches utilized in these studies involve both computational or bioinformatics methodologies and direct experimentation. However, the factors that regulate miRNA biogenesis and function require further clarification. miRNA biogenesis and miRNA machinery are regulated by a number of molecules including Drosha, DICER, and DGCR8 (Macfarlane & Murphy, 2010). All miRNAs are not regulated in a similar manner; individual classes of miRNAs are differentially regulated. Regulation by some of the important miRNA regulatory molecules is described later and is also summarized in Fig. 2.

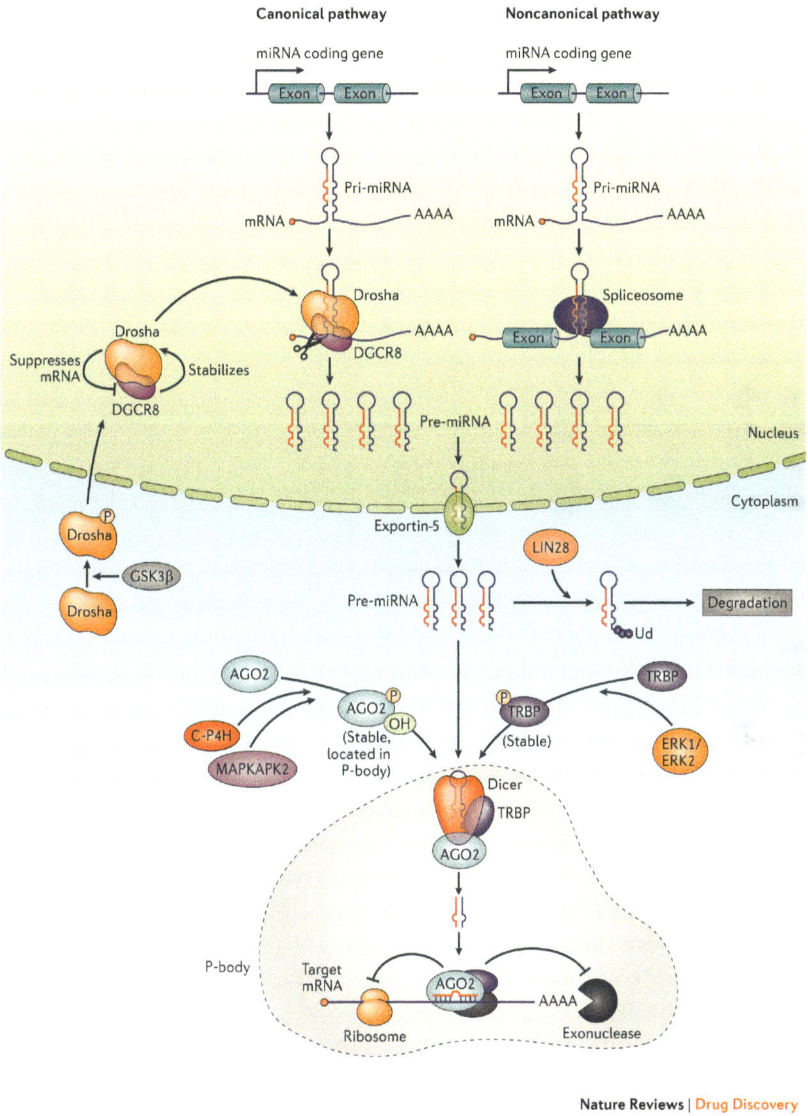

Fig. 1 miRNA biogenesis: RNA PolII transcribes genes encoding miRNAs to produce primary miRNA, which are processed to form the precursor miRNA. Pre-miRNAs are then exported by exportin-5 to the cytoplasm for further processing to form the mature miRNAs. *Adapted from Li, Z., & Rana, T. M. (2014).* Nature Reviews. Drug Discovery, 13, *622–638.*

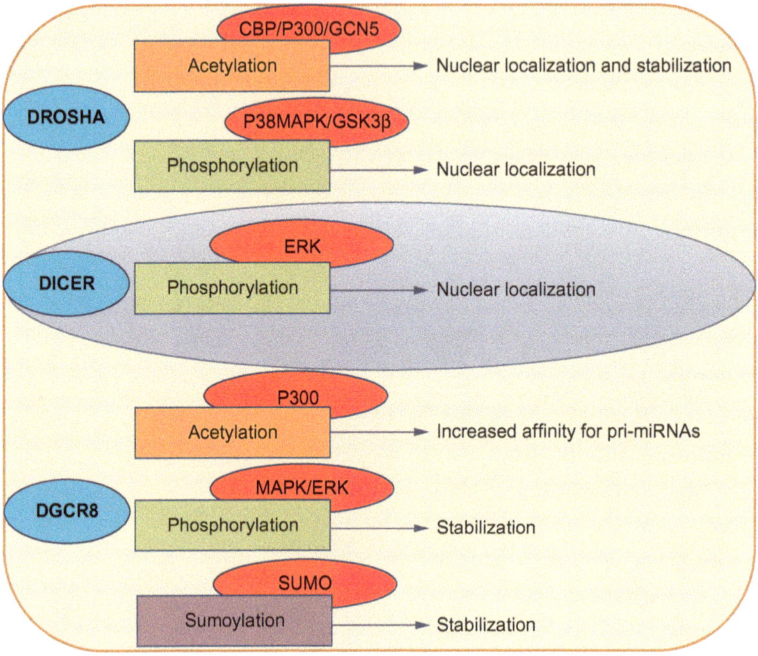

Fig. 2 Schematic representation of the regulation of proteins involved in miRNA biogenesis pathway: Drosha is stabilized by acetylation, and it increases the affinity of DGCR8 toward pri-miRNAs. Phosphorylation helps in the nuclear localization of DICER and DGCR8, which further facilitates nuclear localization of Drosha. Sumoylation is reported to stabilize DGCR8.

4.1 AGO Proteins

Argonaute (AGO) proteins play a critical role in miRNA biogenesis by forming the catalytic engine of the silencing complex (Mallory & Vaucheret, 2010). Phosphorylation of AGO2 protein at Serine 387, which is mediated by MAPK-activated protein kinase 2 (MAPKAPK2), results in its localization to processing bodies (Zeng, Sankala, Zhang, & Graves, 2008) where miRNA-targeted mRNAs are processed and degraded. Akt3 also regulates miRNA processing by phosphorylating AGO2 at Serine 387. This functions as a molecular switch between target mRNA cleavage and translational repression activities of AGO2 (Horman et al., 2013). During hypoxia, EGFR phosphorylates AGO2 at Tyr 393 resulting in its dissociation from DICER. This dissociation leads to a decreased processing of miRNAs from the precursor to mature stage (Shen et al., 2013). AGO1–4 are modified by poly (ADP-ribose) and this modification is enhanced during stress (Leung et al., 2011). Ubiquitination also plays an important role in maintenance of AGO2 stability. Wei et al. found that ubiquitination of Ms_AGO2 is mediated by the mouse homologue of lin-41 (mLin-41), both in vitro and in vivo (Wei et al., 2013). Autophagy is also reported to play a critical role in AGO protein destabilization (Smibert, Yang, Azzam, Liu, & Lai, 2013).

4.2 Drosha Proteins

Drosha, a class 2 ribonuclease enzyme (Filippov, Solovyev, Filippova, & Gill, 2000), also called RNASEN gene (Wu, Xu, Miraglia, & Crooke, 2000), is a core nuclease that initiates miRNA processing in the nucleus. It generates precursor molecules by cleaving primary miRNA transcripts that are then exported into the cytoplasm for further processing (Lee et al., 2003). Acetylation and ubiquitination have opposite effects on the regulation of the Drosha protein (Tang et al., 2013). HDAC inhibitors (HDACI) like Trichostatin A and Nicotinamide increase the protein levels of Drosha without affecting the transcript levels (Tang et al., 2013) and increase miRNA processing. Treatment with MG132 also increases the level of Drosha (Tang et al., 2013). Drosha can be acetylated at its N′-terminus region by CBP, GCN5, or P300, which inhibits its degradation through the proteasome degradation pathway. DGCR8 (DiGeorge syndrome critical region 8) was reported to stabilize Drosha through protein–protein interactions (Han et al., 2009). Under stress conditions, P38MAPK phosphorylates Drosha (Yang et al., 2015). This results in a decreased interaction of Drosha with DGCR8, which promotes its nuclear export and degradation by Calpain (Yang et al., 2015). This phosphorylation-mediated degradation of Drosha promotes cell death under stressful conditions. Drosha is also reported to be phosphorylated by GSK3β (glycogen synthase kinase 3 beta), the function of which is not currently well defined (Tang, Li, Tucker, & Ramratnam, 2011).

4.3 DGCR8 Proteins

DGCR8 is a mandatory component of the RNA interference pathway (Han et al., 2004). It binds to Drosha to form the microprocessor complex. The microprocessor complex cleaves the pri-miRNA to pre-miRNA, which matures to functional miRNA. HDACI increases the miRNA processing by modifying the acetylation of DGCR8 (Wada, Kikuchi, & Furukawa, 2012) by P300 (Wada et al., 2012). MAPK/ERK can phosphorylate DGCR8 (Herbert, Pimienta, DeGregorio, Alexandrov, & Steitz, 2013), and expression of phosphomimetic DGCR8 or inhibition of phosphatases enhances the level of both DGCR8 and Drosha (Herbert et al., 2013). This is attributed to the increased stability of DGCR8 protein rather than increased mRNA stability or localization of protein molecules in the complex (Herbert et al., 2013). Cells expressing the wild-type phosphomimetic showed a progrowth miRNA profile, an increased cellular growth and faster wound closure as compared to cells expressing the mutant form. ABL kinase is also reported to phosphorylate DGCR8 to induce a group of miRNAs specifically after DNA damage (Tu et al., 2015). Interestingly, ABL enhances the level of miR-34c, which is a known regulator of the miRNA biogenesis pathway (Tu et al., 2015).

4.4 DICER Proteins

Drake et al. showed that ERK phosphorylates DICER during oogenesis in *C. elegans* at two conserved residues in its RNase IIIb and double-stranded RNA (dsRNA)-binding domains, and this initiates the translocation of DICER into the nucleus (Drake et al., 2014). Phosphorylation of DICER was found to inhibit its function (Kurzynska-Kokorniak et al., 2015). DICER protein expression is also regulated by different kinds of stress, e.g., reactive oxygen species, dsRNA, or type 1 interferon (Wiesen & Tomasi, 2009).

5. DIRECT (TRANSCRIPTIONAL) REGULATION

5.1 c-MYC

MYC is a helix–loop–helix–leucine zipper transcription factor that regulates approximately 10%–15% of genes that control cell growth and apoptosis (Luscher & Larsson, 1999). c-MYC dysregulation can cause different types of cancer and is associated with activation of oncogenes (Miller, Thomas, Islam, Muench, & Sedoris, 2012). c-MYC also regulates different types of miRNAs (Mogilyansky & Rigoutsos, 2013), both at a transcriptional level and at a posttranscriptional level. c-MYC activates transcription of the miR-17-92 cluster (O'Donnell, Wentzel, Zeller, Dang, & Mendell, 2005), which is specifically activated in different types of cancer. Additionally, c-MYC was shown to activate miR-17, miR-20a, miR-18a, miR-92a, and miR-19a (Mogilyansky & Rigoutsos, 2013). c-MYC represses a group of miRNAs including, let-7, miR-16, and miR-29b (Kawano, Tanaka, Itonaga, Iwasaki, & Tsumura, 2015) and miR-34a, miR-26a, and miR-16 (Zhang et al., 2012). Repression of let-7 by c-MYC is important for cell cycle regulation in different diseases. Interestingly, let-7 transcriptionally downregulates c-MYC (Kim et al., 2009) and this feedback loop is dysregulated in cancer and other diseases.

Experimental evidence suggests that Drosha, DICER, and DGCR8 are the major regulators of miRNA processing. c-MYC regulates miRNA processing indirectly by activating Drosha (Wang, Zhao, Gao, & Wu, 2013). Also, a further study showed that c-MYC transcriptionally regulates Drosha (Wang et al., 2013). c-MYC expression positively correlates with expression of Drosha (Wang et al., 2013), which increases following serum stimulation. Recent studies establish the regulatory role of p53 (Suzuki et al., 2009), p63 (Huang, Jeong, et al., 2012; Su et al., 2010), and Smad (Blahna & Hata, 2012; Davis, Hilyard, Nguyen, Lagna, & Hata, 2010; Davis-Dusenbery & Hata, 2011) on miR regulation. Drosha–miR axis regulation by c-MYC adds a new and interesting dimension in this field.

5.2 E2F1

The E2F1 transcription factor functions as both an oncogene and a tumor suppressor depending on its transcriptional targets (Johnson, 2000). Increased expression of E2F1 causes transformation and tumor development. Overexpression of Rb (retinoblastoma) deregulates E2F1 to cause DNA damage and cancer (Johnson, 2000). Several miRNAs have been shown to regulate E2F1, i.e., miR-330, miR-106a, miR-17, miR-34, miR-223, and miR-15 (Hollern, Honeysett, Cardiff, & Andrechek, 2014).

Interestingly, miRNAs can also be regulated by E2F1. E2F1 directly binds to the promoter of the cluster miRNA, miR-17-92 to induce expression (Lee et al., 2009; Nevins, 2001). This cluster on other hand can downregulate E2F1 thus creating an autoregulatory loop. The miR-17-92 cluster is the first polycistronic miRNA cluster shown to be involved in tumorigenesis. Two other paralogs exist in the human genome; the miR-106a-363 cluster and the miR-106b-25 cluster and E2F1 can regulate both of these paralog (Raghuwanshi et al., 2015) clusters. p53 and c-MYC also play crucial roles in regulating E2F1-miRs (Ji et al., 2011; Tan, Li, Lim, & Tan, 2014). Additionally, E2F1 directly targets miR-449a/b (Polager & Ginsberg, 2009), a miRNA family that structurally resembles the miR-34 family. miR-449 inhibits proliferation and promotes apoptosis through a p53-dependent mechanism (Polager & Ginsberg, 2009). Furthermore, overexpression of E2F1 enhances miR-15 and miR-16 (Wang et al.,

2014), which are known to inhibit cyclin-E: G1/S regulator and E2F1 target genes (Wang et al., 2014). The role of E2F1 as an oncogene/tumor suppressor is well rationalized by the miRs that are regulated by E2F1, and this provides new insights into the molecular *milieu* and functional regulation by E2F1.

5.3 p53

The tumor suppressor p53 plays a critical role in the stability of the genome and it is the most frequently mutated gene in cancer (Hollstein, Sidransky, Vogelstein, & Harris, 1991; Rivlin, Brosh, Oren, & Rotter, 2011). Normal function of p53 is disrupted during the initiation and progression phases of cancer (Muller & Vousden, 2014). Fifty percent of human cancers harbor different mutations of p53 (Olivier, Hollstein, & Hainaut, 2010). In normal humans and other organisms MDM2 inhibits p53 by promoting its degradation mediated by the ubiquitin proteasome pathway. Stress conditions block the p53–MDM2 interaction and activate p53 thereby leading to cell cycle arrest and apoptosis. p53 controls an array of genes through its transcriptional activities. Also, p53 transcription factor controls an assortment of transcriptional networks tested for 750 miRNAs in neuroblastoma. miRNAs regulated by p53 inhibit proliferation of neuroblastoma cells. p53 regulates these miRs through direct or indirect binding to their promoters (Barlev, Sayan, Candi, & Okorokov, 2010; Beckerman & Prives, 2010; He, He, Lowe, & Hannon, 2007).

miR-34a is a direct target of p53 and elicits multiple downstream effects. miR-34 suppresses SIRT1 and HDM4, which are negative regulators of p53 and miR-34 deletion adversely affects the function of p53. Also, miR-34 deficiencies promote tumorigenesis in specific contexts. Knockdown of miR-34 in the mouse exhibits minimal effects on the p53 response. However, in human cells miR-34 has a complex effect on p53 response. In HCT116 cells, overexpression of miR-34a, but not miR-34b or miR-34c, activates p53 (Navarro & Lieberman, 2015). miR-34a inhibits SIRT1 and SIRT1 deacetylates p53 that blocks the function of p53. A positive feedback loop between miR-34 and p53 exists to regulate cell proliferation and apoptosis (Yamakuchi & Lowenstein, 2009). Besides miR-34, several other miRNAs are also regulated by p53 (summarized in Fig. 3).

Genome-wide miRNA expression assays demonstrated that miR-182, miR-34, miR-222, miR-203, and miR-423 are regulated by p53 (Rihani et al., 2015). In response to DNA damage, p53 promotes the processing of

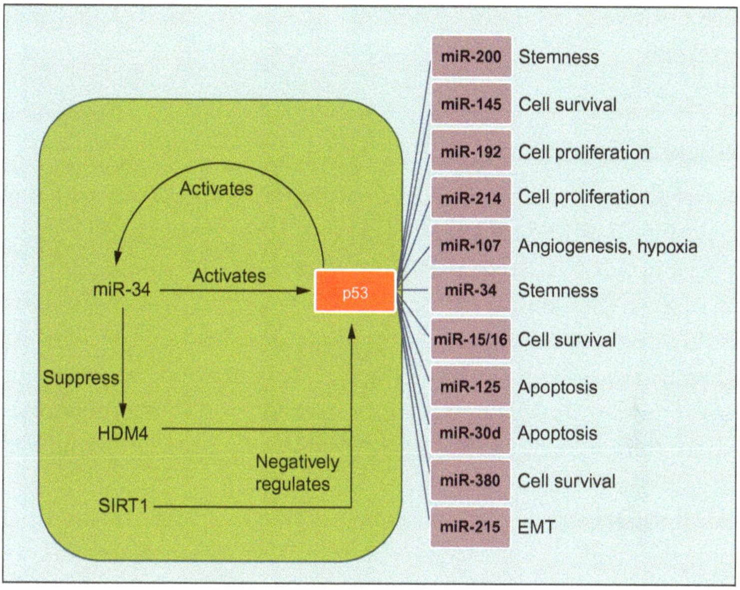

Fig. 3 Regulation of miRNAs by p53: p53 regulates a number of miRNAs involved in cellular proliferation, apoptosis, survival, stemness, angiogenesis, and EMT (epithelial–mesenchymal transition). Also, a feedback loop exists between miR-34–SIRT1–p53.

miR-16, miR-143, and miR-145, which are Drosha mediated, resulting in growth suppressive functions. These miRNAs regulate key mediators of the cell cycle (K-RAS by miR-143, CDKs by miR-16/miR-145). In doxorubicin-treated cells, p53 interacts with Drosha and enhances its function. The mutated version of p53, i.e., p53 R175H or R273H, abrogates this characteristic property of p53, which explains the divergent roles of p53 in cancer contexts. Conversely, miRNAs like miR-34 and miR-504 regulate p53 expression and function (Hu et al., 2010).

Other miRNAs that are regulated by p53 are miR-200a/200b/429 (Tamura et al., 2015) and miR-200c (Schubert & Brabletz, 2011). p53 can also regulate c-MYC (Sachdeva et al., 2009), which can modify a series of miRNAs as described earlier.

6. INDIRECT REGULATION

The expression of miRNAs can be regulated indirectly by a number of factors. In many of these cases the precise mechanism(s) of regulation have not been studied in detail. Some of these regulatory molecules are discussed later.

6.1 Nuclear Receptors

Nuclear receptors are ligand-dependent transcription factors that regulate the transcription of downstream target genes through binding to their promoter regions (Glass & Saijo, 2010). These are a class of protein molecules in cells that sense different kinds of hormones, i.e., steroid or thyroid hormones. In response to different signaling cascades they regulate gene expression, thereby controlling almost all the pathways in organisms. A number of proteins and other molecules belong in the nuclear receptor category, e.g., vitamin D, steroids, thyroid hormones, retinoic acid, and orphan receptors (Olefsky, 2001; Pawlak, Lefebvre, & Staels, 2012).

Additional receptors like androgen receptor, estrogen receptor (ER), progesterone, or glucocorticoid receptors belong to this group. miR-14 targets a nuclear receptor (EcR), which is a steroid hormone receptor (Varghese & Cohen, 2007). The steroid hormone ecdysone and its receptors in insects are critical regulators of development and metamorphosis. Ecdysone receptors and ecdysone signaling contribute as part of a positive regulatory loop that increases the ecdysone receptor to increase the ecdysone pulses. Ecdysone signaling downregulates miRNAs, including miR-14. Also, miR-14 targets EcR thereby resulting in an autoregulatory loop.

Nuclear receptor DAF-12 is a target of the let-7 family (Hammell, Karp, & Ambros, 2009). The DAF-12 and let-7 loop consist of DAF-12 and let-7. Activated DAF-12 represses the let-7 (Hammell et al., 2009) miRNA family. This feedback loop also plays an important role in development. Farnesoid X receptor (FXR) inhibits the miR-34a gene resulting in SIRT1 upregulation (Lee & Kemper, 2010).

A feedback loop exists between the ER and miR-375 (de Souza Rocha Simonini et al., 2010). ER has also been shown to bind to the miR-221/miR-222 promoter and recruit SMRT and NCoR to downregulate miR-221/miR-222 (Di Leva et al., 2010). Estrogen also induces c-MYC, which impacts on miRNA gene regulation (Castellano et al., 2009). Additionally, estrogen regulates other miRNAs including let-7, miR-21, and miR-98 (Bhat-Nakshatri et al., 2009).

Exportin-5 which controls translocation of precursors is upregulated by progesterone and estradiols (Yang & Wang, 2011). Also, DICER is induced by progestins and estradiol (Nothnick, Healy, & Hong, 2010). AGO1 and

Table 1 Regulation of miRNAs by Diverse Receptors

Receptor	miRNA Regulated	References
Androgen receptor	miR-221	Ambs et al. (2008)
Progesterone receptor	miR-200 and miR-320	Williams, Renthal, Condon, Gerard, and Mendelson (2012) and Xia et al. (2010)
HNF4α	miR-122	Li et al. (2011)
PPAR	miR-29 and Let-7	Ye, Hu, Lin, Zhang, and Perez-Polo (2010) and Shah et al. (2007)
LXR	miR-613	Ou et al. (2011)
PXR	miR-31, miR-193, and miR-296	Mahony, Corcoran, Feingold, and Benos (2007) and Saumet et al. (2009)
RXR	miR-23 and miR-210	Saumet et al. (2009)
TLX	miR-9	Zhao, Sun, Li, and Shi (2009)
SHP	miR-433, miR-127, and miR-206	Song and Wang (2008)
Glucocorticoid	miR-223 and miR-15, 16 cluster	Rainer et al. (2009)

AGO2 are downregulated in ER-positive cancers (Cheng, Fu, Alves, & Gerstein, 2009), and estradiol induces AGO2, which is a component of the RISC complex. Several other receptors regulating miRs are listed in Table 1.

6.2 Epigenetic Control

Many of the miRNA genes are regulated and transcribed by a unique RNA polymerase (Schanen & Li, 2011). Therefore, aberrant expression of oncogenes and tumor suppressors regulating RNA polymerase can alter the normal pattern of miRNA. Methylation of promoter DNA of tumor suppressors is common in many cancers, and accordingly several miRNA loci including miR-203, miR-137, and miR-34 are differentially regulated in different cancers by methylation (Davis-Dusenbery & Hata, 2010). Conversely, let-7a-3 is hypomethylated in lung cancer leading to its increased expression (Brueckner et al., 2007).

The promoters of miRNAs are regulated by modification of histones (Handy, Castro, & Loscalzo, 2011). HDAC (histone deacetylase) inhibitors were reported to upregulate a set of miRNAs, e.g., miR-1 (Nasser et al., 2008). However, this regulation is both cancer-specific and cell-type specific. In SKBR3, a HER2-positive breast cancer cell line, many miRNAs are significantly downregulated following treatment with HDACI (Scott, Mattie, Berger, Benz, & Benz, 2006). H3K56 acetylation also regulates several miRNAs in stem cells (Scott et al., 2006).

6.3 TGF-β Signaling

Smads or the signal transducers of TGF-β pathway modulate the expression of miRNAs that in turn regulate several downstream pathways. Chromatin analyses reveal that the promoter elements of several miRNAs are regulated by Smads (Blahna & Hata, 2012). Expression assays also reveal that several miRNAs are regulated by R-Smads. TGF-β induces miR-217 and miR-216 via Smad-binding elements (Kato et al., 2009). miR-29, which is also repressed by Smads, promotes renal fibrosis (Qin et al., 2011).

Recent studies imply that Smad4 knockdown results in dysregulated expression of a number of miRNAs. Smad4 binds to the miR-155 promoter, which is critical for miR-155 induction, resulting in the subsequent downregulation of RhoA. In addition, other miRNAs like miR-192, miR-143, and miR-145 are degraded by Smads (Blahna & Hata, 2012). TGF-β on the other hand increases miR-21 through an indirect regulation, which is dependent on p68 expression (Blahna & Hata, 2012). This documents a critical role of Smads in miRNA processing and biogenesis. These pathways are an important component in attempting to fully comprehend the complexities of gene regulation.

6.4 MDA-7/IL-24

mda-7/IL-24 (melanoma differentiation associated gene-7/interleukin-24) was shown to have anticancer activity in vitro and in vivo in preclinical animal models (Emdad et al., 2009; Fisher, 2005; Jiang et al., 1996; Menezes et al., 2014, 2015; Su et al., 1998). Also a phase I/II clinical trial in advanced cancers involving intratumoral injection of an adenovirus expressing *mda-7/IL-24* (Ad.*mda-7*; INGN 241); demonstrated the therapeutic role of *mda-7/IL-24* without harming normal cells or tissues (Cunningham et al., 2005; Fisher et al., 2003, 2007). *mda-7/IL-24*, a member of the IL-10 cytokine gene family (Dash et al., 2010; Pestka et al., 2004), was cloned using subtraction hybridization and induction of terminal cancer cell differentiation in melanoma cells (Jiang & Fisher, 1993; Jiang, Lin, Su, Goldstein, &

Fisher, 1995). Forced overexpression of *mda-7/IL-24* blocks angiogenesis and sensitizes cancer cells to radiation or chemotherapy (Nishikawa, Ramesh, Munshi, Chada, & Meyn, 2004; Ramesh et al., 2003). Extensive research has established the bystander antitumor activity of *mda-7* (Sauane et al., 2008; Su et al., 2005). MDA-7/IL-24 protein interacts with the chaperone protein BiP/GRP78 to initiate an unfolded protein response specifically in cancer cells leading to apoptosis (Gupta et al., 2006).

A recent study by Pradhan et al. has uncovered a novel and unique role of *mda-7/IL-24* on the regulation patterns of miRNAs (summarized in Fig. 4) (Pradhan et al., 2017). Overexpression of *mda-7/IL-24* in an aggressive triple-negative breast cancer cell line, i.e., MDA-MB-231, identified several differentially regulated miRNAs including miR-200c, let-7c, and miR-320. miR-200c, an important regulator of tumor metastasis and epithelial–mesenchymal transition, was found to be downregulated by *mda-7/IL-24* (Pradhan et al., 2017). miR-17, reported to contribute to regulation of the G1–S transition of cell cycle, was upregulated in these treated cells. miR-185, a tumor suppressor, was upregulated in *mda-7/IL-24*

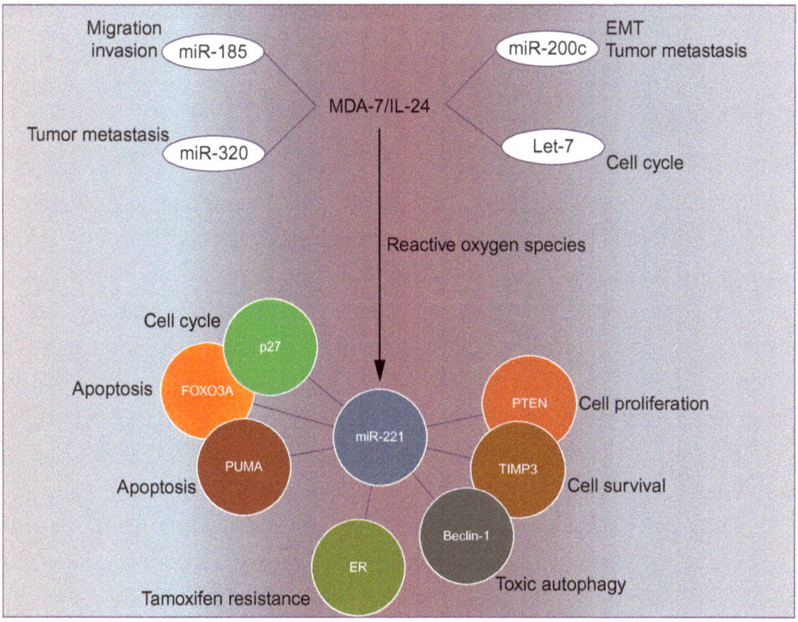

Fig. 4 *mda-7/IL-24* regulates a subset of miRNAs, which are involved in diverse downstream signaling cascades. miR-221 is a key miRNA regulated by *mda-7/IL-24*, involved in a number of oncogenic signaling events as depicted in this figure.

overexpressed cells. The let-7 miRNA families were also found to be deregulated in *mda-7/IL-24*-infected cells. Let-7c was upregulated, while let-7e showed no change with respect to control in *mda-7/IL-24* over-expressing cells (unpublished observation). Additionally, miR-221, a key miRNA in many cancers, was shown to be downregulated by *mda-7/IL-24*. p27 (le Sage et al., 2007), a key modulator of cell cycle, and PUMA (Zhang et al., 2010), a proapoptotic gene, are targeted by miR-221. miR-221 also blocks the action of tamoxifen by targeting the ER in ER-positive breast cancers (Zhao et al., 2008). Other tumor suppressor targets of miR-221 include PTEN (Garofalo et al., 2009) and p57 (Fornari et al., 2008). By regulating these target genes miR-221 plays a critical role in cancer progression. On the other hand, promyelocytic leukemia zinc finger transcription factor functions as a transcriptional repressor of miR-221 (Felicetti et al., 2008).

Pradhan et al. further showed that *mda-7/IL-24* downregulates miR-221, which in turn upregulates beclin-1 and promotes toxic autophagy that switches to apoptosis in breast cancer cells (Pradhan et al., 2017). This study supports a direct regulatory role of *mda-7/IL-24* in miRNA regulation in cancer cells and establishes a novel role of the *mda-7/IL-24*–miR-221–beclin-1 axis in cancer cell-specific death (Pradhan et al., 2017) (summarized in Fig. 5).

6.5 hPNPase

Human PNPase (polynucleotide phosphorylase) is a type 1 IFN-inducible exoribonuclease (Leszczyniecka et al., 2002; Leszczyniecka, Su, Kang, Sarkar, & Fisher, 2003). It degrades specific messenger RNAs and selected noncoding RNAs (Das et al., 2010; Sarkar & Fisher, 2006a, 2006b; Sarkar et al., 2003; Sarkar, Park, & Fisher, 2006). Additionally, previous studies demonstrated that hPNPase can exert an effect on miRNA expression (Das et al., 2011, 2010; Sokhi, Bacolod, et al., 2013). In human melanoma cells infected with an adenovirus expressing hPNPase a number of mRNAs and miRNAs are downregulated (Sokhi, Bacolod, et al., 2013; Sokhi et al., 2014; Sokhi, Das, et al., 2013). One of these miRNAs is miR-221 (Das et al., 2011), which is a key regulator of the cell cycle targeting p27 protein (le Sage et al., 2007). Studies have also shown that exoribonuclease encoding small RNA degrading nuclease (sdn) degrades miRNAs in *Arabidopsis* (Ramachandran & Chen, 2008). Additionally, PNPase also regulates miR-222 and miR-106b without any alteration of miR-25, miR-184, or let-7a, which are in the same cluster of miR-106b (Das et al., 2010). This

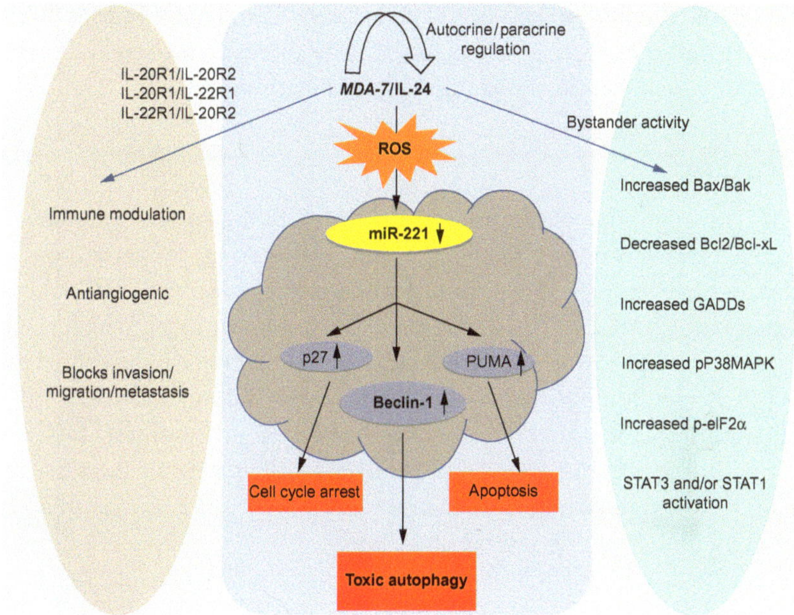

Fig. 5 Schematic representation of *mda-7/IL-24* regulating miR-221: *mda-7/IL-24* downregulates miR-221, which in turn upregulates several downstream genes to block cellular proliferation, induce apoptotic signals, and toxic autophagy. *mda-7/IL-24* also regulates several pathways involved in cancer as shown. *Adapted from Pradhan, A. K., Talukdar, S., Bhoopathi, P., Shen, X. N., Emdad, L., Das, S. K., et al. (2017). mda-7/IL-24 mediates cancer cell-specific death via regulation of miR-221 and the beclin-1 axis.* Cancer Research, *77, 949–959.*

confirms the selective regulation of miRNAs by hPNPase. IFN-β induces expression of hPNPase (Leszczyniecka et al., 2003; Sarkar et al., 2006), and also IFN-β induces downregulation of miR-221 and subsequent upregulation of p27 (Das et al., 2010). Overexpression of miR-221 confers resistance to growth arrest by IFN-β (Das et al., 2010). These studies explore a novel regulation of mature miRNAs by hPNPase through selective protein degradation (Das et al., 2011, 2010). miR-193, miR-29b, and miR-320 are also downregulated by hPNPase (Das et al., 2010).

6.6 hnRNPA1

The heterogeneous nuclear ribonucleoprotein A1 or hnRNPA1 is required for processing of some of the miRNAs including miR-7-92 cluster, miR-18a, miR-101, and let-7a (Slezak-Prochazka, Durmus, Kroesen, & van den

Berg, 2010). Binding of hnRNPA1 structurally rearranges the miRNA to produce a better cleavage site for Drosha. This increases the level of mature miRNAs, which can deregulate downstream pathways.

6.7 ADARs

ADAR or adenosine deaminase acting on RNA catalyzes adenine-to-inosine conversion in dsRNA. The inosine can then be converted to guanosine, which can pair to cytosine. Several primary and pre-miRNAs are substrates of ADARs. miR-142 is edited at the primary level, which results in its suppressed processing by Drosha (Yang et al., 2006). This unedited pri-miRNA is rapidly degraded by the ribonuclease Tudor-SN. miR-151 is another miR that is edited by ADAR and prevents processing by DICER, resulting in the accumulation of pre-miRNAs. This kind of modification can alter the expression level of mature miRNAs, which can change the level of a number of downstream mRNAs.

6.8 Stress

Generation of miRNAs, as described earlier, depends on a number of factors including the processing enzymes DICER. Hence, the level of DICER can regulate the downstream pathways. Wiesen and Tomasi have shown that DICER protein expression is downregulated by multiple stresses like phorbol esters or reactive oxygen species (Wiesen & Tomasi, 2009). Also, dsRNA and interferons (type 1) repress DICER, while IFN-γ induces DICER. This regulation is mostly posttranscriptional.

6.9 Diet and Natural Products

Diet and lifestyle significantly contribute to a number of diseases including cancer, which leads to the introduction and application of natural chemopreventive agents to inhibit different disease processes. Several recent studies indicate the regulatory role of nutritional factors in miRNA regulation (Gavrilas et al., 2016; Ross & Davis, 2011). Vitamin A which is involved in vision and other pathways is reported to regulate a number of miRNAs. Particularly, all-trans retinoic acid or ATRA, the active form of vitamin A, functions as a tumor suppressor by regulating a number of miRNAs (Parasramka et al., 2012). Like vitamin A, vitamin B, vitamin D, and vitamin E are also reported to regulate a number of miRNAs. In addition, fatty acids regulate miRNAs ultimately regulating diverse downstream pathways.

Several other nutrients like curcumin, resveratrol, catechins, and isoflavones are also reported to regulate miRNAs (Parasramka et al., 2012).

7. CONCLUSIONS AND FUTURE PERSPECTIVES

Cancer is a complex multigenic disorder involving multiple abnormal genetic and epigenetic changes, which lead to transformation of normal cells into cancer cells (Friedmann, 1992; Welch & Fisher, 2016). This process is often a continuum that can progress culminating in cancer cells with the capacity to invade into the bloodstream and migrate and colonize other sites in the body, the process of metastasis (Hurst & Welch, 2011; Welch & Fisher, 2016). It is ultimately this metastatic process that is responsible for an estimated 90% of patient deaths. Changes promoting the cancerous state may occur in the coding region or in the noncoding region of genes. miRNAs regulate a vast array of physiologically relevant biological processes depending upon the genes they target and in cancer they function as both oncogenes and tumor suppressors. miRNAs are involved in the initiation and progression of cancer to metastasis and different subsets of miRNAs are often associated with specific cancers (Davis-Dusenbery & Hata, 2010; Ma, 2016). Previous and current research on miRNAs have significantly expanded our understanding of the biogenesis of miRNAs, and their functional regulation of target genes, however, more research is necessary to fully comprehend the functional nuances of these interesting molecules. miRNAs or noncoding RNAs can be useful as gene therapies for cancer. However, the applications of miRNAs as therapeutics (either tumor targeted or nontargeted) present major challenges in the clinic. Immune responses and off-target effects leading to cytotoxicity must be considered and can compromise the efficacy of miRNAs as therapeutics. Further research into the regulatory pathways, both upstream and downstream, will help identify new potential targets and facilitate movement of these unique gene regulatory molecules from bench to bedside. Once appropriate targets are identified, medicinal chemistry approaches can also be used to develop pharmacological inhibitors to block cancer-promoting targets, including those regulating the processes of invasion and metastasis, thereby suppressing or preventing cancer cell pathogenesis (Kegelman et al., 2017).

ACKNOWLEDGMENTS

Support for our research endeavors was provided in part by NIH, NCI SPORE P50 CA058236, R01 CA097318, R01 CA134721, and P01 CA104177, the Samuel Waxman

Cancer Research Foundation (SWCRF) and the National Foundation for Cancer Research (NFCR).

Conflict of Interest: P.B.F. is cofounder of serves as a consultant to and has ownership interest in Cancer Targeting Systems (CTS), Inc. Virginia Commonwealth University, Johns Hopkins University, and Columbia University have ownership interest in CTS, Inc. Competing financial interests had no role in study designs, data collections and interpretations, or the decision to submit this work for publication. The other authors declare that no competing interests exist.

REFERENCES

Ambs, S., Prueitt, R. L., Yi, M., Hudson, R. S., Howe, T. M., Petrocca, F., et al. (2008). Genomic profiling of microRNA and messenger RNA reveals deregulated microRNA expression in prostate cancer. *Cancer Research, 68*, 6162–6170.

Ardekani, A. M., & Naeini, M. M. (2010). The role of MicroRNAs in human diseases. *Avicenna Journal of Medical Biotechnology, 2*, 161–179.

Bagga, S., Bracht, J., Hunter, S., Massirer, K., Holtz, J., Eachus, R., et al. (2005). Regulation by let-7 and lin-4 miRNAs results in target mRNA degradation. *Cell, 122*, 553–563.

Barlev, N. A., Sayan, B. S., Candi, E., & Okorokov, A. L. (2010). The microRNA and p53 families join forces against cancer. *Cell Death & Differentiation, 17*, 373–375.

Beckerman, R., & Prives, C. (2010). Transcriptional regulation by p53. *Cold Spring Harbor Perspectives in Biology, 2*. a000935.

Bhat-Nakshatri, P., Wang, G., Collins, N. R., Thomson, M. J., Geistlinger, T. R., Carroll, J. S., et al. (2009). Estradiol-regulated microRNAs control estradiol response in breast cancer cells. *Nucleic Acids Research, 37*, 4850–4861.

Blahna, M. T., & Hata, A. (2012). Smad-mediated regulation of microRNA biosynthesis. *FEBS Letters, 586*, 1906–1912.

Brueckner, B., Stresemann, C., Kuner, R., Mund, C., Musch, T., Meister, M., et al. (2007). The human let-7a-3 locus contains an epigenetically regulated microRNA gene with oncogenic function. *Cancer Research, 67*, 1419–1423.

Cammaerts, S., Strazisar, M., De Rijk, P., & Del Favero, J. (2015). Genetic variants in microRNA genes: Impact on microRNA expression, function, and disease. *Frontiers in Genetics, 6*, 186.

Castellano, L., Giamas, G., Jacob, J., Coombes, R. C., Lucchesi, W., Thiruchelvam, P., et al. (2009). The estrogen receptor-alpha-induced microRNA signature regulates itself and its transcriptional response. *Proceedings of the National Academy of Sciences of the United States of America, 106*, 15732–15737.

Cheng, C., Fu, X., Alves, P., & Gerstein, M. (2009). mRNA expression profiles show differential regulatory effects of microRNAs between estrogen receptor-positive and estrogen receptor-negative breast cancer. *Genome Biology, 10*, R90.

Cunningham, C. C., Chada, S., Merritt, J. A., Tong, A., Senzer, N., Zhang, Y., et al. (2005). Clinical and local biological effects of an intratumoral injection of mda-7 (IL24; INGN 241) in patients with advanced carcinoma: A phase I study. *Molecular Therapy, 11*, 149–159.

Das, S. K., Bhutia, S. K., Sokhi, U. K., Dash, R., Azab, B., Sarkar, D., et al. (2011). Human polynucleotide phosphorylase (hPNPase(old-35)): An evolutionary conserved gene with an expanding repertoire of RNA degradation functions. *Oncogene, 30*, 1733–1743.

Das, S. K., Sokhi, U. K., Bhutia, S. K., Azab, B., Su, Z. Z., Sarkar, D., et al. (2010). Human polynucleotide phosphorylase selectively and preferentially degrades microRNA-221 in human melanoma cells. *Proceedings of the National Academy of Sciences of the United States of America, 107*, 11948–11953.

Dash, R., Bhutia, S. K., Azab, B., Su, Z. Z., Quinn, B. A., Kegelmen, T. P., et al. (2010). Mda-7/IL-24: A unique member of the IL-10 gene family promoting cancer-targeted toxicity. *Cytokine & Growth Factor Reviews*, *21*, 381–391.

Davis, B. N., Hilyard, A. C., Nguyen, P. H., Lagna, G., & Hata, A. (2010). Smad proteins bind a conserved RNA sequence to promote microRNA maturation by Drosha. *Molecular Cell*, *39*, 373–384.

Davis-Dusenbery, B. N., & Hata, A. (2010). MicroRNA in cancer: The involvement of aberrant MicroRNA biogenesis regulatory pathways. *Genes & Cancer*, *1*, 1100–1114.

Davis-Dusenbery, B. N., & Hata, A. (2011). Smad-mediated miRNA processing: A critical role for a conserved RNA sequence. *RNA Biology*, *8*, 71–76.

de Souza Rocha Simonini, P., Breiling, A., Gupta, N., Malekpour, M., Youns, M., Omranipour, R., et al. (2010). Epigenetically deregulated microRNA-375 is involved in a positive feedback loop with estrogen receptor alpha in breast cancer cells. *Cancer Research*, *70*, 9175–9184.

Di Leva, G., Gasparini, P., Piovan, C., Ngankeu, A., Garofalo, M., Taccioli, C., et al. (2010). MicroRNA cluster 221-222 and estrogen receptor alpha interactions in breast cancer. *Journal of the National Cancer Institute*, *102*, 706–721.

Drake, M., Furuta, T., Suen, K. M., Gonzalez, G., Liu, B., Kalia, A., et al. (2014). A requirement for ERK-dependent Dicer phosphorylation in coordinating oocyte-to-embryo transition in *C. elegans*. *Developmental Cell*, *31*, 614–628.

Emdad, L., Lebedeva, I. V., Su, Z. Z., Gupta, P., Sauane, M., Dash, R., et al. (2009). Historical perspective and recent insights into our understanding of the molecular and biochemical basis of the antitumor properties of mda-7/IL-24. *Cancer Biology & Therapy*, *8*, 391–400.

Felicetti, F., Errico, M. C., Bottero, L., Segnalini, P., Stoppacciaro, A., Biffoni, M., et al. (2008). The promyelocytic leukemia zinc finger-microRNA-221/-222 pathway controls melanoma progression through multiple oncogenic mechanisms. *Cancer Research*, *68*, 2745–2754.

Filippov, V., Solovyev, V., Filippova, M., & Gill, S. S. (2000). A novel type of RNase III family proteins in eukaryotes. *Gene*, *245*, 213–221.

Fisher, P. B. (2005). Is *mda-7/IL-24* a magic bullet for cancer? *Cancer Research*, *65*, 10128–10138.

Fisher, P. B., Gopalkrishnan, R. V., Chada, S., Ramesh, R., Grimm, E. A., Rosenfeld, M. R., et al. (2003). mda-7/IL-24, a novel cancer selective apoptosis inducing cytokine gene: From the laboratory into the clinic. *Cancer Biology & Therapy*, *2*, S23–S37.

Fisher, P. B., Sarkar, D., Lebedeva, I. V., Emdad, L., Gupta, P., Sauane, M., et al. (2007). Melanoma differentiation associated gene-7/interleukin-24 (*mda*-7/IL-24): Novel gene therapeutic for metastatic melanoma. *Toxicology and Applied Pharmacology*, *224*, 300–307.

Fornari, F., Gramantieri, L., Ferracin, M., Veronese, A., Sabbioni, S., Calin, G. A., et al. (2008). MiR-221 controls CDKN1C/p57 and CDKN1B/p27 expression in human hepatocellular carcinoma. *Oncogene*, *27*, 5651–5661.

Friedman, R. C., Farh, K. K., Burge, C. B., & Bartel, D. P. (2009). Most mammalian mRNAs are conserved targets of microRNAs. *Genome Research*, *19*, 92–105.

Friedmann, T. (1992). Approaches to gene therapy of complex multigenic diseases: Cancer as a model and implications for cardiovascular disease and diabetes. *Annals of Medicine*, *24*, 411–417.

Garofalo, M., Di Leva, G., Romano, G., Nuovo, G., Suh, S. S., Ngankeu, A., et al. (2009). miR-221&222 regulate TRAIL resistance and enhance tumorigenicity through PTEN and TIMP3 downregulation. *Cancer Cell*, *16*, 498–509.

Gavrilas, L. I., Ionescu, C., Tudoran, O., Lisencu, C., Balacescu, O., & Miere, D. (2016). The role of bioactive dietary components in modulating miRNA expression in colorectal cancer. *Nutrients*, *8*, 590.

Glass, C. K., & Saijo, K. (2010). Nuclear receptor transrepression pathways that regulate inflammation in macrophages and T cells. *Nature Reviews Immunology*, *10*, 365–376.

Gupta, P., Walter, M. R., Su, Z. Z., Lebedeva, I. V., Emdad, L., Randolph, A., et al. (2006). BiP/GRP78 is an intracellular target for MDA-7/IL-24 induction of cancer-specific apoptosis. *Cancer Research*, *66*, 8182–8191.

Hammell, C. M., Karp, X., & Ambros, V. (2009). A feedback circuit involving let-7-family miRNAs and DAF-12 integrates environmental signals and developmental timing in Caenorhabditis elegans. *Proceedings of the National Academy of Sciences of the United States of America*, *106*, 18668–18673.

Han, J., Lee, Y., Yeom, K. H., Kim, Y. K., Jin, H., & Kim, V. N. (2004). The Drosha-DGCR8 complex in primary microRNA processing. *Genes & Development*, *18*, 3016–3027.

Han, J., Pedersen, J. S., Kwon, S. C., Belair, C. D., Kim, Y. K., Yeom, K. H., et al. (2009). Posttranscriptional crossregulation between Drosha and DGCR8. *Cell*, *136*, 75–84.

Handy, D. E., Castro, R., & Loscalzo, J. (2011). Epigenetic modifications: Basic mechanisms and role in cardiovascular disease. *Circulation*, *123*, 2145–2156.

He, L., & Hannon, G. J. (2004). MicroRNAs: Small RNAs with a big role in gene regulation. *Nature Reviews Genetics*, *5*, 522–531.

He, L., He, X., Lowe, S. W., & Hannon, G. J. (2007). MicroRNAs join the p53 network—Another piece in the tumour-suppression puzzle. *Nature Reviews Cancer*, *7*, 819–822.

Herbert, K. M., Pimienta, G., DeGregorio, S. J., Alexandrov, A., & Steitz, J. A. (2013). Phosphorylation of DGCR8 increases its intracellular stability and induces a progrowth miRNA profile. *Cell Reports*, *5*, 1070–1081.

Hill, C. G., Jabbari, N., Matyunina, L. V., & McDonald, J. F. (2014). Functional and evolutionary significance of human microRNA seed region mutations. *PLoS One*, *9*, e115241.

Hollern, D. P., Honeysett, J., Cardiff, R. D., & Andrechek, E. R. (2014). The E2F transcription factors regulate tumor development and metastasis in a mouse model of metastatic breast cancer. *Molecular & Cellular Biology*, *34*, 3229–3243.

Hollstein, M., Sidransky, D., Vogelstein, B., & Harris, C. C. (1991). p53 mutations in human cancers. *Science*, *253*, 49–53.

Holohan, K. N., Lahiri, D. K., Schneider, B. P., Foroud, T., & Saykin, A. J. (2012). Functional microRNAs in Alzheimer's disease and cancer: Differential regulation of common mechanisms and pathways. *Frontiers in Genetics*, *3*, 323.

Horman, S. R., Janas, M. M., Litterst, C., Wang, B., MacRae, I. J., Sever, M. J., et al. (2013). Akt-mediated phosphorylation of argonaute 2 downregulates cleavage and upregulates translational repression of microRNA targets. *Molecular Cell*, *50*, 356–367.

Hsu, P. W., Huang, H. D., Hsu, S. D., Lin, L. Z., Tsou, A. P., Tseng, C. P., et al. (2006). miRNAMap: Genomic maps of microRNA genes and their target genes in mammalian genomes. *Nucleic Acids Research*, *34*, D135–D139.

Hu, W., Chan, C. S., Wu, R., Zhang, C., Sun, Y., Song, J. S., et al. (2010). Negative regulation of tumor suppressor p53 by microRNA miR-504. *Molecular Cell*, *38*, 689–699.

Huang, Y., Jeong, J. S., Okamura, J., Sook-Kim, M., Zhu, H., Guerrero-Preston, R., et al. (2012). Global tumor protein p53/p63 interactome: Making a case for cisplatin chemoresistance. *Cell Cycle*, *11*, 2367–2379.

Huang, P. J., Lin, W. C., Chen, S. C., Lin, Y. H., Sun, C. H., Lyu, P. C., et al. (2012). Identification of putative miRNAs from the deep-branching unicellular flagellates. *Genomics*, *99*, 101–107.

Hurst, D. R., & Welch, D. R. (2011). Metastasis suppressor genes at the interface between the environment and tumor cell growth. *International Review of Cell and Molecular Biology*, *286*, 107–180.

Ji, M., Rao, E., Ramachandrareddy, H., Shen, Y., Jiang, C., Chen, J., et al. (2011). The miR-17-92 microRNA cluster is regulated by multiple mechanisms in B-cell malignancies. *The American Journal of Pathology, 179*, 1645–1656.

Jiang, H., & Fisher, P. B. (1993). Use of a sensitive and efficient subtraction hybridization protocol for the identification of genes differentially regulated during the induction of differentiation in human melanoma cells. *Molecular and Cellular Differentiation, 1*, 285–299.

Jiang, H., Lin, J. J., Su, Z. Z., Goldstein, N. I., & Fisher, P. B. (1995). Subtraction hybridization identifies a novel melanoma differentiation associated gene, mda-7, modulated during human melanoma differentiation, growth and progression. *Oncogene, 11*, 2477–2486.

Jiang, H., Su, Z. Z., Lin, J. J., Goldstein, N. I., Young, C. S. H., & Fisher, P. B. (1996). The melanoma differentiation associated gene *mda*-7 suppresses cancer cell growth. *Proceedings of the National Academy of Sciences of the United States of America, 93*, 9160–9165.

Johnson, D. G. (2000). The paradox of E2F1: Oncogene and tumor suppressor gene. *Molecular Carcinogenesis, 27*, 151–157.

Kato, M., Putta, S., Wang, M., Yuan, H., Lanting, L., Nair, I., et al. (2009). TGF-beta activates Akt kinase through a microRNA-dependent amplifying circuit targeting PTEN. *Nature Cell Biology, 11*, 881–889.

Kawano, M., Tanaka, K., Itonaga, I., Iwasaki, T., & Tsumura, H. (2015). c-Myc represses tumor-suppressive microRNAs, let-7a, miR-16 and miR-29b, and induces cyclin D2-mediated cell proliferation in Ewing's sarcoma cell line. *PLoS One, 10*, e0138560.

Kegelman, T. P., Wu, B., Das, S. K., Talukdar, S., Beckta, J. M., Hu, B., et al. (2017). Inhibition of radiation-induced glioblastoma invasion by genetic and pharmacological targeting of MDA-9/syntenin. *Proceedings of the National Academy of Sciences of the United States of America, 114*, 370–375.

Kim, H. H., Kuwano, Y., Srikantan, S., Lee, E. K., Martindale, J. L., & Gorospe, M. (2009). HuR recruits let-7/RISC to repress c-Myc expression. *Genes & Development, 23*, 1743–1748.

Kurzynska-Kokorniak, A., Koralewska, N., Pokornowska, M., Urbanowicz, A., Tworak, A., Mickiewicz, A., et al. (2015). The many faces of Dicer: The complexity of the mechanisms regulating Dicer gene expression and enzyme activities. *Nucleic Acids Research, 43*, 4365–4380.

Lai, E. C., Tomancak, P., Williams, R. W., & Rubin, G. M. (2003). Computational identification of Drosophila microRNA genes. *Genome Biology, 4*, R42.

le Sage, C., Nagel, R., Egan, D. A., Schrier, M., Mesman, E., Mangiola, A., et al. (2007). Regulation of the p27(Kip1) tumor suppressor by miR-221 and miR-222 promotes cancer cell proliferation. *The EMBO Journal, 26*, 3699–3708.

Lee, Y., Ahn, C., Han, J., Choi, H., Kim, J., Yim, J., et al. (2003). The nuclear RNase III Drosha initiates microRNA processing. *Nature, 425*, 415–419.

Lee, K. H., Chen, Y. L., Yeh, S. D., Hsiao, M., Lin, J. T., Goan, Y. G., et al. (2009). MicroRNA-330 acts as tumor suppressor and induces apoptosis of prostate cancer cells through E2F1-mediated suppression of Akt phosphorylation. *Oncogene, 28*, 3360–3370.

Lee, J., & Kemper, J. K. (2010). Controlling SIRT1 expression by microRNAs in health and metabolic disease. *Aging, 2*, 527–534.

Leszczyniecka, M., Kang, D. C., Sarkar, D., Su, Z. Z., Holmes, M., Valerie, K., et al. (2002). Identification and cloning of human polynucleotide phosphorylase, hPNPase old-35, in the context of terminal differentiation and cellular senescence. *Proceedings of the National Academy of Sciences of the United States of America, 99*, 16636–16641.

Leszczyniecka, M., Su, Z. Z., Kang, D. C., Sarkar, D., & Fisher, P. B. (2003). Expression regulation and genomic organization of human polynucleotide phosphorylase, hPNPase(old-35), a type I interferon inducible early response gene. *Gene, 316*, 143–156.

Leung, A. K., Vyas, S., Rood, J. E., Bhutkar, A., Sharp, P. A., & Chang, P. (2011). Poly (ADP-ribose) regulates stress responses and microRNA activity in the cytoplasm. *Molecular Cell*, *42*, 489–499.

Lewis, M. A., & Steel, K. P. (2010). MicroRNAs in mouse development and disease. *Seminars in Cell & Developmental Biology*, *21*, 774–780.

Li, Z. Y., Xi, Y., Zhu, W. N., Zeng, C., Zhang, Z. Q., Guo, Z. C., et al. (2011). Positive regulation of hepatic miR-122 expression by HNF4alpha. *Journal of Hepatology*, *55*, 602–611.

Lin, S. L., Miller, J. D., & Ying, S. Y. (2006). Intronic microRNA (miRNA). *Journal of Biomedicine and Biotechnology*, *2006*. 26818.

Luscher, B., & Larsson, L. G. (1999). The basic region/helix-loop-helix/leucine zipper domain of Myc proto-oncoproteins: Function and regulation. *Oncogene*, *18*, 2955–2966.

Ma, L. (2016). MicroRNA and metastasis. *Advances in Cancer Research*, *132*, 165–207.

Macfarlane, L. A., & Murphy, P. R. (2010). MicroRNA: Biogenesis, function and role in cancer. *Current Genomics*, *11*, 537–561.

Mahony, S., Corcoran, D. L., Feingold, E., & Benos, P. V. (2007). Regulatory conservation of protein coding and microRNA genes in vertebrates: Lessons from the opossum genome. *Genome Biology*, *8*, R84.

Mallory, A., & Vaucheret, H. (2010). Form, function, and regulation of ARGONAUTE proteins. *Plant Cell*, *22*, 3879–3889.

Menezes, M. E., Bhatia, S., Bhoopathi, P., Emdad, L., Dasgupta, S., Das, S. K., et al. (2014). MDA-7/IL-24: Multifunctional cancer killing cytokine. *Advances in Experimental Medicine and Biology*, *818*, 127–153.

Menezes, M. E., Shen, X. N., Das, S. K., Emdad, L., Guo, C., Yuan, F., et al. (2015). MDA-7/IL-24 functions as a tumor suppressor gene in vivo in transgenic mouse models of breast cancer. *Oncotarget*, *6*, 36928–36942.

Miller, D. M., Thomas, S. D., Islam, A., Muench, D., & Sedoris, K. (2012). c-Myc and cancer metabolism. *Clinical Cancer Research*, *18*, 5546–5553.

Miska, E. A., Alvarez-Saavedra, E., Abbott, A. L., Lau, N. C., Hellman, A. B., McGonagle, S. M., et al. (2007). Most Caenorhabditis elegans microRNAs are individually not essential for development or viability. *PLoS Genetics*, *3*, e215.

Mogilyansky, E., & Rigoutsos, I. (2013). The miR-17/92 cluster: A comprehensive update on its genomics, genetics, functions and increasingly important and numerous roles in health and disease. *Cell Death & Differentiation*, *20*, 1603–1614.

Mott, J. L. (2009). MicroRNAs involved in tumor suppressor and oncogene pathways: Implications for hepatobiliary neoplasia. *Hepatology*, *50*, 630–637.

Muller, P. A., & Vousden, K. H. (2014). Mutant p53 in cancer: New functions and therapeutic opportunities. *Cancer Cell*, *25*, 304–317.

Nasser, M. W., Datta, J., Nuovo, G., Kutay, H., Motiwala, T., Majumder, S., et al. (2008). Down-regulation of micro-RNA-1 (miR-1) in lung cancer. Suppression of tumorigenic property of lung cancer cells and their sensitization to doxorubicin-induced apoptosis by miR-1. *The Journal of Biological Chemistry*, *283*, 33394–33405.

Navarro, F., & Lieberman, J. (2015). miR-34 and p53: New insights into a complex functional relationship. *PLoS One*, *10*, e0132767.

Nevins, J. R. (2001). The Rb/E2F pathway and cancer. *Human Molecular Genetics*, *10*, 699–703.

Nishikawa, T., Ramesh, R., Munshi, A., Chada, S., & Meyn, R. E. (2004). Adenovirus-mediated mda-7 (IL24) gene therapy suppresses angiogenesis and sensitizes NSCLC xenograft tumors to radiation. *Molecular Therapy*, *9*, 818–828.

Nothnick, W. B., Healy, C., & Hong, X. (2010). Steroidal regulation of uterine miRNAs is associated with modulation of the miRNA biogenesis components exportin-5 and Dicer1. *Endocrine*, *37*, 265–273.

O'Donnell, K. A., Wentzel, E. A., Zeller, K. I., Dang, C. V., & Mendell, J. T. (2005). c-Myc-regulated microRNAs modulate E2F1 expression. *Nature, 435*, 839–843.

Olefsky, J. M. (2001). Nuclear receptor minireview series. *The Journal of Biological Chemistry, 276*, 36863–36864.

Olivier, M., Hollstein, M., & Hainaut, P. (2010). TP53 mutations in human cancers: Origins, consequences, and clinical use. *Cold Spring Harbor Perspectives in Biology, 2*. a001008.

Ou, Z., Wada, T., Gramignoli, R., Li, S., Strom, S. C., Huang, M., et al. (2011). MicroRNA hsa-miR-613 targets the human LXRalpha gene and mediates a feedback loop of LXRalpha autoregulation. *Molecular Endocrinology, 25*, 584–596.

Parasramka, M. A., Dashwood, W. M., Wang, R., Abdelli, A., Bailey, G. S., Williams, D. E., et al. (2012). MicroRNA profiling of carcinogen-induced rat colon tumors and the influence of dietary spinach. *Molecular Nutrition & Food Research, 56*, 1259–1269.

Pawlak, M., Lefebvre, P., & Staels, B. (2012). General molecular biology and architecture of nuclear receptors. *Current Topics in Medicinal Chemistry, 12*, 486–504.

Pestka, S., Krause, C. D., Sarkar, D., Walter, M. R., Shi, Y., & Fisher, P. B. (2004). Interleukin-10 and related cytokines and receptors. *Annual Reviews of Immunology, 22*, 929–979.

Polager, S., & Ginsberg, D. (2009). p53 and E2f: Partners in life and death. *Nature Reviews Cancer, 9*, 738–748.

Pradhan, A. K., Talukdar, S., Bhoopathi, P., Shen, X. N., Emdad, L., Das, S. K., et al. (2017). mda-7/IL-24 mediates cancer cell-specific death via regulation of miR-221 and the beclin-1 axis. *Cancer Research, 77*, 949–959.

Qin, W., Chung, A. C., Huang, X. R., Meng, X. M., Hui, D. S., Yu, C. M., et al. (2011). TGF-beta/Smad3 signaling promotes renal fibrosis by inhibiting miR-29. *Journal of the American Society of Nephrology: JASN, 22*, 1462–1474.

Raghuwanshi, S., Karnati, H. K., Sarvothaman, S., Gutti, U., Saladi, R. G., Tummala, P. R., et al. (2015). microRNAs: Key players in hematopoiesis. *Advances in Experimental Medicine and Biology, 887*, 171–211.

Rainer, J., Ploner, C., Jesacher, S., Ploner, A., Eduardoff, M., Mansha, M., et al. (2009). Glucocorticoid-regulated microRNAs and mirtrons in acute lymphoblastic leukemia. *Leukemia, 23*, 746–752.

Ramachandran, V., & Chen, X. (2008). Degradation of microRNAs by a family of exoribonucleases in Arabidopsis. *Science, 321*, 1490–1492.

Ramesh, R., Mhashilkar, A. M., Tanaka, F., Saito, Y., Branch, C. D., Sieger, K., et al. (2003). Melanoma differentiation-associated gene 7/interleukin (IL)-24 is a novel ligand that regulates angiogenesis via the IL-22 receptor. *Cancer Research, 63*, 5105–5113.

Reddy, K. B. (2015). MicroRNA (miRNA) in cancer. *Cancer Cell International, 15*, 38.

Rihani, A., Van Goethem, A., Ongenaert, M., De Brouwer, S., Volders, P. J., Agarwal, S., et al. (2015). Genome wide expression profiling of p53 regulated miRNAs in neuroblastoma. *Scientific Reports, 5*, 9027.

Rinn, J. L., & Chang, H. Y. (2012). Genome regulation by long noncoding RNAs. *Annual Review of Biochemistry, 81*, 145–166.

Rivlin, N., Brosh, R., Oren, M., & Rotter, V. (2011). Mutations in the p53 tumor suppressor gene: Important milestones at the various steps of tumorigenesis. *Genes & Cancer, 2*, 466–474.

Ross, S. A., & Davis, C. D. (2011). MicroRNA, nutrition, and cancer prevention. *Advances in Nutrition, 2*, 472–485.

Sachdeva, M., Zhu, S., Wu, F., Wu, H., Walia, V., Kumar, S., et al. (2009). p53 represses c-Myc through induction of the tumor suppressor miR-145. *Proceedings of the National Academy of Sciences of the United States of America, 106*, 3207–3212.

Sarkar, D., & Fisher, P. B. (2006a). Human polynucleotide phosphorylase (hPNPase old-35): An RNA degradation enzyme with pleiotrophic biological effects. *Cell Cycle, 5*, 1080–1084.

Sarkar, D., & Fisher, P. B. (2006b). Polynucleotide phosphorylase: An evolutionary conserved gene with an expanding repertoire of functions. *Pharmacology & Therapeutics, 112*, 243–263.

Sarkar, D., Leszczyniecka, M., Kang, D. C., Lebedeva, I. V., Valerie, K., Dhar, S., et al. (2003). Downregulation of Myc as a potential target for growth arrest induced by human polynucleotide phosphorylase (*hPNPase*$^{old-35}$) in human melanoma cells. *The Journal of Biological Chemistry, 278*, 24542–24551.

Sarkar, D., Park, E. S., & Fisher, P. B. (2006). Defining the mechanism by which IFN-beta dowregulates c-Myc expression in human melanoma cells: Pivotal role for human polynucleotide phosphorylase (hPNPaseold-35). *Cell Death & Differentiation, 13*, 1541–1553.

Sauane, M., Su, Z. Z., Gupta, P., Lebedeva, I. V., Dent, P., Sarkar, D., et al. (2008). Autocrine regulation of mda-7/IL-24 mediates cancer-specific apoptosis. *Proceedings of the National Academy of Sciences of the United States of America, 105*, 9763–9768.

Saumet, A., Vetter, G., Bouttier, M., Portales-Casamar, E., Wasserman, W. W., Maurin, T., et al. (2009). Transcriptional repression of microRNA genes by PML-RARA increases expression of key cancer proteins in acute promyelocytic leukemia. *Blood, 113*, 412–421.

Sayed, D., & Abdellatif, M. (2011). MicroRNAs in development and disease. *Physiological Reviews, 91*, 827–887.

Schanen, B. C., & Li, X. (2011). Transcriptional regulation of mammalian miRNA genes. *Genomics, 97*, 1–6.

Schubert, J., & Brabletz, T. (2011). p53 spreads out further: Suppression of EMT and stemness by activating miR-200c expression. *Cell Research, 21*, 705–707.

Scott, G. K., Mattie, M. D., Berger, C. E., Benz, S. C., & Benz, C. C. (2006). Rapid alteration of microRNA levels by histone deacetylase inhibition. *Cancer Research, 66*, 1277–1281.

Shabalina, S. A., & Koonin, E. V. (2008). Origins and evolution of eukaryotic RNA interference. *Trends in Ecology & Evolution, 23*, 578–587.

Shah, Y. M., Morimura, K., Yang, Q., Tanabe, T., Takagi, M., & Gonzalez, F. J. (2007). Peroxisome proliferator-activated receptor alpha regulates a microRNA-mediated signaling cascade responsible for hepatocellular proliferation. *Molecular and Cellular Biology, 27*, 4238–4247.

Shen, J., Xia, W., Khotskaya, Y. B., Huo, L., Nakanishi, K., Lim, S. O., et al. (2013). EGFR modulates microRNA maturation in response to hypoxia through phosphorylation of AGO2. *Nature, 497*, 383–387.

Skalsky, R. L., & Cullen, B. R. (2010). Viruses, microRNAs, and host interactions. *Annual Reviews of Microbiology, 64*, 123–141.

Slezak-Prochazka, I., Durmus, S., Kroesen, B. J., & van den Berg, A. (2010). MicroRNAs, macrocontrol: Regulation of miRNA processing. *RNA, 16*, 1087–1095.

Smibert, P., Yang, J. S., Azzam, G., Liu, J. L., & Lai, E. C. (2013). Homeostatic control of Argonaute stability by microRNA availability. *Nature Structural & Molecular Biology, 20*, 789–795.

Sokhi, U. K., Bacolod, M. D., Dasgupta, S., Emdad, L., Das, S. K., Dumur, C. I., et al. (2013). Identification of genes potentially regulated by human polynucleotide phosphorylase (hPNPase old-35) using melanoma as a model. *PLoS One, 8*, e76284.

Sokhi, U. K., Bacolod, M. D., Emdad, L., Das, S. K., Dumur, C. I., Miles, M. F., et al. (2014). Analysis of global changes in gene expression induced by human polynucleotide phosphorylase (hPNPase(old-35)). *Journal of Cellular Physiology, 229*, 1952–1962.

Sokhi, U. K., Das, S. K., Dasgupta, S., Emdad, L., Shiang, R., DeSalle, R., et al. (2013). Human polynucleotide phosphorylase (hPNPaseold-35): Should I eat you or not—That is the question? *Advances in Cancer Research, 119*, 161–190.

Song, G., & Wang, L. (2008). Transcriptional mechanism for the paired miR-433 and miR-127 genes by nuclear receptors SHP and ERRgamma. *Nucleic Acids Research, 36*, 5727–5735.

Su, X., Chakravarti, D., Cho, M. S., Liu, L., Gi, Y. J., Lin, Y. L., et al. (2010). TAp63 suppresses metastasis through coordinate regulation of Dicer and miRNAs. *Nature, 467*, 986–990.
Su, Z. Z., Emdad, L., Sauane, M., Lebedeva, I. V., Sarkar, D., Gupta, P., et al. (2005). Unique aspects of mda-7/IL-24 antitumor bystander activity: Establishing a role for secretion of MDA-7/IL-24 protein by normal cells. *Oncogene, 24*, 7552–7566.
Su, Z. Z., Madireddi, M. T., Lin, J. J., Young, C. S. H., Kitada, S., Reed, J. C., et al. (1998). The cancer growth suppressor gene *mda*-7 selectively induces apoptosis in human breast cancer cells and inhibits tumor growth in nude mice. *Proceedings of the National Academy of Sciences of the United States of America, 95*, 14400–14405.
Suzuki, H. I., Yamagata, K., Sugimoto, K., Iwamoto, T., Kato, S., & Miyazono, K. (2009). Modulation of microRNA processing by p53. *Nature, 460*, 529–533.
Svoronos, A. A., Engelman, D. M., & Slack, F. J. (2016). OncomiR or tumor suppressor? The duplicity of MicroRNAs in cancer. *Cancer Research, 76*, 3666–3670.
Tamura, M., Sasaki, Y., Kobashi, K., Takeda, K., Nakagaki, T., Idogawa, M., et al. (2015). CRKL oncogene is downregulated by p53 through miR-200s. *Cancer Science, 106*, 1033–1040.
Tan, W., Li, Y., Lim, S. G., & Tan, T. M. (2014). miR-106b-25/miR-17-92 clusters: Polycistrons with oncogenic roles in hepatocellular carcinoma. *World Journal of Gastroenterology, 20*, 5962–5972.
Tang, X., Li, M., Tucker, L., & Ramratnam, B. (2011). Glycogen synthase kinase 3 beta (GSK3beta) phosphorylates the RNAase III enzyme Drosha at S300 and S302. *PLoS One, 6*, e20391.
Tang, X., Wen, S., Zheng, D., Tucker, L., Cao, L., Pantazatos, D., et al. (2013). Acetylation of Drosha on the N-terminus inhibits its degradation by ubiquitination. *PLoS One, 8*, e72503.
Tu, C. C., Zhong, Y., Nguyen, L., Tsai, A., Sridevi, P., Tarn, W. Y., et al. (2015). The kinase ABL phosphorylates the microprocessor subunit DGCR8 to stimulate primary microRNA processing in response to DNA damage. *Science Signaling, 8*, ra64.
Varghese, J., & Cohen, S. M. (2007). MicroRNA miR-14 acts to modulate a positive autoregulatory loop controlling steroid hormone signaling in Drosophila. *Genes & Development, 21*, 2277–2282.
Wada, T., Kikuchi, J., & Furukawa, Y. (2012). Histone deacetylase 1 enhances microRNA processing via deacetylation of DGCR8. *EMBO Reports, 13*, 142–149.
Wang, B., Hsu, S. H., Wang, X., Kutay, H., Bid, H. K., Yu, J., et al. (2014). Reciprocal regulation of microRNA-122 and c-Myc in hepatocellular cancer: Role of E2F1 and transcription factor dimerization partner 2. *Hepatology, 59*, 555–566.
Wang, X., Zhao, X., Gao, P., & Wu, M. (2013). c-Myc modulates microRNA processing via the transcriptional regulation of Drosha. *Scientific Reports, 3*, 1942.
Wei, K., Wu, L., Chen, Y., Lin, Y., Wang, Y., Liu, X., et al. (2013). Argonaute protein as a linker to command center of physiological processes. *Chinese Journal of Cancer Research = Chung-Kuo Yen Cheng Yen Chiu, 25*, 430–441.
Welch, D. R., & Fisher, P. B. (2016). Preface. *Advances in Cancer Research, 132*, xi–xiv.
Wiesen, J. L., & Tomasi, T. B. (2009). Dicer is regulated by cellular stresses and interferons. *Molecular Immunology, 46*, 1222–1228.
Williams, K. C., Renthal, N. E., Condon, J. C., Gerard, R. D., & Mendelson, C. R. (2012). MicroRNA-200a serves a key role in the decline of progesterone receptor function leading to term and preterm labor. *Proceedings of the National Academy of Sciences of the United States of America, 109*, 7529–7534.
Witkos, T. M., Koscianska, E., & Krzyzosiak, W. J. (2011). Practical aspects of microRNA target prediction. *Current Molecular Medicine, 11*, 93–109.
Wu, H., Xu, H., Miraglia, L. J., & Crooke, S. T. (2000). Human RNase III is a 160-kDa protein involved in preribosomal RNA processing. *The Journal of Biological Chemistry, 275*, 36957–36965.

Xia, H. F., Jin, X. H., Song, P. P., Cui, Y., Liu, C. M., & Ma, X. (2010). Temporal and spatial regulation of miR-320 in the uterus during embryo implantation in the rat. *International Journal of Molecular Sciences*, *11*, 719–730.

Yamakuchi, M., & Lowenstein, C. J. (2009). MiR-34, SIRT1 and p53: The feedback loop. *Cell Cycle*, *8*, 712–715.

Yang, W., Chendrimada, T. P., Wang, Q., Higuchi, M., Seeburg, P. H., Shiekhattar, R., et al. (2006). Modulation of microRNA processing and expression through RNA editing by ADAR deaminases. *Nature Structural & Molecular Biology*, *13*, 13–21.

Yang, Q., Li, W., She, H., Dou, J., Duong, D. M., Du, Y., et al. (2015). Stress induces p38 MAPK-mediated phosphorylation and inhibition of Drosha-dependent cell survival. *Molecular Cell*, *57*, 721–734.

Yang, Z., & Wang, L. (2011). Regulation of microRNA expression and function by nuclear receptor signaling. *Cell & Bioscience*, *1*, 31.

Ye, Y., Hu, Z., Lin, Y., Zhang, C., & Perez-Polo, J. R. (2010). Downregulation of microRNA-29 by antisense inhibitors and a PPAR-gamma agonist protects against myocardial ischaemia-reperfusion injury. *Cardiovascular Research*, *87*, 535–544.

Zeng, Y., Sankala, H., Zhang, X., & Graves, P. R. (2008). Phosphorylation of Argonaute 2 at serine-387 facilitates its localization to processing bodies. *Biochemical Journal*, *413*, 429–436.

Zhang, X., Chen, X., Lin, J., Lwin, T., Wright, G., Moscinski, L. C., et al. (2012). Myc represses miR-15a/miR-16-1 expression through recruitment of HDAC3 in mantle cell and other non-Hodgkin B-cell lymphomas. *Oncogene*, *31*, 3002–3008.

Zhang, B., Pan, X., Cobb, G. P., & Anderson, T. A. (2007). MicroRNAs as oncogenes and tumor suppressors. *Developmental Biology*, *302*, 1–12.

Zhang, B., Wang, Q., & Pan, X. (2007). MicroRNAs and their regulatory roles in animals and plants. *Journal of Cellular Physiology*, *210*, 279–289.

Zhang, C. Z., Zhang, J. X., Zhang, A. L., Shi, Z. D., Han, L., Jia, Z. F., et al. (2010). MiR-221 and miR-222 target PUMA to induce cell survival in glioblastoma. *Molecular Cancer*, *9*, 229.

Zhao, J. J., Lin, J., Yang, H., Kong, W., He, L., Ma, X., et al. (2008). MicroRNA-221/222 negatively regulates estrogen receptor alpha and is associated with tamoxifen resistance in breast cancer. *The Journal of Biological Chemistry*, *283*, 31079–31086.

Zhao, C., Sun, G., Li, S., & Shi, Y. (2009). A feedback regulatory loop involving microRNA-9 and nuclear receptor TLX in neural stem cell fate determination. *Nature Structural & Molecular Biology*, *16*, 365–371.

CHAPTER THREE

Animal Models to Study MicroRNA Function

Arpita S. Pal*, Andrea L. Kasinski†,1

*PULSe Graduate Program, Purdue University, West Lafayette, IN, United States
†Purdue Center for Cancer Research, Purdue University, West Lafayette, IN, United States
[1]Corresponding author: e-mail address: akasinski@purdue.edu

Contents

1. Introduction 54
 1.1 MicroRNA Biogenesis, Mechanism of Action and Function 55
 1.2 MiRNA Function and Relevance in Cancer 64
 1.3 Tools That Advanced MiRNA Research 67
2. Generation of Model Organisms and Their Use in MiRNA Functional Studies 68
 2.1 *Caenorhabditis elegans* 68
 2.2 *Drosophila melanogaster* (Fruit Flies) 75
 2.3 *Danio rerio* (Zebrafish) 81
 2.4 *Mus musculus* (Mouse) 87
3. Conclusion 97
References 105

Abstract

The discovery of the microRNAs, *lin-4* and *let-7* as critical mediators of normal development in *Caenorhabditis elegans* and their conservation throughout evolution has spearheaded research toward identifying novel roles of microRNAs in other cellular processes. To accurately elucidate these fundamental functions, especially in the context of an intact organism, various microRNA transgenic models have been generated and evaluated. Transgenic *C. elegans* (worms), *Drosophila melanogaster* (flies), *Danio rerio* (zebrafish), and *Mus musculus* (mouse) have contributed immensely toward uncovering the roles of multiple microRNAs in cellular processes such as proliferation, differentiation, and apoptosis, pathways that are severely altered in human diseases such as cancer. The simple model organisms, *C. elegans*, *D. melanogaster*, and *D. rerio*, do not develop cancers but have proved to be convenient systesm in microRNA research, especially in characterizing the microRNA biogenesis machinery which is often dysregulated during human tumorigenesis. The microRNA-dependent events delineated via these simple in vivo systems have been further verified in vitro, and in more complex models of cancers, such as *M. musculus*. The focus of this review is to provide an overview of the important contributions made in the microRNA field using model organisms. The simple model systems provided the basis for the importance of microRNAs in normal cellular physiology, while the more complex animal systems provided evidence for the role

Advances in Cancer Research, Volume 135
ISSN 0065-230X
http://dx.doi.org/10.1016/bs.acr.2017.06.006

of microRNAs dysregulation in cancers. Highlights include an overview of the various strategies used to generate transgenic organisms and a review of the use of transgenic mice for evaluating preclinical efficacy of microRNA-based cancer therapeutics.

1. INTRODUCTION

The central dogma of molecular biology, laid down by Francis Crick in 1958, stated that the fundamental role of RNA molecules is to transmit the genetic code into proteins (Crick, 1958, 1970). However, with the characterization of the first transfer RNA (tRNA) in 1965 (Holley, Apgar, Everett, et al., 1965), additional RNAs emerged that violated the central dogma. For example, ribosomal RNAs (rRNAs) (Fox & Woese, 1975), small nuclear RNAs (snRNAs) (Lerner & Steitz, 1979), and small nucleolar RNAs (snoRNAs) (Filipowicz & Kiss, 1993) are not translated into protein products like a messenger RNAs (mRNAs) but indirectly influence the process of protein synthesis. Such RNA molecules were collectively termed "noncoding RNAs" (review: Cech & Steitz, 2014). While the novel roles performed by some noncoding RNAs were being assimilated as imperative molecular mechanisms, a study conducted to identify genes in heterochronic signaling incidentally led to the discovery of an additional noncoding RNA, a 22-nucleotide RNA molecule, *lin-4* (Lee, Feinbaum, & Ambros, 1993). The discovery of *lin-4* further defied the central dogma of molecular biology via an unprecedented mechanism and led to the establishment of a new class of small noncoding RNAs called "microRNAs" (Lagos-Quintana, Rauhut, Lendeckel, & Tuschl, 2001; Lau, Lim, Weinstein, & Bartel, 2001; Lee & Ambros, 2001).

Lin-4 was the first microRNA (miRNA) identified, which was determined to be indispensable for the normal development of *Caenorhabditis elegans* (Lee et al., 1993). Functionally, *lin-4* interacts with the 3′-untranslated region (3′-UTR) of the mRNA transcript *lin-14*, resulting in a marked repression of the lin-14 protein (Wightman, Ha, & Ruvkun, 1993). Unfortunately, due to limited knowledge in RNA biology at the time, *lin-4* and its peculiar role were overlooked to be a worm-specific phenomenon. Seven years later, a second *C. elegans* miRNA, *let-7*, was discovered which encouraged further miRNA investigations (Reinhart, Slack, Basson, et al., 2000). It became apparent that *let-7* was not only critical for the development of *C. elegans* but is also evolutionarily conserved in other organisms, including humans (Pasquinelli, Reinhart, Slack, et al., 2000). Currently ~2500 human-encoded miRNAs have been identified, which are listed in a miRNA database, miRBase (http://www.mirbase.org/, Release 21) (Kozomara & Griffiths-Jones, 2014). In addition to their identification, biochemical and molecular studies have determined that the canonical function of miRNAs is to posttranscriptionally regulate a repertoire of protein-coding mRNA transcripts, whereas a few miRNAs perform unanticipated or "noncanonical" functions (reviews: Cipolla, 2014; Hausser & Zavolan, 2014).

Following the identification of these first two miRNAs, *lin-4* and *let-7*, our understanding of miRNAs in normal physiology and diseased states, such as cancer (reviews: Adams, Kasinski, & Slack, 2014; Orellana & Kasinski, 2015), has advanced remarkably. Advancements in the field have been possible due to state-of-the-art technologies such as high-throughput screening and deep sequencing, but majorly due to the development of appropriate in vivo model systems (review: Steinkraus, Toegel, & Fulga, 2016). Therefore, the focus of this review will be on the various model systems that have been instrumental in elucidating the roles of miRNAs in cancers and the technologies that have been extensively applied to generate these animal model systems. Briefly, the utility of in vivo models for evaluating the potential of miRNAs as therapeutic agents or targets for treatment of various cancers will also be touched upon.

1.1 MicroRNA Biogenesis, Mechanism of Action and Function

1.1.1 Biogenesis

1.1.1.1 Expression of MiRNA Genes

The transcription of miRNA genes is regulated by multiple mechanisms eventually dictating the level of expression of a particular miRNA in normal or diseased states (reviews: Kasinski & Slack, 2011; Lin & Gregory, 2015; Rupaimoole & Slack, 2017).

1.1.1.1.1 Regulation Mediated by Availability of Transcription Factors A transcription factor can enhance or repress the expression of a miRNA gene depending on the availability of the particular factor (reviews: Calin & Fabbri, 2010; Sato, Tsuchiya, Meltzer, & Shimizu, 2011). The prominent tumor suppressor p53, which functions as a transcription factor for several genes, also enhances the transcription of miRNA genes. Examples of miRNAs that are directly induced by p53 include *mir-34a* and *b/c* (Fabbri, Bottoni, Shimizu, et al., 2011; Hermeking, 2007; Raver-Shapira,

Marciano, Meiri, et al., 2007; Wei, Wu, Vega, et al., 2006) and two miR-200 subfamilies, *mir-200c/141* and *mir-200a/200b/429* (Chang, Chao, Xia, et al., 2011; Kim, Veronese, Pichiorri, et al., 2011). In the case of *mir-15a~16-1* and mir-107 a p53-indirect effect leads to upregulation of the miRNAs through activation of the host genes *DLEU2* (Fabbri et al., 2011) and *PANK1* (Bohlig, Friedrich, & Engeland, 2011), respectively. Conversely, MYC, a well-studied oncogenic transcription factor, negatively regulates the tumor-suppressive miRNA, *let-7a*-1 (Chang, Yu, Lee, et al., 2008; Wang, Lin, Li, et al., 2011).

1.1.1.1.2 Regulation Mediated by Genomic Location of MiRNA Genes

1.1.1.1.2.1 Location in the Epigenome Transcriptional activation or inactivation of specific miRNA genes is largely influenced by epigenetics. Such epigenetic regulation includes the proximity of the miRNA gene promoter to a CpG island, various histone modifications to the chromatin, and availability of factors that maintain and regulate expression from the epigenome. The expression of *mir-127*, a miRNA located near a CpG island is dependent on the methylation status of the promoter, implying epigenetic control on the expression of miRNAs (Lujambio, Ropero, Ballestar, et al., 2007; Saito, Liang, Egger, et al., 2006). MiRNAs also undergo massive upregulation when the DNA methylatransferases 1 and 3b (DNMT1, DNMT3) are downregulated (Lujambio et al., 2007), lending further support to the role of DNA methylation in regulating miRNA expression (reviews: Calin & Fabbri, 2010; Sato et al., 2011).

1.1.1.1.2.2 Location Relative to Host Genes The origin of a miRNA gene within a specific chromosomal location controls the expression of the miRNA. In the context of other genes, miRNAs genes are either intragenic where they are embedded within a host gene or intergenic if they are located between two genes on a chromosome. Expression of an intragenic miRNA is dependent on the expression of the host gene (reviews: Calin & Fabbri, 2010; Sato et al., 2011). MiR-126 is one such miRNA whose expression is concomitantly controlled by epigenetic regulation of its host gene *EGFL7* (Saito et al., 2009). Intragenic miRNAs are also regulated by canonical mechanisms that influence host gene expression such as transcription factor occupancy at the promoter of the host gene (Bohlig et al., 2011; Fabbri et al., 2011; review Zhang & Zeng, 2010). MiRNAs that are not directly regulated by a host gene are still subject to nearby epigenetic influence. For example, *let-7a-3* and miR-129-1 expression are dependent on a nearby region of the genome that is prone to altered methylation states during the onset of cancer. The proximity to this differentially methylated region severely impacts their expression (Bandres, Agirre, Bitarte, et al., 2009; Brueckner, Stresemann, Kuner, et al., 2007; Sato et al., 2011).

1.1.1.1.2.3 Regulation by MiRNA Copy Number A single mature miRNA can be expressed and processed from multiple loci in the genome. For example, three individual genes encoding human *mir-7* produce an identical mature miRNA product (Li & Carthew, 2005). Conversely, miR-21 is generated from a single genomic locus (Buscaglia & Li, 2011; Hirata, Murai, Yanaihara, et al., 2014). The advantages of miRNAs originating from various loci relative to a miRNA originating from a single locus are discussed in a later section.

1.1.1.1.2.4 Cancer-Associated Genomic Regions Specific regions in the human genome that are prone to amplification or loss upon the onset of cancers are referred to as cancer-associated genomic regions (CAGRs). CAGRs contain amplified or deleted miRNA and/or protein-coding genes. Many of these genetic aberrations are required for tumorigenesis. MiRNAs that are lost are frequently located in either fragile sites of the genome or regions susceptible to loss of heterozygosity (LOH). For example, the *mir-15a∼16-1* cluster located in a fragile region of the genome at 13q14.3 is frequently deleted in chronic lymphocytic leukemia (CLL) patients (Calin, Sevignani, Dumitru, et al., 2004; Fabbri et al., 2011; Lerner, Harada, Loven, et al., 2009). Whereas other miRNAs are commonly amplified in multiple cancers due to their location in fragile regions. For example, the 17q23–25 chromosomal region containing *mir-21* gene is an amplified CAGR (reviews: Calin & Fabbri, 2010; Sato et al., 2011) which results in overexpression of miR-21, a feature commonly observed in multiple cancers (Hirata et al., 2014)

1.1.1.2 Process of Biogenesis

The primary miRNA (pri-miRNA) transcript produced as a result of RNA polymerase II/III-dependent transcription containing a single miRNA or as a cluster of miRNAs produces a monocistronic or polycistronic pri-miRNA transcript, respectively (Lee, Kim, Han, et al., 2004). Pri-miRNA transcripts form stem–loop structures flanked by single-stranded (ss) RNA ends. For RNA polymerase II transcripts, the ends contain a canonical 5′ 7-methylguanosine cap and a polyadenylation signaling at the 3′-end. The size of a typical pri-miRNA can range from a 100 to a few kilobases in length and can originate from either intragenic or intergenic miRNA genes (Bartel, 2004; Lee et al., 2004) (Fig. 1).

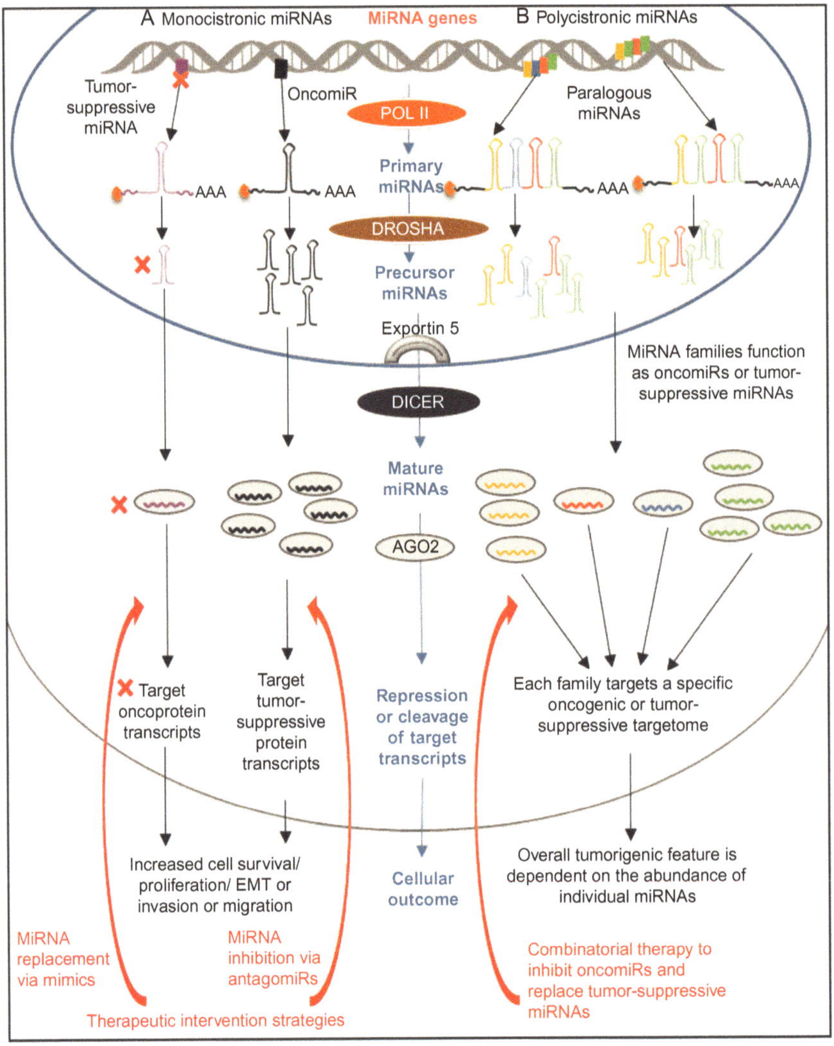

Fig. 1 Overview of oncomiRs or tumor-suppressive miRNAs encoded as monocistronic or polycistronic genes, their involvement in tumorigenesis, and potential use as miRNA-based cancer therapeutics. (A) A monocistronic miRNA gene encodes a transcript containing a single primary miRNA. In cancers, one mechanism to alter the abundance of a mature miRNAs is through changes in transcription of the primary miRNA, where the expression of a tumor-suppressive miRNA is downregulated, while that of an oncomiR is enhanced. A tumor-suppressive miRNA typically targets transcripts encoding oncogenic proteins; therefore, miRNA replacement therapies using tumor-suppressive miRNA mimics are currently being tested. OncomiRs on the other hand target tumor suppressor protein transcripts, and hence, their inhibition via antagomiRs is also a potential miRNA-based therapeutic strategy. (B) Transcription of a polycistronic miRNA

Processing of most pri-miRNAs begins with the association of the RNaseIII enzyme DROSHA and its cofactor Di George Syndrome Critical Region 8 (DGCR8) forming the microprocessor complex (Gregory, Yan, Amuthan, et al., 2004; Han, Lee, Yeom, et al., 2006). The microprocessor complex recognizes the ssRNA regions of the pri-miRNA sequence flanking the stem–loop and cleaves the ends. The resultant ~60–80 nucleotide long hairpin structure is referred to as a precursor miRNA (pre--miRNA) (Gregory et al., 2004; Han et al., 2006). The pre-miRNA is translocated into the cytoplasm via Exportin-5 where another RNase III enzyme, DICER1, performs additional processing (Park, Heo, Tian, et al., 2011; Yi, Qin, Macara, & Cullen, 2003). DICER1 cleaves the pre-miRNA to generate a ~22-nucleotide duplex molecule containing the guide and the passenger miRNA strands. Following cleavage, the DICER1-miRNA duplex associates with transactivation-responsive RNA-binding protein (TRBP) that mediates a stable transfer of the miRNA duplex into an Argonaute protein (AGO) (Chendrimada, Gregory, Kumaraswamy, et al., 2005; Kawamata, Yoda, & Tomari, 2011). Selective incorporation of the miRNA duplex into either AGO1, AGO2, AGO3, or AGO4 is dictated by the presence of bulges or mismatches in nucleotides 9–12 of the duplex (Cenik & Zamore, 2011; Zhang, Niu, Carbonell, et al., 2014). Incorporation of a miRNA duplex in AGO2, an AGO protein with endonuclease activity, results in selective cleavage of the passenger strand. The ssRNA guide strand is retained, and with AGO forms the mature miRNA-induced silencing complex (miRISC) (Burroughs, Ando, de Hoon, et al., 2011; Liu, Carmell, Rivas, et al., 2004). On the other hand, the endonuclease activity-deficient AGOs, AGO1, AGO3, and AGO4, generate a functional miRISC by binding to the guide strand and separating the passenger strand based on thermodynamic instability. The released passenger strand is shunted for further degradation (Dueck, Ziegler,

gene or a miRNA cluster results in a primary miRNA transcript containing multiple miRNAs. The duplication of a cluster, and expression of a more or less intact cluster from multiple genomic loci, generates paralogous miRNAs. The resultant miRNAs from paralogues can be predominantly tumor-suppressive or oncogenic; however, their function is often largely context dependent—i.e., tissue specific, temporally regulated, etc. The potential therapeutic strategy targeting miRNAs expressed from clusters depends on the abundance of the individual tumor-suppressive or oncogenic miRNAs. Combinatorial miRNA therapeutics is a potential strategy currently being evaluated to combat tumorigenesis, where the altered ratio between oncogenic/tumor-suppressive miRNAs drives cancer development.

Eichner, Berezikov, & Meister, 2012; Biogenesis reviews: Bartel, 2004; Lin & Gregory, 2015; Rupaimoole & Slack, 2017; Argonautes reviews: Cenik & Zamore, 2011; Nakanishi, 2016).

1.1.2 Mechanisms of Action and Functions of MiRNAs

1.1.2.1 Incorporation Into miRISC and Targeting

The well-established role of functionally active miRISC is to negatively regulate transcription of target protein-coding transcripts. The canonical mechanism by which miRISC performs its function depends on the extent of complementary binding between the 5′-end "seed region" of the miRNA, the 3′-UTR of the target mRNA transcript, and the enzymatic activity of the AGO protein (Bartel, 2009; Cenik & Zamore, 2011; Cipolla, 2014; Lee, Ajay, Yook, et al., 2009). Perfect complementarity between the seed sequence, nucleotides 2–7 of the guide miRNA strand, and the target results in either degradation or translational repression of the target. The fate of the target transcript is dictated by whether the incorporated AGO displays catalytic activity or not, and whether additional complementarity occurs between the target and the miRNA. If a catalytically active AGO is incorporated into miRISC and the binding between the miRNA and the target is complementary between nucleotides 9–12 of the miRNA, then target cleavage will occur (Doench, Petersen, & Sharp, 2003; Liu et al., 2004). Translational inhibition in the absence of target degradation occurs when the miRNA binds to its target via partial complementarity (reviews: Bartel, 2009; Cenik & Zamore, 2011; Cipolla, 2014) or if an endonuclease-deficient AGO is included in miRISC. The partial complementarity between the miRNA and its target is highly conserved across species, providing the basis for a combinatorial interactome. A combinatorial interactome is the mechanism by which a single miRNA regulates multiple targets, thus simultaneously exerting its regulatory effects on various signaling pathways. For example, the very well-studied miRNA miR-21 simultaneously targets transcripts of proteins that regulate cell division and apoptosis, such as phosphatase and tensin homolog (PTEN) (Zhang et al., 2010), and programmed cell death 4 (PDCD4) (Lu, Liu, Stribinskis, et al., 2008) to drive the process of tumorigenesis (review: Buscaglia & Li, 2011). Partial complementarity between the miRNA and the target also facilitates targeting of a single transcript mRNA by multiple miRNAs resulting in enhanced repression of the target. This resulting moderate-to-severe downregulation of target transcripts via a miRISC is the canonical mode of action of miRNAs (reviews: Bartel, 2004, 2009; Ha & Kim, 2014).

1.1.2.2 The Role of Family Members in Expression and Targeting

The mechanism of action of miRNAs originating from a single locus or a single mature miRNA originating from multiple loci remains largely unchanged. In these instances the same cohort of target mRNA transcripts is repressed (Buscaglia & Li, 2011; Kefas, Godlewski, Comeau, et al., 2008; Zhang et al., 2010; Zhao, Tao, Zhou, et al., 2015). However, miRNAs originating from several loci that contain subtle variations in their mature sequences (Bartel, 2004; Roush & Slack, 2008) can exert their repressive functions on a larger repertoire of target transcripts. These miRNA families have acquired an evolutionary advantage relative to miRNAs with a single mature miRNA sequence. In addition to an increased pool of potential targets, the presence of multiple miRNA family members across the genome may allow at least one of the family members to evade transcriptional or epigenomic regulation. Therefore, the presence of multiple genetically distinct miRNA family members may prevent the depletion of an entire pool of a specific mature miRNA during the onset of a diseased state. For example, transcription of the 12 *let-7* miRNA genes produces 9 unique mature miRNA sequences that differ by at most 3 nucleotides (reviews: Lee, Han, Kwon, & Lee, 2016; Roush & Slack, 2008). These minor nucleic acid changes can potentially alter the targeting affinity of the various family members (Fig. 1 depicts miRNA family members). It has also become apparent that the promoter of *let-7a3* resides in a heavily methylated region of the genome in normal cells resulting in low levels of *let-7a3* in a normal cell. This is however not the case for the other *let-7* family members which are highly expressed under normal conditions leading to a stably differentiated state of the cell. Nonetheless, upon the onset of tumorigenesis, the methylation state of cells become severely disrupted, and except for *let-7a3* all the other *let-7* isoforms become repressed (Brueckner et al., 2007; Roush & Slack, 2008; Sato et al., 2011). Thus, the presence of multiple genetic loci encoding miRNA family members and slight variations in sequence between members adds an additional layer of complexity in the regulation of miRNAs in adverse cellular conditions.

Similar to the *let-7* family, another well-studied family of miRNAs is the miR-34 family. The three canonical miR-34 family members include miR-34a that arises from a monocistronic locus, and miR-34b/c, which are expressed from a polycistronic transcript (Hermeking, 2007; Lujambio et al., 2007; Raver-Shapira et al., 2007) (Fig. 1 depicts mono- and polycistronic miRNA genes). The function of miR-34 in normal physiology is well established as an inducer of cellular senescence and cell cycle arrest (Vogt, Munding, Gruner, et al., 2011; review: Hermeking, 2007). Nevertheless, the advantage of multiloci encoding miR-34 family members is that miR-34a and miR-34b/c can be differentially regulated in a tissue-specific context (Balca-Silva, Sousa Neves, Goncalves, et al., 2012; Nadal, Chen, Gallegos, et al., 2013; Tanaka, Toyooka, Soh, et al., 2012).Recent reports suggest that the tissue-specific expression of the miR-34 paralogues miR-449a/b/c adds an additional level of complexity to the control of cancer cell proliferation, invasion, and migration (Bou Kheir, Futoma-Kazmierczak, Jacobsen, et al., 2011; Lize, Herr, Klimke, Bals, & Dobbelstein, 2010; Mao, Zhao, Zhou, et al., 2016). Indeed, it was not until the paralogue *mir-449a/b/c* cluster was deleted in mouse models that the *mir-34a*, *mir-34b/c* double mutant displayed a phenotype (Comazzetto, Di Giacomo, Rasmussen, et al., 2014; Concepcion, Han, Mu, et al., 2012; Song, Walentek, Sponer, et al., 2014).

1.1.2.3 The Role of MiRNA Clusters and Paralogous in Targeting

Analogous to the overlapping role that miRNA family members have on gene expression, some paralogous clusters can also have overlapping roles, while others have gained novel functions. A paralogous miRNA cluster is generated when a cluster undergoes duplication and translocates to another area of the genome (Fig. 1 depicts miRNA paralogues). The resultant paralogue may express miRNAs similar to the parent cluster, located in relatively analogous positions (Hertel, Lindemeyer, Missal, et al., 2006; Khuu, Utheim, & Sehic, 2016; Sun, Gao, Zhou, et al., 2013). One such miRNA cluster, miR-17∼92 has been extensively studied due to its implication in the human developmental syndrome, Feingold disease. Loss of *mir-17∼92* results in severe skeletal abnormalities, and learning and developmental disabilities associated with Feingold disease (Mogilyansky & Rigoutsos, 2013). However, similar developmental defects were not observed following the knock-out of two *mir-17∼92* paralogous clusters—*mir-106b∼25* and *mir-106a∼363* (Ventura, Young, Winslow, et al., 2008). Additionally, the presence of a single wild-type *mir-17∼92* allele was capable of mitigating the deleterious effects of the loss of *mir-17∼92*, despite the absence of its paralogues (Mu, Han, Betel, et al., 2009). Collectively the three paralogous clusters encode a total of 15 miRNAs that can be subclassified into 4 miRNA families that are presumed to target analogous target transcripts.

However, in this case it can be inferred that alterations in certain nucleotides of the paralogues may have ceased their ability to compensate for *mir-17~92* deletion (Khuu et al., 2016; Mu et al., 2009; Ventura et al., 2008). Therefore, in order to dissect the function of each miRNA in a family of miRNAs or within paralogues generation of appropriate model systems are essential.

1.1.2.4 Use of Bioinformatics to Elucidate MiRNA Function

Recently, there has been a surge in the development of bioinformatic tools to precisely predict targets of a miRNA, or predict miRNAs that target a particular transcript (Steinkraus et al., 2016). Several computationally predicated miRNA–target pairs based on complementarity between the miRNA seed sequence and the 3′-UTR of transcripts have been experimentally validated (Bartel, 2004, 2009; Cho, Jang, Jun, et al., 2013; Johnson, Grosshans, Shingara, et al., 2005; Lee et al., 1993; Lewis, Burge, & Bartel, 2005; Roush & Slack, 2008). Thus, implying that the developed algorithms are powerful in predicting relevant targets of novel miRNAs with unknown functions based on the canonical mode of action of miRNAs. However, there is increasing evidence that demonstrates some noncanonical mechanisms of actions of miRNAs (Cipolla, 2014; Ha & Kim, 2014; Seok, Ham, Jang, & Chi, 2016). For example, a passenger strand (also known as miRNA*) may not always be released for degradation once the miRNA:miRNA* duplex is incorporated into AGO. The miRNA* strand may have an equal or a higher potential of becoming incorporated in an active miRISC (Chang, Kao, YH, et al., 2013; Cipolla, 2014). Moreover, certain miRNAs modulate the expression of targets either positively or negatively by physically interacting with the 5′-UTR of transcripts (Helwak, Kudla, Dudnakova, & Tollervey, 2013; Lee et al., 2009), the coding sequences of transcripts (Forman, Legesse-Miller, & Coller, 2008; Helwak et al., 2013), or with gene promoters (Kim, Saetrom, Snove, & Rossi, 2008). Mature miRNAs have also been experimentally validated to interact with other noncoding RNAs such as other miRNAs, long noncoding RNAs (lncRNAs) (Helwak et al., 2013; Yoon, Abdelmohsen, & Gorospe, 2014), or circular RNAs (circRNAs) (Hansen, Jensen, Clausen, et al., 2013; Helwak et al., 2013). The canonical "seed-sequence" binding hypothesis has also been challenged as miRNAs can bind some targets independently of the seed sequence (Loeb, Khan, Canner, et al., 2012; Martin, Wani, Steptoe, et al., 2014). Data from these studies, along with the identification that certain miRNAs are contained in extracellular vesicles such as exosomes, have added an additional level of complexity in the mechanism by which miRNAs function, including nonautonomous mechanisms (Fabbri, 2012).

1.2 MiRNA Function and Relevance in Cancer

MiRNAs are important players in the normal developmental processes of animal species. As such, disruption in the normal physiological levels of certain miRNAs can lead to the development of multiple diseases, including cancers.

Detailed characterization of various miRNAs has revealed many important properties of these powerful posttranscriptional modulators in both normal and diseased states. In the context of cancers, certain miRNAs have been identified as functional "drivers of cancer," whereas others are regarded as mere "passengers" in the tumorigenic process. A few known miRNA drivers of cancer become upregulated, while others are severely downregulated or lost. The miRNAs that promote hallmarks of cancer are referred to as oncogenic miRNAs (oncomiRs). Those that prevent or reduce tumorigenesis are collectively called tumor-suppressive miRNAs (Fig. 1).

1.2.1 OncomiRs

OncomiR coding genes are frequently located in regions of the genome that are aberrantly amplified or are subject to increased expression (Esquela-Kerscher & Slack, 2006). Increased expression of an oncomiR can be attributed to enhanced transcription of the oncomiR gene due to (i) availability of transcription factors, (ii) hypomethylation of its promoter, or (iii) its location in an intra- or intergenic region that is subject to increased expression in cancer via other mechanisms. OncomiRs can also be upregulated due to defects in biogenesis and/or stability of the mature miRNA (Bartel, 2004; Calin & Fabbri, 2010; Ha & Kim, 2014; Lin & Gregory, 2015; Mu et al., 2009). The way by which an oncomiR typically functions is through targeting tumor-suppressive protein-coding transcripts via canonical mechanisms, or through other less understood noncanonical mechanisms.

The first oncomiR to be validated was the miR-17∼92 cluster (oncomiR-1). Overexpression of the cluster led to the development of lymphoproliferative and autoimmune diseases in mice via targeting of BIM, a proapoptotic protein (Xiao, Srinivasan, Calado, et al., 2008). Other targets of miR-17∼92 that support the oncogenic role for this cluster include

PTEN, *E2Fs*, and *MYC*. More detailed analysis of this cluster confirmed that the cell-type and context-specific processing of individual miRNAs from the cluster adds an additional level of complexity to the function of the oncomiR (Khuu et al., 2016; Olive, Jiang, & He, 2010). Cooperatively the individual miRNAs processed from miR-17∼92 functions as an oncomiR. However, miR-92 alone can antagonize an additional cluster member, miR-19, and also negatively regulates the oncogenic effects of c-Myc (Mu et al., 2009; Olive et al., 2010; Olive, Sabio, Bennett, et al., 2013). Because miR-19 alone can recapitulate the oncogenic role of the intact mir-17∼92 cluster (Olive, Bennett, Walker, et al., 2009; Olive et al., 2013), negative regulation by miR-92 suggest that miR-92 may be functioning as a tumor suppressor. The function of the miR-17∼92 miRNA cluster is extremely intriguing and is currently under active investigation. Specifically, molecular roles and tissue-specific effects of individual miRNAs of the miR-17∼92 cluster are being determined in appropriate model systems (Du, Wang, Sliz, & Gregory, 2015; Mu et al., 2009; Olive et al., 2009, 2013; Sandhu, Fassan, Volinia, et al., 2013; Ventura et al., 2008). These positive findings highlight the importance of carefully dissecting individual miRNAs from clusters so as to accurately identify the functions of each of the miRNAs contained within them.

Other miRNAs that have been well established as oncomiRs due to their implication in multiple solid tumors, and hematological malignancies are miR-21 and miR-155 (Babar, Cheng, Booth, et al., 2012; Costinean, Zanesi, Pekarsky, et al., 2006; Medina, Nolde, & Slack, 2010). Independent studies determined that overexpression of individual miRNAs such as miR-21 and miR-155 are sufficient to cause lymphoproliferative diseases. The mechanism by which miR-155 initiates cancer is not well understood; however, in leukemic mouse models, it was determined that miR-155 promotes cancer progression, perhaps through gradual downregulation of its targets, SHIP and C/EBP (Costinean, Sandhu, Pedersen, et al., 2009). In miR-21-dependent mouse models of lung cancer or pre-B-lymphoma, downregulation of the miR-21 targets PTEN and PDCD4 (negative regulators of cell death and cell cycle, respectively) contributed to enhanced proliferation and growth (Buscaglia & Li, 2011; Lu et al., 2008; Zhang et al., 2010).

1.2.2 Tumor-Suppressive MiRNA

About 50% of the miRNAs involved in repressing oncogenic protein-coding genes are located in or are close to fragile regions of the genome that are frequently deleted in cancer. Additional mechanisms elicited by cancer cells to repress tumor-suppressive miRNAs include LOH, hypermethylation of the promoter, or the activation of transcriptional repressors that specifically downregulate the expression of the miRNA gene (Calin & Fabbri, 2010; Calin et al., 2004; Raver-Shapira et al., 2007; Vogt et al., 2011). In the case of most tumor-suppressive miRNA genes, identification of their role in development preceded their characterization as tumor suppressors. The most striking example of such a tumor-suppressive miRNA is *let-7*. *Let-7* was identified as a crucial differentiation factor in *C. elegans* prior to its identified role in tumorigenesis. Indeed, the development of cancer requires a reversal of a well-differentiated state to an undifferentiated state; thus, it is perhaps not surprising that downregulation or loss of *let-7* family members is common in tumorigenesis (Lee, Han, et al., 2016; Roush & Slack, 2008).

High levels of *let-7* expressed from multiple genomic loci are expected in normal fully differentiated cells (Lee, Han, et al., 2016; Roush & Slack, 2008). This results in repression of *let-7* targets which are important oncogenes, such as *KRAS*, *NRAS* (Johnson et al., 2005; Kumar, Erkeland, Pester, et al., 2008), HMGA2 (Kumar et al., 2008; Yu, Yao, Zhu, et al., 2007), *LIN28* (Heo et al., 2008; Newman, Thomson, & Hammond, 2008; Piskounova, Viswanathan, Janas, et al., 2008; Viswanathan, Daley, & Gregory, 2008), and *MYC* (Sampson, Rong, Han, et al., 2007). A candid tumor-suppressive miRNA, such as *let-7*, has multiple loci of origin in order to maintain an appropriate level of the tumor-suppressive miRNA as a defense mechanism against developing cancers (Misso, Di Martino, De Rosa, et al., 2014; Roush & Slack, 2008). However, since most *let-7* isoforms are located in regions of the genome frequently deleted in cancer, *let-7* is severely downregulated in multiple cancers (Calin et al., 2004). One anomaly to this rule is the expression of *let-7a-3* gene. In lung adenocarcinoma, epigenetic regulation of the gene encoding *let-7-a3* results in hypomethylation of the promoter enhancing the accumulation of the *pre-let-7a-3* transcript in lung cancer cells, and subsequently its potential oncogenic effects (Brueckner et al., 2007; Sato et al., 2011).

Additional miRNAs that have been well established as tumor suppressors include miR-15a and miR-16-1, which were among the first miRNAs identified that demonstrated a negative correlation with the development of cancers (Calin, Dumitru, Shimizu, et al., 2002; Calin et al., 2004). *Mir-15a∼16-1* is located in 13q14.3, a region that is homozygously or hemizygously lost in over 50% of CLL cases (Calin et al., 2002). Early reports correlated loss of *mir-15a∼16-1* with an increase in expression of *BCL2*, a prosurvival factor that normally prevents cell death (Bonci, Coppola, Musumeci, et al., 2008; Calin, Cimmino, Fabbri, et al., 2008; Cimmino, Calin, Fabbri, et al., 2005; Tsujimoto, Finger, Yunis, Nowell, & Croce, 1984). In addition to targeting *BCL-2*, targets of miR-15a~16-1 include several cell-regulatory proteins, such as MCL1 (Calin et al., 2008; Pekarsky & Croce, 2015), another BCL2-family member, CCND1 (Bonci et al., 2008), a cell cycle regulator, and WNT3A (Bonci et al., 2008), a protein that induces several tumorigenic features including survival, proliferation, and invasion. Hence, it is speculated that the simultaneous overexpression of these prosurvival oncoproteins, as a result of the loss of *mir-15a~16-1* cluster, may synergistically contribute to the development of cancers (Aqeilan, Calin, & Croce, 2010; Calin et al., 2002). To further evaluate the tumor-suppressive potential of each miRNA in the *miR-15a~16-1* cluster, modeling CLL in more sophisticated model systems is required.

1.3 Tools That Advanced MiRNA Research

Research in the miRNA field exponentially increased following the discovery of the second miRNA in *C. elegans*, *let-7*. *Let-7* was experimentally found to be homologous in a few species such as *Drosophila melanogaster* (fruit fly) and *Danio rerio* (zebrafish) (Pasquinelli et al., 2000; Roush & Slack, 2008). However, additional computational evidence showed that *let-7* was conserved further throughout evolution in all metazoans evaluated, and therefore, the discovery of *let-7* was regarded as one of the most significant breakthroughs in the history of miRNA research (Orellana & Kasinski, 2015; Roush & Slack, 2008). Moreover, bioinformatic analysis corroborated with molecular studies also established that the *let-7* target, *lin-41*, is conserved across species (Lagos-Quintana, Rauhut, Borkhardt, & Tuschl, 2003; Lee & Ambros, 2001; Pasquinelli et al., 2000; Wightman et al., 1993). These studies were fundamental in establishing the proposed global mechanism of action of miRNAs that miRNAs negatively regulate protein-coding gene expression through miRNA:target interactions. The more recent utility of computational target prediction algorithms in miRNA research has further advanced the field in two ways: (i) a miRNA and its targets can be predicted in silico such that a biologically relevant target is putatively identified prior to in depth biochemical and molecular analysis, and (ii) computational methods spread awareness about similarities between various species. This knowledge can be used to identify appropriate, and perhaps simpler organisms that can be used as an adequate model system in miRNA research. In silico analysis combined with biochemical and molecular studies conducted in in vivo models, and in vitro from cells isolated from the in vivo models (Kasinski & Slack, 2013), or from human samples, have remarkably advanced our understanding of miRNAs in development and disease.

While the functional significance of miRNAs in various developmental stages of *C. elegans* was emerging and the correlation between miRNA levels and disease such as cancer was being reported, a spontaneously occurring mouse model of CLL, the New Zealand Black (NZB) mouse was discovered. This discovery incidentally highlighted the importance of mouse models in delineating the role of miRNAs in cancers. It was determined that the endogenous loss of the *mir-15a∼16-1* cluster in this mouse led to the spontaneous development of CLL as the mouse aged (Raveche, Salerno, Scaglione, et al., 2007; Scaglione, Salerno, Balan, et al., 2007). This discovery followed by bioinformatic developments that determined that ∼60% of miRNA loci remain conserved from mouse to humans spearheaded the use of murine models in miRNA research (Griffiths-Jones, Saini, van Dongen, & Enright, 2008; Lagos-Quintana et al., 2003). Indeed, from worms to mice many miRNAs and the components of the miRNA biogenesis machinery are conserved. The benefits of using simple model organisms such as worms, zebrafish, and Drosophila in miRNA-based biochemical studies have indirectly paved the path toward characterization of miRNAs that have relevance in cancer (Figs. 4 and 5). Validation assays and clinically relevant studies become possible due to the increased use of very robust and durable mouse model systems (Figs. 4 and 5). Therefore, in this review, a short background on the historical perspective of using various model systems will be followed by a detailed explanation on the current technology used to generate appropriate model systems. Some of the most successful models that have faithfully aided in uncovering the roles of various tumor-suppressive or oncogenic miRNAs involved in cancer will be discussed. In closing, the pros and cons of the most widely used model systems in miRNA functional studies in cancer will be elaborated upon.

2. GENERATION OF MODEL ORGANISMS AND THEIR USE IN MiRNA FUNCTIONAL STUDIES

2.1 *Caenorhabditis elegans*

Although *C. elegans* do not develop cancer, they have been extensively used as a model organism to identify the functions of molecules and delineate pathways involved in normal cellular processes that are severely impaired in cancer, such as cell proliferation, differentiation, metabolism, and death (Kato & Slack, 2008; Kirienko, Mani, & Fay, 2010; Kyriakakis, Markaki, & Tavernarakis, 2015). The completely sequenced *C. elegans* genome revealed that ~60% of its miRNAs have a human orthologue (Ibáñez-Ventoso, Vora, & Driscoll, 2008). However, since *C. elegans* have a reduced number of miRNA family members studying miRNA function in *C. elegans* excludes redundancy as a hurdle to overcome. Moreover, *C. elegans* are self-fertilizing hermaphrodites that can produce a large number of genetically identical offspring. Additionally, their visually traceable, well-organized transparent body makes *C. elegans* an excellent model system (Kato & Slack, 2008; Kirienko et al., 2010; Kyriakakis et al., 2015). Phenotypic and genetic screens, application of molecular techniques, and development of transgenic *C. elegans* have identified a few critical miRNAs, including their mechanism of action. Studies in *C. elegans* have also been an instrumental in understanding the molecular basis of miRNA biogenesis (Bracht, Hunter, Eachus, Weeks, & Pasquinelli, 2004; Grishok, Pasquinelli, Conte, et al., 2001; Ketting et al., 2001). For example, the *C. elegans* ortholog of DICER, *dcr-1*, involved in RNA-mediated silencing, was identified as a critical component necessary for the processing of mature *let-7* from its precursor molecule (Ketting et al., 2001).

2.1.1 Generation of Transgenic C. elegans for Use in MiRNA Functional Studies

The most common mechanism used to generate transgenic *C. elegans* is transformation via either (i) microinjection or (ii) DNA bombardment. Transformation is widely used to ectopically introduce a transgene or fragment of DNA of interest into the animals to rescue a mutant phenotype or to overexpress or silence a gene. The DNA is typically codelivered with a scorable marker to determine successful transformation. For example, a scorable marker such as a promoter-driven gfp::transgene allows for the selection of GFP-positive worms when the promoter is positively regulated (Evans, 2006).

Microinjection: Microinjection is a precise mechanism of introducing DNAs into the distal end of the worm gonad, which is composed of a syncytium of cells sharing cytoplasmic material. The transgene is usually a plasmid, cosmid, phage, Yeast artificial chromosome (YAC), or PCR product coinjected into the gonad with a scorable marker. Injected DNAs undergo efficient homologous recombination with each other to generate large extrachromosomal arrays. The extrachromosomal arrays contain multiple copies of the transgenic DNA that do not usually integrate into the genome but can become inheritable by a fraction of the F1 generation. However, when integration is essential, random incorporation of the extrachromosomal DNA can be induced using radiation (gamma or UV) or through the use of a single DNA oligonucleotide that facilitates random integration and suppresses array formation (Evans, 2006; Mello, Kramer, Stinchcomb, & Ambros, 1991).

DNA bombardment: DNA bombardment is a specialized technique where the transgene and coinjected transformation plasmid DNA mix is coated onto a gold microparticle and is bombarded into the worm using a gene gun. The advantage of this technique is that it produces a considerable number of nonhomologous integrants posttransformation (Berezikov, Bargmann, & Plasterk, 2004; Evans, 2006; Praitis, Casey, Collar, & Austin, 2001).

Apart from the above-mentioned techniques, additional newer strategies are now being applied toward generating transgenic C. elegans, such as the CRISPR–Cas9 system, discussed in a later section.

2.1.2 C. elegans *as a Model System for Studying the Function of MiRNAs in Cancer*

The first discovered miRNA, *lin-4*, was identified in *C. elegans* via a conventional method of forward genetic mutagenesis screen, with the intent to mutate heterochronic genes resulting in phenotypic developmental defects. *Lin-4 lof* worms reiterate early phases of developmental fates (L1) at later stages of development, resulting in the absence of well-differentiated adult phenotypes such as the adult cuticle, and a developed vulva. Development into adulthood in *lin-4* null mutants was rescued by microinjection of *lin-4* PCR products confirming that *lin-4 lof* was responsible for the mutant heterochronic phenotype. The phenotype of *lin-4 lof* mutants is completely opposite to that of *lin-14 lof* mutants attributed to the fact that the *lin-4* miRNA negatively regulates the *lin-14* mRNA transcript by binding to several complementary sequences in the *lin-14* 3′-UTR. When the *lin-4* complementary sequences were mutated abnormally high lin-14 protein levels led to worms with retarded developmental phenotypes at late developmental stages (Lee et al., 1993). In addition to suppressing *lin-14*, *lin-4* also posttranscriptionally regulates the heterochronic genes, *lin28* and *hbl-1*. Thus, *lin-4* is regarded as a critical switch in nematodes for the development of well-differentiated adult structures (Lin, Johnson, Abraham, et al., 2003; Moss, Lee, & Ambros, 1997). Nevertheless, its homologs remained unidentified in higher organisms for many years, and *lin-4* was presumed to have been lost during the course of evolution. But recent advances in bioinformatics have provided researchers with the tools needed to identify the human homolog for *lin-4*, miR-125. Analogous to the role of *lin-4* in worm development, miR-125 in human cells targets LIN28 resulting in the acquisition of a differentiated state in normal cells (Liang, Wong, Ying, et al., 2010; Wu & Belasco, 2005). In multiple human cancers, the two miR-125 family members, miR-125a/b, are severely underexpressed, specifically in leukemia (Sonoki, Iwanaga, Mitsuya, & Asou, 2005) and melanoma (Zhang et al., 2016), and ovarian (Lee, Kim, & Jeon, 2016), breast (Iorio, Ferracin, Liu, et al., 2005), oral (Henson, Bhattacharjee, O'Dee, Feingold, & Gollin, 2009), and thyroid (Visone, Pallante, Vecchione, et al., 2007) carcinomas. In concert, ectopic miR-125 prevents cellular proliferation and migration in bladder cancer (Zhao, He, Li, et al., 2015), inhibits epithelial–mesenchymal transition (EMT) of triple-negative breast cancer cells (Hong, Pan, Jiang, et al., 2016), and induces radiosensitivity and chemosensitivity in breast cancer and osteosarcoma, respectively (Metheetrairut, Adams, Nallur, Weidhaas, & Slack, 2017; Wang, Yu, Liu, et al., 2016).

Seven years after the identification of *lin-4*, the second miRNA, *let*-7, was subsequently discovered, again through genetic analysis of the heterochronic pathway in *C. elegans* (Lee et al., 1993; Reinhart et al., 2000). Mutants with severely retarded developmental phenotypes were identified and used to map the sequence of *let*-7. *Let*-7 is expressed in the later stages of worm development and regulates the transition of L4 larval stage worms to adulthood. A striking phenotype of *let*-7 *lof* worms is lethality as they fail to transition from larval to adults, at nonpermissive temperatures. Most worms die due to bursting of the vulva. However, microinjecting the worms with PCR fragments containing the *let*-7 sequence rescued the progenies. Further analysis via Northern blotting verified that *let*-7 did not encode a protein but instead encoded a 22-nucleotide RNA molecule in the rescued progenies. The offspring were selectively scored via coinjecting a GFP reporter plasmid, goa-1::GFP. The microinjected worms developed normally, supporting a role for the *let*-7 RNA product in inhibiting the bursting vulva phenotype. Due to the in silico finding that *let*-7 is complementary to the 3′-UTR of lin-41, target validations that *let*-7 regulates lin-41 were performed. A lacZ reporter gene was fused to the 3′-UTR of lin-41 and was coinjected with goa-1::GFP in a *let*-7 wild-type organism. Reduced luciferase levels and subsequent validation studies confirmed that *let*-7 exerts negative posttranscriptional regulation of *lin-41*. Moreover, overexpression of *let-7* was shown to be implicated in premature development of *C. elegans*, therefore, acting as a critical developmental switch in worms (Reinhart et al., 2000).

After the discovery of the first two miRNAs in *C. elegans*, it was speculated that nematodes had invented a novel mechanism to sequentially control their developmental course. This notion was nonetheless challenged by simple bioinformatic analysis conducted to investigate if *let-7* was conserved in other organisms. Sequence analysis in *D. melanogaster* (fruit fly), *D. rerio* (zebrafish), *Mus musculus* (mouse), *Gallus gallus* (chicken), and *Homo sapiens* (humans) confirmed that *let-7* had been preserved throughout evolution (Lagos-Quintana et al., 2003; Pasquinelli et al., 2000). Similarly, the *let-7* target, *lin-41*, was also conserved (Pasquinelli et al., 2000). Moreover, other detectable *let-7* family members, miR-48, miR-84, and miR-241, were also identified as heterochronic miRNAs crucial for the temporal patterning of development in *C. elegans* (Abbott, Alvarez-Saavedra, Miska, et al., 2005). Both bioinformatic and biochemical analysis of *let-7* family members revealed a considerable redundancy in target specificity between miRNA family members, suggesting a complex mechanism by which family members function. Although miRs-48, -84, and -241 lack human homologs, several *let-7* loci are present in humans generating nine mature *let-7* family members; therefore, there is a need to dissect the individual functions or functional redundancy between human-encoded *let-7* family members (reviews: Lee, Han, et al., 2016; Roush & Slack, 2008).

Despite the widespread acceptance of the role of *let-7* in development, not much is known about *let-7* in human embryonic development due to ethical constraints. However, shortly after the finding that miR-15a/*16-1* is a *bona fide* tumor-suppressive miRNA cluster, investigators evaluated the potential of *let-7* as a tumor-suppressive miRNA. *Lof* of *let-7* in seam cells of *C. elegans*, leading to the inability of the seam cells to exit the cell cycle and become terminally differentiated, indicated that *let-7* may play a role in maintaining the balance between cell differentiation and proliferation (Reinhart et al., 2000; Roush & Slack, 2008). In human cancers *let-7* expression is often reduced (Takamizawa, Konishi, Yanagisawa, et al., 2004), confirmed by the finding that multiple *let-7* family members are located in fragile regions of the genome that are often lost in various malignancies (Calin et al., 2004), signifying that *let-7* is a tumor-suppressive miRNA.

To delineate the molecular mechanism of *let-7* in cancers, additional *let-7* targets were computationally predicted in worms which identified *let-60* as a putative target. Reporter plasmids containing the 3′-UTR of *let-60* with or without the putative *let-7* binding sites verified that *let-60* was indeed a *let-7* target. Moreover, *let-7* mutant worms that usually die at nonpermissive temperatures, when fed with silencing RNAs (RNAi) directed to *let-60* survived. This was the first experimental evidence suggesting a novel role for *let-7* in negatively regulating *let-60* via a posttranscriptional mechanism. This breakthrough suggested that additional miRNAs may be dysregulated in cancer, other than the initially discovered mir-15 ~ 16 cluster lost in CLL. Indeed, *let-60* is the human homolog of the RAS protooncogenes, proteins that are amplified and constitutively activated in multiple human cancers (Johnson et al., 2005). Multiple studies have since validated the negative regulation of RAS exerted by *let-7* in various model systems and in cells derived from human patients, which has led to the advancement of *let-7* family members as potential cancer therapeutics to target such oncogenes (Esquela-Kerscher, Trang, Wiggins, et al., 2008; Kasinski, Kelnar, Stahlhut, et al., 2015; Kumar et al., 2008; Trang, Medina, Wiggins, et al., 2010; Trang, Wiggins, Daige, et al., 2011; Kasinski & Slack, 2010; Kim, Kasinski, & Slack, 2011).

With the emergence of the fact that miRNAs exist not merely as key developmental switches in nematodes, but are also crucial for normal cellular behavior in multiple species, including humans, severely dysregulated miRNAs are in the limelight for their roles in cancer initiation, development, and progression. One such miRNA, miR-34, frequently lost in various cancers, has been associated with stress-response in normal cells, and with radio- and chemotherapeutic response in human cancer cells (Hermeking, 2007; Hummel, Hussey, & Haier, 2010; Lacombe & Zenhausern, 2017; Vogt et al., 2011). In vitro studies in wild-type and p53-mutated mouse and human cells determined that miR-34 was capable of suppressing the cell cycle via a p53-mediated pathway (Raver-Shapira et al., 2007). However, the effects of miR-34 modulation in response to radio- and chemotherapies in vivo were unknown.

To determine the biochemical changes imposed by miR-34 and to record miR-34 dependent molecular observations in vivo, a mir-34-promoter::gfp transgene was constructed and microinjected in *C. elegans* (Kato, Paranjape, Muller, et al., 2009). GFP signals from miR-34 transgenic animals were invariably identified in somatic tissues, including the vulval cells. Although, miR-34 expression was detected in the vulva in late larval stages and in the adult worm, miR-34 *lof* did not affect the development of vulva indicating that miR-34 was not a heterochronic miRNA (Kato et al., 2009). Moreover, although miR-34 was upregulated in wild-type transgenic worms following exposure to radiation, miR-34 was not transcriptionally activated by the p53 worm homolog *cep-1* (Kato et al., 2009). This observation in worms was contradictory to that predicted from human cell lines and may be attributed to evolution of higher organisms (Kato et al., 2009; Lacombe & Zenhausern, 2017). Nevertheless, since miR-34 was upregulated in worms following exposure to radiation, the next big question was how does miR-34 affect apoptotic and nonapoptotic pathways postradiation exposure? *C. elegans* have been characterized as an excellent model to study both apoptosis in germline cells, and necrosis (nonapoptotic pathway) in vulva cells (Gartner, Milstein, Ahmed, Hodgkin, & Hengartner, 2000; Weidhaas, Eisenmann, Holub, & Nallur, 2006). Upon exposure of miR-34 mutant worms to radiation it was observed that the germline cells of the worms were unaffected, yet the vulva cells were radiosensitized. Thus, miR-34 is an essential miRNA for induction of apoptosis but is a suppressor of necrosis in nematodes. The data also provided evidence that the single miRNA, miR-34, can independently function as an apoptotic inducer, irrespective of p53 status (Kato et al., 2009). Additionally, *C. elegans* express only one isoform of miR-34, while miR-34 in humans is encoded by four family members miR-34a/b/c and mir-449a (Bou Kheir et al., 2011; Lize et al., 2010; Mao et al., 2016), indicating a possible evolutionary advantage toward enhancing miR-34-mediated tumor-suppressive function. This notion requires further evaluation in an appropriate model system.

Conventionally, transformation has been utilized to generate transgenic worms; however, with new sophisticated techniques, such as transgenesis via the CRISPR–Cas9 system, targeting an endogenous locus has now become the preferred approach. The use of CRISPR–Cas9 system to knock out *let-7* in *C. elegans* validated the role of *let-7* in maintaining vulval integrity via the regulation of lin-41. This study exemplified the importance of ablating a miRNA to holistically understand its effects in normal physiology via reverse genetics (Ecsedi, Rausch, & Grosshans, 2015).

C. elegans have provided a platform for investigators to biochemically delineate the functions of some important miRNAs that are also crucial players in human cancers. However, during the time that experiments were being conducted in *C. elegans*, the use of computational knowledge to elucidate *let-7* homology between species opened up the possibility of using additional simple organisms to identify the intricate functions of miRNAs. Conceptually, gathering functional knowledge for a single conserved miRNA in various species will aid in appropriately documenting the function of the miRNA, ultimately enhancing our understanding of the molecular biology of human diseases, such as cancer.

2.2 *Drosophila melanogaster* (Fruit Flies)

Fruit flies are a simple yet useful model system in cancer research, since cellular alterations in flies leads to the development of a few of the hallmarks of cancers, unlike in *C. elegans*. Although *D. melanogaster* lacks an intact angiogenic system, they have been very useful models to study cell survival and proliferation, cell death and apoptosis, and invasion and metastasis in the presence of normal cells. Additional advantages of using fruit flies as a model organism include (i) the large brood size, one fly can lay ~100 eggs per day for about 20 days, (ii) the series of conspicuous developmental stages that flies go through from embryogenesis through adulthood over the course of 10 days, and (iii) the fact that genotypic changes are easily phenotypically tractable (Venken & Bellen, 2007).

2.2.1 Generation of Transgenic D. melanogaster *to Model Cancer Phenotypes*

Transgenesis procedures in flies rely on the flies' endogenous transposition mechanism, P-element-mediated transposition. The two-vector transposition system includes a helper vector referred to as the P-transposase vector, and the P-element transposon backbone containing the transgene and a scorable marker or a reporter vector (Fig. 2). The transgene can be constitutive or inducible, or tissue-specific depending on the promoter (Fig. 3). Simultaneously, the scorable marker that is expressed may be a gene that is easily detectable as a physical change—for example, white eye color, small wings, GFP wings, or lethality. The two vectors are coinjected into a generation zero (G0) embryo after which the transgene and the P-transposase are randomly incorporated into the genome. Using P-elements flies have been generated to express Gal4, a yeast-derived transcription factor gene, in a tissue-specific manner. Gal4 associates with the UAS response elements to drive transcription of genes containing them. As such, UAS-driven transgenic lines have also be generated. Crossing UAS-transgene expressing lines with Gal4 expressing flies yield flies that express the transgene in the pattern of the Gal4 activator, i.e., expressed conditionally and in a tissue-specific manner (reviews: (Brumby & Richardson, 2005; Venken & Bellen, 2007) (Fig. 3).

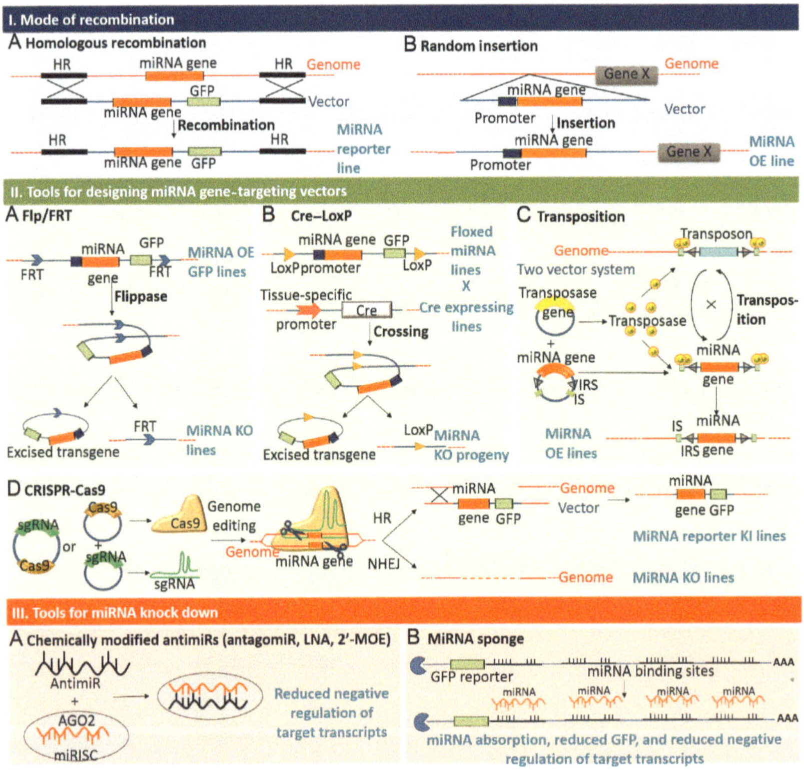

Fig. 2 Strategies for the generation of targeting vector to knock-out miRNAs, and tools employed to knock-down miRNAs. (I) Modes of recombination that govern the genomic location of incorporation of the transgene. (A) Homologous recombination (HR) allows a site-specific insertion of the transgene via crossing-over of the specific HR sites between the genomic site and the vector, in the presence of the enzyme recombinase. (B) Random insertion results in incorporation of the transgene at a random site in the genome. (II) Various gene-editing used for the generation of gene-targeting vectors. (A) Flp/FRT system: Exogenously added or endogenously expressed Flippase (Flp) recombinase allows site-specific recombination with Flp recombinase target (FRT) sites flanking the transgenic miRNA gene. This results in knocking out the targeted miRNA. (B) Cre–LoxP system: This system is analogous to the Flp/FRT system. The Cre-recombinase catalyzes site-specific recombination between two LoxP sites flanking the miRNA of interest, resulting in its excision and miRNA knock-out. (C) Transposition is the mode of transgenic vector incorporation widely used in *D. melanogaster*. This mode of insertion of transgene utilizes multiple mechanisms to insert the transgene at a specific transposon site on the genome. The mechanism here explains a two-vector system. Transposase, the enzyme that facilitates transposition, is encoded by the transposase vector, and the miRNA gene to be transposed at the transposon site in the genome is encoded by the second vector, flanked by inverted repeat sequences (IRS) and insertional sites (IS), that mediate the transgene exchange with a random transposon. (D) CRISPR–Cas9 system: Cas9-vector and an sgRNA vector are expressed in cells. The

Although the conventional method for generating transgenic flies via transposition has not become obsolete, newer strategies for more efficient transgenesis are gaining attention, such as the Flp–FRT, Cre–LoxP (Venken & Bellen, 2007), and CRISPR–Cas9 systems (Kondo & Ueda, 2013) (Fig. 2). Such innovative and novel technologies have significantly advanced the development of transgenic animal models in miRNA functional studies, especially in the context of elucidating the function of miRNAs in various cancer-related events.

2.2.2 D. melanogaster *as a Model for Studying the Functions of MiRNAs in Cancer*

After the discovery of *let-7* in *C. elegans*, *let-7* research was extended to *D. melanogaster* since flies only express a single *let-7*. In flies, *let-7* is produced as a polycistronic pri-miRNA encoding miR-100, *let-7*, and miR-125, also conserved in humans (Caygill & Johnston, 2008; Chawla, Deosthale, Childress, Wu, & Sokol, 2016; Henson et al., 2009; Sokol, Xu, Jan, & Ambros, 2008). To evaluate the role of *let-7* in the progression of juvenile flies into adulthood, the *let-7* locus was replaced by Gal4 downstream of the *let-7* promoter. The absence of *let-7* did not affect morphogenesis of pupae into adult flies; however, the mutant adults displayed severe flight, motility, and fertility defects (Sokol et al., 2008). This study demonstrated that although loss of *let-7* in juvenile flies did not phenocopy the dramatic larval lethality observed in *C. elegans*, *let-7 lof* resulted in severe detrimental effects in the developing flies.

Cas9 endonuclease associates with the expressed sgRNA, which guides Cas9 to a homologous region in the genome to generate a double strand (ds) break. The ds break is repaired in an error-prone manner using nonhomologous end joining (NHEJ), or a targeting vector is inserted at the breakpoint via homologous recombination to generate either a miRNA knock-out or a knock-in of a miRNA and/or a reporter vector, respectively. (III) Strategies to exogenously or endogenously knock-down a mature miRNA (A and B). (A) Various chemical modifications on small miRNA-complementary oligonucleotides, double-stranded or single-stranded, have successfully been generated, to sequester functional mature miRNAs and inhibit their function. AntagomiRs are ssRNAs conjugated with cholesterol. Locked-nucleic acid (LNA) are generated via the formation of a 2′,4′methylene bridge in the ribose resulting in a stable bicyclic nucleotide. 2′-MOE are 2′-*O*-methoxyethyl phosphorothioate-modified oligonucleotides. (B) A miRNA sponge depicted here contains multiple binding sites (Cech & Steitz, 2014; Filipowicz & Kiss, 1993; Lee et al., 1993) for a specific miRNA in the 3′-UTR of a reporter vector. Sequestration of the miRNA results in negative regulation of the reporter and reduced regulation of the endogenous miRNA targets.

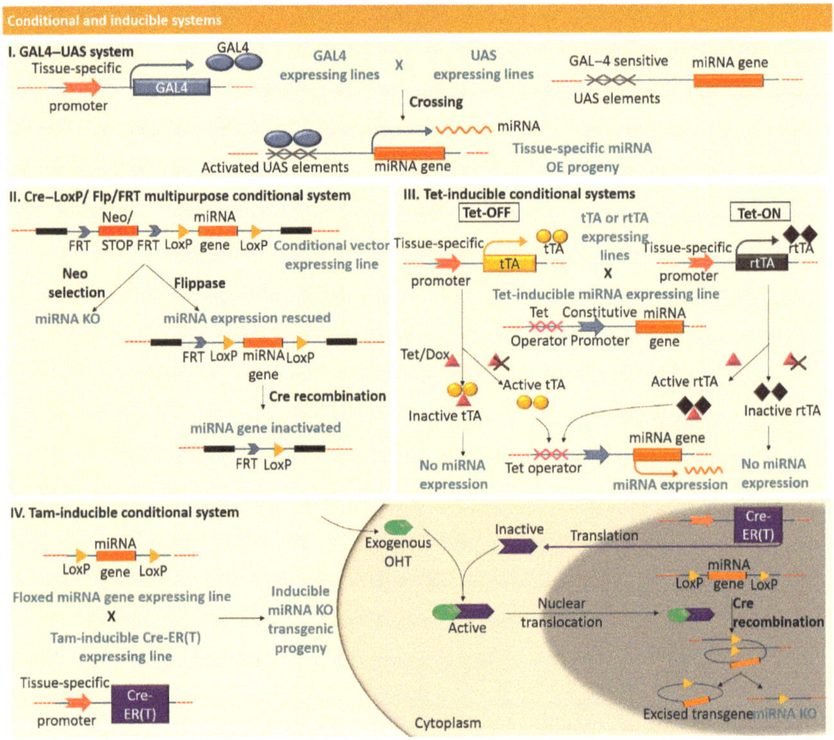

Fig. 3 Conditional and inducible systems. (I) The GAL4/upstream activating sequence (UAS) system (GAL4–UAS) is an inducible system has been utilized in the generation of transgenic flies and zebrafish models. Tissue-specifically expressed GAL4 lines are crossed with a line constitutively expressing the transgene-encoded downstream of a UAS element, allowing GAL4-mediated activation of UAS in a tissue-specific manner. Specific binding of GAL4 to UAS element allows the transcription of the miRNA gene, resulting in a tissue-specific overexpression (OE) of the miRNA in the offspring. (II) A combination of Cre–LoxP and Flp/FRT is a powerful tool to generate a multipurpose conditional and inducible targeting vector. In this case, expression of the neomycin (Neo)/STOP cassette generates a knock-out first vector, inhibiting the expression of the downstream miRNA gene. However, expression of Flp leads to excision of the STOP cassette through recombination of the two FRT sites, rescuing the miRNA gene expression. This system allows miRNA functional studies first in the absence of miRNA expression, following which the effects of rescuing the miRNA can be evaluated. Finally, the effects of the loss of miRNA can be confirmed upon complete inactivation of the miRNA gene, achieved via expression of Cre. (III) The tetracycline-inducible systems (Tet-OFF and Tet-ON) have proved to be very versatile in the generation of transgenic model systems. Tet-OFF is mediated via the expression of the Tet transactivator (tTA), whereas the Tet-ON system is dependent on the expression of the reverse tTA (rtTA). Lines expressing tTa or rtTA in a tissue-specific manner are crossed with a transgenic strain expressing the miRNA gene under the control of a constitutive promoter incorporated downstream of a tetracycline activated element, the Tet Operator (TetO). tTa binds to

An important contribution to the miRNA field made through the use of *D. melanogaster* was achieved using the yeast Flp–FRT system to delineate the consequence of knocking out a miRNA via homologous end recombination (Caygill & Johnston, 2008; Sokol et al., 2008). Using this system, the individual miRNAs from the *let*-7/mir100/mir-125 cluster were knocked out, resulting in data that supported that *let*-7 was sufficient for the normal development of *D. melanogaster*, whereas miR-100 and miR-125 were dispensable (Sokol et al., 2008). In humans the cluster exists as three paralogs, miR-100/*let-7a-2*/miR-125b-1, miR-99a/*let-7c*/miR-125b-2, and miR-99b/*let-7e*/miR-125a. These paralogues exist on chromosomal regions that are frequently lost in multiple cancers (Calin et al., 2004; Roush & Slack, 2008). In a few recent studies, one or more of the individual miRNAs in each cluster has been confirmed to be negatively associated with cancers due to their potent tumor-suppressive functions (Rane, Erb, Nappo, et al., 2016; Roush & Slack, 2008; Sun, Zhang, & Ma, 2015; Wang, Yu, Guan, Zhang, & Cheng, 2017) (Fig. 4).

Until recently, transgenic *D. melanogaster* generation was accomplished using technologies such as the Flp–FRT or the GAL4–UAS systems that mediated successful knock-in or knock-out of miRNA genes in a spatiotemporal pattern. The transgenic *D. melanogaster* models so generated have contributed immensely toward identifying the roles of novel miRNAs in cellular functions of flies, and miRNAs that have a potential role in human cancers. However, the state-of-the-art technology for modeling human

the TetO in the absence of ttetracycline (Tet) or Doxycycline (Dox), leading to the constitutive expression of the transgenic miRNA gene, while the rtTA remains inactive and unable to bind to TetO in the absence of Tet/Dox inhibiting the expression of the miRNA gene. Upon the addition of Tet/Dox to the Tet-OFF system, tTa binds to Tet/Dox and the miRNA gene expression is turned off, whereas in the case of Tet-ON system, Tet/Dox binds to rtTa enabling it to induce the expression of the miRNA gene via direct interaction with TetO. (IV) Tamoxifen (Tam)-inducible conditional system is an extensively used inducible system in the generation of transgenic model organisms. A strain containing a floxed miRNA gene is crossed to a Tam-inducible Cre-ER(T) expressing line, generating an inducible miRNA knock-out strain. Cre-ER(T) is the estrogen receptor (ER) ligand-binding domain fused to Cre recombinase, which remains inactive due to sequestration in the cytoplasm. However, upon exogenous addition of hydroxytamoxifen (OHT), the OHT–Cre-ER(T) complex translocates into the nucleus and actively allows Cre-mediated recombination of the two LoxP sites to occur. The resulting Cre–LoxP recombination knocks out the miRNA gene from the specific tissue expressing the Tam-inducible Cre-ER(T) vector.

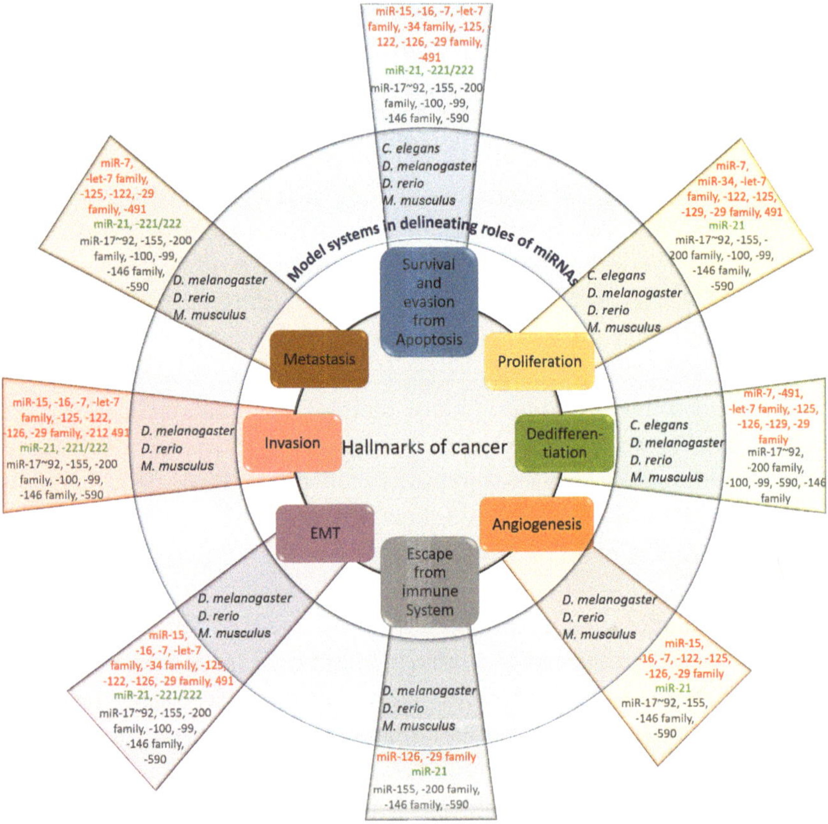

Fig. 4 Functions of miRNAs in regulation of the hallmarks of cancer, identified via the use of in vivo model organisms. Hallmarks of cancer are the cellular processes that become severely dysregulated upon the onset of a cancer. The various model organisms, owing to their endogenous properties, have been utilized to delineate the functions of the enlisted miRNAs that mediate the specific cancerous feature. MiRNAs represented in *red* are *bona fide* tumor-suppressive miRNAs, in *green* are oncomiRs, while *gray* represents miRNAs that, depending on their context, can function as either a tumor-suppressive or oncogenic miRNA.

cancers in vivo is the CRISPR–Cas9 system, which in a recent study was successfully used to knock-out miR-219 and miR-315 in *D. melanogaster* (Kondo & Ueda, 2013). Although the roles of miR-219 and miR-315 were not evaluated biochemically in the mutant flies, other investigations have reported that miR-219 is an essential neurodifferentiation factor (Dugas, Cuellar, Scholze, et al., 2010; Hudish, Blasky, & Appel, 2013; Zhao, He, Han, et al., 2010), and is suppressed in several human cancers, of which the reduced expression ultimately drives the acquisition of tumorigenic properties via diverse mechanisms (Lahdaoui, Delpu, Vincent, et al., 2015; Lei, Zou, Li, et al., 2013; Shi, DL, Huang, & Tan, 2014; Xiong, Zhang, Xiao, et al., 2015). Regardless, this pioneering study was the first to highlight the power of the CRISPR–Cas9 technology in developing transgenic *D. melanogaster* models to study miRNA *lof*, which will likely lead to new and innovative miRNA functional studies.

D. melanogaster have proven to be an instrumental models in understanding the fundamentals of miRNA biogenesis including studies identifying the molecular mechanisms of components such as dicer (Hutvagner et al., 2001) and *locquacious* (TRBP homolog) (Saito, Ishizuka, Siomi, & Siomi, 2005). Indeed, studies in *D. melanogaster* validated that certain miRNAs remained conserved across evolutionary history, such as *let*-7 (Ruby, Jan, & Bartel, 2007). However, due to the failure to find orthologous miRNAs for bantam (Brennecke, Hipfner, Stark, Russell, & Cohen, 2003) and miR-14 (Xu, Vernooy, Guo, & Hay, 2003), and the lack of common targets or overlapping pathways regulated by miRNAs such as miR-7, current miRNA studies are mostly being conducted in other model systems discussed in this review. This difference in miRNA conservation among flies and humans may be a result of evolutionary canalization. Evolutionary canalization suggests that *D. melanogaster*-encoded miRNAs are constantly evolving, acquiring distinct properties, and gaining robustness in their conspicuous functionalities (Kalinowski, Brown, Ganda, et al., 2014; Kefas et al., 2008; Li & Carthew, 2005; Li, Cassidy, Reinke, Fischboeck, & Carthew, 2009; Zhao, Tao, et al., 2015; review: Hornstein & Shomron, 2006).

2.3 *Danio rerio* (Zebrafish)

A model system that has been and continues to be a major contributor in elucidating the functional role of miRNAs in both normal and cancerous cells is zebrafish. Ever since in silico predictions identified *let*-7 as a conserved miRNA across species, additional miRNAs initially identified in zebrafish have displayed a striking homology in composition and function to miRNAs encoded by humans and other vertebrates (Lagos-Quintana et al., 2003; Pasquinelli et al., 2000). *D. rerio* has been a useful model to conduct miRNA-based studies because of the homology with human-encoded miRNAs and additional features of zebrafish such as (i) the small size, an adult zebrafish is about 2–3 cm in length, (ii) the large brood size, a female lays about 100 eggs every 2–3 days, (iii) the ability to easily visualize the forming embryos since fertilization is external, and (iv) the short

3-month generation time of the progeny. Additionally, due to an intact angiogenic and immune system, and a well-developed organ system, successful engraftment of human cancer cells into zebrafish embryos has enabled tumor growth in the host microenvironment. Thus, zebrafish have also demonstrated to be successful model systems to assess tumoral response to anticancer treatments in vivo (Geiger, Fu, & Kao, 2008; Liu & Leach, 2011). Taken together, although zebrafish is a simplistic model system, the compelling features of *D. rerio* have contributed toward the use of this organism to better understand the effects of aberrantly altering the endogenous levels of certain miRNAs (Geiger et al., 2008; Liu & Leach, 2011). Without question zebrafish has proven to be a successful model system that has led to the identification of multiple miRNAs via basic biochemical and molecular studies and holds immense potential to be utilized as a model to identify clinically relevant miRNAs (Fig. 5).

2.3.1 Generation of Transgenic D. rerio *for Use in MiRNA Functional Research*

There are multiple ways to generate transgenic zebrafish via microinjection at the one- or two-cell stage. In addition to its use in flies, the previously described Gal4–UAS system via P-element-mediated transposition has been used to conditionally express transgenes in zebrafish (Brumby & Richardson, 2005; Liu & Leach, 2011; Venken & Bellen, 2007; Xiong, Wei, Zhu, & Sun, 2013) (Figs. 2 and 3). With regard to miRNA overexpressing lines, injection of a plasmid or a linearized transgene inclusive of the miRNA driven by a constitutive or conditional promoter (Xiong et al., 2013) and injection of miRNA mimics (Giraldez, Cinalli, Glasner, et al., 2005) are common strategies used to overexpress a miRNA.

However, most of the zebrafish lines generated for studying the function of a miRNA have been created using a reverse genetics approach via knock-out or knock-down strategies. The tools used to knock-out miRNAs include (i) transcription activator-like effector nuclease (Liu et al., 2013) or (ii) the CRISPR–Cas9 system (Xiao, Wang, Hu, et al., 2013). Whereas, most reported knock-down studies make use of (i) synthetic antisense RNA analogues called morpholinos (Krutzfeldt, Rajewsky, Braich, et al., 2005), or (ii) heavily modified antisense RNA oligonucleotides, locked-nucleic acids (LNAs) (Wienholds, Kloosterman, Miska, et al., 2005; review: Liu & Leach, 2011). Additional approaches include combinations of the

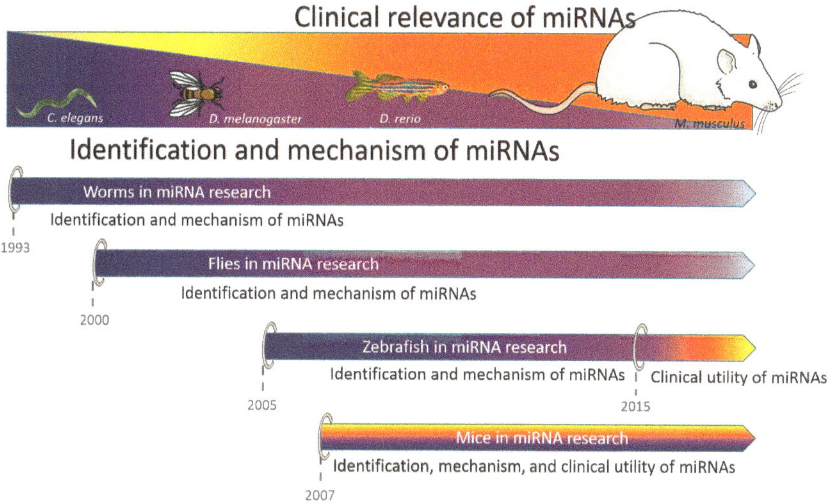

Fig. 5 Utility of model systems in various aspects of miRNA research. The model systems highlighted in this review include *C. elegans* (worms), *D. melanogaster* (flies), *D. rerio* (fish), and *M. musculus* (mice). All the enlisted model systems have contributed immensely toward the identification of evolutionarily conserved miRNAs and delineating their mechanism of action in normal cells. The simple organisms, worms and flies, do not develop cancers; hence, they have mainly been utilized to delineate miRNA biogenesis and function. However, the more complex systems, fish and mice, with an intact immune and angiogenic system have shown immense robustness in identifying the normal functions of miRNAs, and their roles in driving tumorigenesis. Since the discovery of the first miRNA, *lin-4*, in 1993, these in vivo systems have come a long way and have demonstrated their applicability as preclinical model organisms to predict therapeutic relevance of certain miRNAs. Illustration created using graphics from Library of Science & Medical Illustrations (http://www.somersault1824.com/science-illustrations/) and Servier Medical Art (http://www.servier.fr/servier-medical-art).

above-mentioned strategies incorporating inducible vectors, or the use of Cre–LoxP and Flp–FRT systems (Figs. 2 and 3).

2.3.2 D. rerio *as a Model for Studying the Function of MiRNA in Cancer*

Zebrafish serve as a model that is simple for understanding the basic mechanisms of miRNAs in human diseases such as cancer (Pasquinelli et al., 2000; Roush & Slack, 2008), similar to *C. elegans* and *D. melanogaster*. However, *D. rerio* is evolutionarily closer to humans (Pasquinelli et al., 2000; Roush & Slack, 2008). One of the founding miRNAs, *let-7*, is absent in zebrafish embryos during the first ~48 h postfertilization. Overexpressing *let-7* during this critical time in zebrafish embryos causes severe developmental defects.

However, upon countering the overexpressed *let-7* with morpholinos, the defects are reversed. The absence of *let-7* during the first ~48 h of development, but continuous expression until adulthood, describes the heterochronic nature of *let-7* (Kloosterman, Wienholds, Ketting, & Plasterk, 2004). This observation made in invertebrate models is also typical in the case of vertebrates (Kloosterman et al., 2004; Wienholds et al., 2005). However, due to ethical concerns, similar experiments cannot be conducted in human embryos; therefore, the temporal nature of *let-7* in humans has yet to be validated. Importantly, *let-7* family members are highly conserved between zebrafish and mammals. There are 11 mature *let-7* miRNAs expressed in zebrafish, while in humans, there are 9. Hence, zebrafish is an excellent model to begin to dissect the individual contributions of miRNA family members. This observed conservation also suggests that *let-7* family members may display redundancy in activities during vertebral development (Pasquinelli et al., 2000; Roush & Slack, 2008). Although *let-7* family members are implicated as tumor suppressors, functional redundancy due to the presence and expression of other family members presents a challenge toward precisely discerning the function of each miRNA in the *let-7* family. Currently, researchers using the CRISPR–Cas9 system have successfully generated knock-outs of each of the *let-7* family members in zebrafish. Since the knock-outs are viable, these zebrafish transgenics may be useful to uncover the functions of individual members of the *let-7* family. However, detailed biochemical characterization of the individual knock-outs has yet to be performed in vivo (Narayanan, Hill-Teran, Moro, et al., 2016).

The finding that *let-7* is conserved in zebrafish, but that *let-7* does not have a role in early zebrafish development has prompted researchers to identify other miRNAs that contribute to zebrafish developmental. One groundbreaking contribution to the miRNA field using zebrafish was the elucidation that miRNAs are dispensable for cell-fate determination, despite their indispensable roles in highly related cell-fate specification, tissue, and organ formation (Wienholds et al., 2005). Using microarray analysis conducted on whole organisms at different stages of development, it was determined that most miRNAs are not expressed during the first 12 h postfertilization; however, heightened expression is observed postorganogenesis at 96 h. More detailed in situ analysis showed that during development many miRNAs are expressed in a tissue-specific manner (Wienholds et al., 2005). For example, in *Dicer* mutant fish with global miNRA downregulation, the importance of miR-430 in brain morphogenesis was demonstrated. Mimics of the miR-430 family were injected into one-cell stage *Dicer* mutant embryos, which successfully rescued the defective brain morphology of *Dicer* mutant fish (Giraldez et al., 2005). This investigation also highlighted that mature miRNA mimics can form functional miRISC complexes in the absence of active Dicer, shedding light on an unknown mechanism of miRNAs biogenesis with immense applicability in therapeutics. Another significant conclusion drawn from this study was that miRNA expression can be discriminated based on specialized cell types within a specific organ. For example, miR-217 and miR-7 are highly expressed in exocrine and endocrine cells of the pancreas, respectively (Correa-Medina, Bravo-Egana, Rosero, et al., 2009; Goodwin, Rosenzweig, Zhang, et al., 2014; Wienholds et al., 2005). This observation solidified the notion that miRNAs are not always required for cell-fate determination but may also be crucial for cellular differentiation, tissue formation, and maintenance of tissue-identity in a whole organism.

With the findings from zebrafish research that miRNAs are involved in differentiation, tissue formation, and maintenance of tissue identity, zebrafish have indirectly contributed to the characterization of tumor-suppressive miRNAs that are frequently lost in poorly differentiated human cancer cells originating from a specific organ. For example, developmental studies performed in zebrafish demonstrated that miR-122 is a liver-specific miRNA that is only expressed during organogenesis (Wienholds et al., 2005). Analysis of miR-122 in mouse models of liver cancer and in human hepatocellular carcinoma (HCC) patients verified that miR-122 is liver specific and functions as a tumor suppressor; loss of miR-122 correlates with aggressive HCC and poor prognosis (Girard, Jacquemin, Munnich, Lyonnet, & Henrion-Caude, 2008). Similar, although slightly more controversial results were found for miR-126. MiR-126 was shown to be specifically expressed in differentiating endothelial cells during zebrafish organogenesis (Wienholds et al., 2005). To dissect the function of miR-126 in endothelial cell biology, morpholinos knock-down of miR-126 confirmed that miR-126 is essential for several aspects of endothelial cell biology including cell survival, migration, tissue organization, and vascular integrity and stability (Fish, Santoro, Morton, et al., 2008). This study suggests that enhancing the expression of miR-126 in endothelial cells may inhibit migration and invasion of tumor cells through a well-integrated endothelium by enhancing endothelial cell differentiation. However, additional contradictory reports in multiple human cancers also suggest that overexpression of miR-126 in cancer cells may induce proliferation of tumor cells via increased vascularization of tumors (Meister & Schmidt, 2010). Therefore, more comprehensive studies are required to further delineate the role of miR-126 in human cancers.

An additional miRNA identified in zebrafish that showed tissue specificity was miR-200 (Wienholds et al., 2005). MiR-200 was determined to be involved in the development of sensory organs of epithelial origin in both zebrafish and mouse (Choi, Zakhary, Choi, et al., 2008). Functionally, loss of miR-200 in zebrafish during organogenesis results in the generation of embryos with underdeveloped olfactory neurons, due to terminal differentiation of olfactory progenitor cells. This finding identified that expression of miR-200 is critical for the development of sensory epithelial in zebrafish through preventing differentiation (Choi et al., 2008). Consequently, several studies confirmed that loss of expression of miR-200 family members is responsible for EMT of multiple human cancer cells (Korpal & Kang, 2008).

In a more clinical context, a recent study used zebrafish to assess the emerging role of exosomes as vehicles for drugs delivery. Exosomes are an integral mode of cellular communication, and a mechanism that is often hijacked by cancer cells. Recent literature suggests that one of the major macromolecules contained within exosomes that mediate the cancerous phenotypes that exosomes promote are miRNAs (An, Qin, Xu, et al., 2015; Milane, Singh, Mattheolabakis, Suresh, & Amiji, 2015). Therefore, current preclinical research is being directed toward exploring the potential use of exosomes as in vivo drug delivery vehicles for conventional therapeutics and for delivery of therapeutic miRNAs. In this pioneering study, DiD-labeled human brain cancer cells were xenotransplanted in the zebrafish brain ventricle to generate a model for primary glioblastoma–astrocytoma. Following which, exosomes derived from mouse brain endothelial cells were loaded with a fluorescently labeled drug—doxorubicin, and the loaded exosomes were injected into the cardinal vein of zebrafish embryos. In vivo fluorescent imaging confirmed penetration of the blood–brain barrier. Moreover, the therapeutic effects of doxorubicin were confirmed by a reduction in vascular endothelial growth factor mRNA levels, and a dramatic reduction in size of the xenotransplanted brain cancer cells (Yang, Martin, Fogarty, et al., 2015). These studies provide evidence that exosomes can be used as efficient drug delivery systems, at least in a simple model system, and that the exogenously added nonself-exosomes do not produce an inflammatory response. This research also opens avenues to assess exosome-mediated delivery of miRNAs that have therapeutic potential, specifically as anticancer drugs using zebrafish as a model system.

Zebrafish is by far the simplest model system that contains a well-developed blood circulatory system that can help to recapitulate the presence of the tumor microenvironment. Therefore, *D. rerio* has an immense potential to contribute toward a better understanding of the role of miRNAs and exosomes in influencing the tumor microenvironment and their potential altered behavior in a model with an intact immune system. Thus, based on the contributions that zebrafish research had in unfolding the functions of a few crucial miRNAs in cancer, it can be accepted that zebrafish is a powerful model organism, and further studies using zebrafish will likely result in more breakthroughs in the field.

2.4 *Mus musculus* (Mouse)

Considerable progress in miRNA research can be attributed to the use of the previously described in vivo models. However, the ability to closely recapitulate human cancers in mouse models and the conservation between humans and mice suggest that studies in mice provide the most meaningful insights on the role of miRNAs in the molecular pathogenesis of human cancers. Evolutionarily, ~60% of mouse miRNA loci are conserved in humans (Griffiths-Jones et al., 2008; Lagos-Quintana et al., 2003). Additionally, mice are widely used because (i) they are smaller in size than other mammals that are closer in evolutionary history to humans, (ii) they are relatively inexpensive and easy to maintain, and (iii) they produce a fairly large number of offspring in a reasonable amount of time. The use of transgenic mouse models to evaluate the contribution of miRNAs in cancers has proved to be a robust and experimentally tractable system. Recent developments in the field have resulted in the generation of new mouse models that better recapitulate the clinical outcomes of patients treated with various therapeutics than previously used conventional mouse models (reviews: Hidalgo, Amant, Biankin, et al., 2014; Kersten, de Visser, van Miltenburg, & Jonkers, 2017), and therefore, current efforts are also being directed toward the development of useful mouse models to evaluate miRNA therapeutics preclinically.

2.4.1 *Generation of Transgenic* M. musculus *for Use in MiRNA Functional Research*

2.4.1.1 Genetically Engineered Mouse Models Generated via Transgenesis

The most simple and straightforward method to generate a genetically engineered mouse model (GEMM) entails microinjecting a transgene into the male pronucleus of a fertilized egg, followed by transplanting the fertilized egg into a pseudopregnant female to generate offspring expressing the randomly incorporated transgene, at variable copy number. Littermates that have successfully incorporated the transgene into the germline are screened and crossed to generate homozygous mice. Generation of transgenic mice using this method represents the first generation of GEMMs expressing a transgene that is expressed from an exogenous promoter or an enhancer element, resulting in constitutive or tissue-specific overexpression of the transgene (Hanahan, Wagner, & Palmiter, 2007). Knock-out models using this method are accomplished using DNA cassettes, or knock-down of an endogenous protein-coding or miRNA gene via transgenesis of shRNAs or miRNA sponges, respectively (reviews: Cheon & Orsulic, 2011; Coumoul & Deng, 2006; Kersten et al., 2017) (Fig. 2).

2.4.1.2 GEMMs Generated via Homologous Recombination

To exchange an endogenous gene via site-specific homologous recombination, gene-targeting vectors containing the gene of interest and a selectable marker, flanked by homologous DNA sequences of insertion, are transfected in vitro into embryonic stem (ES) cells isolated from a blastocyst. ES cells are utilized to generate GEMMs because they are pluripotent and thus retain the capacity to generate into any cell type, including cells of the germline. Additionally, ES cells maintain a normal karyotype in culture post in vitro gene manipulation and exhibit a higher rate of homologous recombination. The genetically engineered ES cells are then reimplanted into the blastocyst of a surrogate female to generate chimeric mice. Chimeric appearance of the animals' coat is indicative of successful incorporation of the gene-targeting vector. Chimeras are bred to generate germline-transmitted GEMMs. GEMMs with knock-in, knock-out, or conditionally overexpressed genes are successfully generated with this method (reviews: Cheon & Orsulic, 2011; Kersten et al., 2017; Zhang et al., 2012).

2.4.2 Strategies Utilized to Generate Gene-Targeting Vectors

A variety of strategies have been employed to fine-tune the design of gene-targeting vectors that are used to recapitulate human cancers in multiple model systems. The application of gene-targeting vectors used to uncover the in vivo functions of miRNAs involved in the development of multiple human cancers has undeniably advanced this field of research. The most common and current technologies used to design gene-targeting vectors for generating GEMMs via homologous recombination of these vectors into ES cells are described in this section.

2.4.2.1 Cre–LoxP System

The Cre–LoxP system was adapted from the bacteriophage, Coliphase P1. Cre, cyclization recombinase, is a 38-kDa site-specific DNA recombinase that specifically recognizes the 34-bp sites of LoxP, locus of X-over of P1, enabling site-specific recombination. The interaction of Cre with paired LoxP sites results in excision or inversion of a DNA fragment, depending on the same or opposite orientations of the LoxP sites, respectively. LoxP sites can be located in *cis* surrounding a specific gene, or can be located in *trans*, where the two LoxP sites are located in separate areas of the genome. One of the advantages of using the Cre–LoxP system is the ability to manipulate the transgene in vivo. The Cre–LoxP system can be utilized to knock-out or knock-in a transgene constitutively, or conditionally with both spatial and temporal control. Incorporation or excision of a reporter gene cassette, such as LacZ or GFP, can be included to indicate successful loss or gain of function of the gene. To generate a tissue-specific conditionally expressing transgenic model either a tissue-specific Cre expressing line is crossed with a constitutively floxed line, i.e., LoxP-transgene-LoxP line, or Cre is expressed in the specific tissue of the floxed line via administration of Cre expressing lentivirus or adenovirus (reviews: Bouabe & Okkenhaug, 2013; Cheon & Orsulic, 2011; Kersten et al., 2017; Xiong et al., 2013; Zhang et al., 2012) (Fig. 3).

2.4.2.2 Flp–FRT System

The Flp–FRT system uses the yeast *Saccharomyces cerevisiae*-derived flippase (Flp) recombinase that allows site-specific recombination with a pair of 34-bp Flp recombinase target (FRT) sites that flank the transgene or a reporter gene cassette. This system is analogous to the Cre–LoxP system and has made marked contributions toward generation of GEMMs (reviews: Coumoul & Deng, 2006; Deng, 2014; Zhang et al., 2012) (Fig. 3).

2.4.2.3 Inducible Systems

An inducible system is a tremendously powerful technology that enables a researcher to precisely control the expression of a transgene that can allow for a more accurate recapitulation of human cancer development and progression. Moreover, knock-out GEMMs generated using inducible systems as opposed to the conventional knock-out of a developmentally essential gene, bypass the potential lethal consequences that may be observed at early developmental stages. Inducible systems also allow for evaluating oncogene addiction in tumor maintenance, and effects of oncogene ablation on tumor progression. An inducible transgene can be generated by multiple mechanisms; however, in this review, a few of the widely used inducible systems to generate GEMMs will be focused on.

2.4.2.3.1 Tetracycline ON/OFF System The tetracycline (Tet)-inducible system is typically used to turn a transgene on or off. The Tet technology is a binary system that includes Tet-controlled transcription factors (Tet transactivator (tTA) or reverse tTA (rtTA)) and an operator sequences of the bacterial Tet Operon (TetO). TetO is fused upstream of the transgene, and when crossed with a tTA expressing line, a Tet-OFF circuit is generated such that the tTA is unable to bind TetO in the presence of Tet, or the less toxic derivative doxycycline (Dox). On the contrary, the Tet-ON system is generated when a TetO-regulated transgene expressing line is crossed with a rtTA line. In this case the transgene function only occurs in the presence of Tet or Dox. Dox-bound rTtA binds to the TetO inducing transgene expression. When Tet or Dox is withdrawn from the diet rtTA ceases to bind to TetO, terminating the downstream transgene expression (a similar but converse mechanism occurs with the Tet-OFF system). Therefore, the reversible nature of the Tet-ON/OFF systems under the control of a tissue-specific promoter and/or in conjugation with Cre–LoxP or Flp–FRT has been especially instrumental in modeling spatially and temporally controlled gene expression in various model organisms (review: Cheon & Orsulic, 2011; Zhang et al., 2012) (Fig. 3).

2.4.2.3.2 Cre-ER(T) System Another powerful tool that has wide applications in reversibly controlling transgene expression is the Cre-ER(T) system. The Cre-ER(T) system uses a mutated estrogen receptor (ER) ligand-binding domain fused to Cre recombinase. The Cre-ER(T) fusion protein is expressed constitutively but remains sequestered in the cytoplasm unless hydroxytamoxifen (OHT) is added. OTH allows Cre-ER(T) to translocate into the nucleus where it acts on the transgenes. Thus, the Cre-ER(T) system can allow for gene expression in either a tissue-specific manner or constitutively via local or systemic administration of OTH, respectively. When a Cre-ER(T) line is bred with a line containing a LoxP-flanked gene of interest, the Cre-ER(T) fusion protein can be temporally induced via OHT resulting in translocation of Cre-ER(T) into the nucleus to exert homologous recombination at that specific site (review: Cheon & Orsulic, 2011; Zhang et al., 2012) (Fig. 3).

The above-mentioned Cre–LoxP and Flp–FRT systems have been created in complex combinations with the inducible systems to generate gene-editing vectors that are not only spatially and temporally controlled but also enable reversible expression of the gene of interest (Fig. 3). A few examples of mouse models that have been used to study the function of miRNAs in cancers have successfully been generated using these technologies. Specific use of these systems is highlighted in the following section.

2.4.3 Using M. musculus *for MiRNA Functional Studies in Cancer*

The first evidence of a mouse that spontaneously developed a cancer homologous to humans was the New Zealand Black (NZB) mouse strain that developed CLL. Similar to the molecular alterations occurring in human CLL patients, these NZB mice had lost the *mir-15∼16 cluster* (Aqeilan et al., 2010; Calin et al., 2004; Klein, Lia, Crespo, et al., 2010; Raveche et al., 2007). This discovery was the founding premise of miRNA involvement in cancer and demonstrated the power in using murine models to gain a better understanding of the contribution of miRNAs in carcinogenesis. Moreover, using this naturally occurring model of CLL, it was determined that exogenous delivery of miR-15∼16 to NZB-derived malignant CLL cell lines could reverse CLL phenotypes through the induction of apoptosis (Aqeilan et al., 2010; Calin et al., 2004; Klein et al., 2010; Raveche et al., 2007; Scaglione et al., 2007). This investigation shed light on the importance of modeling cancers in appropriate organisms and on the use of mouse models to evaluate miRNAs with potential therapeutic application.

Discovering that loss of the miR-15∼16 miRNA cluster in both mice and human patients was driving CLL resulted in an exponential increase in the identification of miRNAs that are misregulated in cancer. This fueled in vivo studies to determine if these misregulated miRNAs had a significant role in promoting or maintaining disease. Initially focusing on the miR-15∼16 cluster, both constitutive and conditional knock-out mouse models for *mir-15a∼16-1* were generated. A GEMM containing a floxed *mir-15a∼16-1* locus was crossed to mice constitutively expressing Cre to generate *mir-15a∼16-1*$^{+/-}$ chimeric mice. To conditionally knock-out *mir-15a∼16-1* exclusively from B-cells, *mir-15a∼16-1*$^{flox/+}$ mice were crossed with CD19-Cre transgenic mice. Intercrossing the F1 chimeras generated constitutively null mice, or mice with *mir-15a∼16-1* deleted only in B-cells (Klein et al., 2010). Both models demonstrated manifestations of CLL phenotypes, and cellular and molecular alterations in the models displayed a striking resemblance to human CLL, depicting the accuracy of transgenic mouse models to study the development and stage of miRNA-mediated cancer progression.

Following the generation of robust models used to evaluate the loss of expression of tumor-suppressive miRNAs, the oncogenic potential of the first speculated oncomiR, *oncomiR-1*, was explored. *OncomiR-1* is a polycistronic gene that gives rise to a single transcript containing seven miRNA precursors, commonly known as the miR-17∼92 cluster (Olive et al., 2010). In patients suffering from various hematologic cancers, the genomic locus of *mir-17∼92* was observed to be amplified and has since been extensively investigated via bioinformatic and biochemical approaches to confirm its oncogenic property (He, Thomson, Hemann, et al., 2005; Lu, Getz, Miska, et al., 2005; Mu et al., 2009; Ota, Tagawa, Karnan, et al., 2004; Xiao et al., 2008). Although initially identified as an oncomiR in hematopoietic malignancies, miR-17∼92 is also involved in the development of solid cancers such as medulloblastoma (Murphy, Obad, Bihannic, et al., 2013), and hepatocellular (Zhu, Han, & Wu, 2015) and malignancies of the lung (Hayashita, Osada, Tatematsu, et al., 2005) and breast (Kim, Chadalapaka, Lee, et al., 2012). Due to its pleotropic role in various malignancies, modeling the cluster in mouse models was a high priority. Since the knock-out model of *mir-17∼92* produced via crossing floxed *mir-17∼92* with Actin-Cre resulted in mice that suffered postnatal lethality due to severe birth defects (Ventura et al., 2008), multiple conditional knock-outs and overexpression GEMMs were generated to investigate the oncogenic role of miR-17∼92 (Mu et al., 2009; Sandhu et al., 2013; Xiao et al., 2008). Targeted overexpression of the entire miR-17∼92 cluster, specifically in B-cells using B-cell specific Ig-heavy chain promoter Eμ-enhancer, resulted in severe B-cell lymphomas and leukemias (Sandhu et al., 2013). To elucidate roles for the individual miRNAs in the cluster, transgenic mice were generated that overexpressed each individual miRNA in the miR-17∼92 polycistron. The results revealed that miR-19 is sufficient to exert the oncogenic potential of *oncomiR-1* (Mu et al., 2009). MiR-92 overexpression, on the other hand, counters the effects of miR-19 in oncogenesis via a feedback mechanism (Olive et al., 2009, 2013). Through the extensive use of mouse models, it is now evident that there exists a complex but delicate balance between miRNAs that are simultaneously expressed, but that may function individually as either potent oncomiRs or tumor-suppressive miRNAs. The balance in expression of such antagonistically functioning miRNAs likely plays a very critical role in maintenance of normal cellular physiology.

Very few phenotypes have been observed in mouse models following altered expression of a single miRNA. Indeed, overexpression of the miR-17∼92 cluster or loss of the miR-15∼16 cluster can both promote tumorigenesis, but in both cases, multiple miRNAs were altered. The miRNA field was further revolutionized following evidence that overexpression of either miR-21 or miR-155 is sufficient to induce tumorigenesis without the contribution of other oncogenic alterations (Babar et al., 2012; Medina et al., 2010). For the first time, it was proven that hematologic cancer maintenance is dependent on a single oncomiRs, such that reducing expression of the miRNA results in cancer regression. In these studies, a transgenic vector expressing the pre-miRNA was placed downstream of a floxed STOP cassette under the control of a Tet promoter, i.e., $miR\text{-}21^{LSLtTA}$. In the absence of Dox, the miRNA is not expressed. To conditionally express the transgene in the lymphoid tissues the miRNA lines were crossed to Nestin-Cre mice resulting in STOP cassette excision, overexpression of the miRNAs, and development of pre-B-cell lymphomas. However, upon impregnating the mouse chow with Dox, there was a rapid regression of pre-B-cell lymphomas and an increase in survival. While the miR-21 study was the first to show that tumors can be addicted to changes in miRNA expression, the miR-155 report reconfirmed the role of miR-155 in lymphoma as initially reported by the Croce group (Babar et al., 2012; Costinean et al., 2006; Medina et al., 2010).

While these aforementioned studies focused on overexpressing or knocking out miRNAs independently of other genetic alterations, additional in vivo evidence has determined that miRNAs cooperate with both protein-coding genes and with each other and thus suggests that miRNA-based therapeutics might have a place in the clinic. For example, miR-21 was found to enhance lung cancer in the inducible autochthonous model driven by the protooncogene, Kras, $Kras^{Lox\text{-}Stop\text{-}Lox(LSL)\text{-}G12D/+}$. In the absence of $Kras^{G12D}$ expression, miR-21 was insufficient to induce oncogenesis in the mice. However, following $Kras^{G12D}$ transgene expression, significantly more tumors were observed, compared to the $Kras^{G12D/+}$ control mice (Hatley, Patrick, Garcia, et al., 2010). Similarly, in the pancreatic autochthonous $Kras^{G12D/+}$ model it was suggested that miR-21 may be involved in pancreatic cancer development via a multistep process (du Rieu, Torrisani, Selves, et al., 2010). Confirmation that miR-21 was essential for pancreatic cancer maintenance was verified in animals that were orthotopically transplanted with pancreatic ductal adenocarcinoma cells. A single dose of intratumoral administration of lentiviral vectors expressing antisense-miR-21 severely impaired tumor cell growth via onset of necrosis due to miR-21 depletion (Sicard, Gayral, Lulka, Buscail, & Cordelier, 2013).

The use of protein-coding transgenic models has also contributed to the miRNA field, specifically to evaluate miRNA-based therapeutics. For example, the *Kras*$^{G12D/+}$ autochthonous lung cancer model was used to evaluate the tumor-suppressive roles of miR-34a and/or family members of *let*-7, and to assess their therapeutic efficacies. To this end, *Kras*$^{G12D/+}$ mice were tail-vein injected with synthetic formulations of each of the miRNAs (miR-34a or *let-7b*), or orthotopically administered viral-encoded *let-7a*. The resulting regressed tumors demonstrated that each of the individual tumor-suppressive miRNAs could act as potential therapeutic agents (Esquela-Kerscher et al., 2008; Kasinski & Slack, 2012; Trang et al., 2011). Following this, additional autochthonous NSCLC mouse models have confirmed that combinatorial RNA therapeutics produce even greater effects than treatment with an individual miRNA, some of these studies include the use of miR-34 combined with *let-7b* (Kasinski et al., 2015), or miR-34 and an siRNA-targeting *Kras* (Xue, Dahlman, Tammela, et al., 2014). More recently, the *Kras*$^{LSL\text{-}G12D/+}$; *p53*$^{fox/flx}$ model was used to asses the utility of delivering a therapeutically relevant miRNA using a novel vehicle-free delivery approach. In this study, progression of murine NSCLC was reduced following systemic administration of miR-34a directly conjugated to a targeting ligand, folate, highlighting the ability to deliver unprotected miRNAs in vivo (Orellana et al., 2017).

The above studies proved the tumor-suppressive role of *let*-7 through the use of exogenous *let*-7; however, the endogenous tumor-suppressive activity of the *let*-7 family in a transgenic model was only recently demonstrated. To delineate if *let*-7 suppresses the Myc-driven tumorigenesis process, *let-7g* and Myc were simultaneously overexpressed using a triple transgenic, liver-specific, Tet-off system. *Let-7g* was successfully overexpressed despite the high levels of Lin28 by innovatively cloning the mature *let-7g* sequence into a miR-21 stem–loop, retaining the miR-21 loop, which prevents Lin28B-mediated processing inhibition of *let*-7 family members. Myc overexpression led to tumorigenesis in the absence of *let-7g* expression; however, proliferation and growth were markedly reduced when *let-7g* was overexpressed. This study also showed that *let*-7 is transiently repressed in tissues undergoing repair and regeneration as the cells require enhanced proliferation. Further, through evaluating a conditional liver-specific knock-out of *let-7b* and *let-7c2*, investigators proved that reduced *let*-7 levels resulted in higher liver mass relative to control, due to increased mTOR signaling activity. This study demonstrates that *let*-7 is expressed in liver tissue, likely to suppress the development of liver cancer, but that controlled balance of *let*-7 levels is the key to maintaining the regenerative capacity of liver (Wu, Nguyen, Zhou, et al., 2015).

Multiple studies have identified *let-7* as a potent tumor suppressor, and the consequences of *lof* of *let-7* results in loss of cellular differentiation, increased proliferation, and tumorigenesis. However, with the growing evidence that *let-7* is an important tumor suppressor, the mechanisms involved in *lof* of *let-7* have become a subject of interest in the field. Early studies evaluating *let-7* led to the intriguing finding that loss of mature *let-7* is not always associated with changes in transcription of the *let-7* gene, which encouraged characterizing *let-7* at the posttranscriptional level. With the groundbreaking finding that RNA-binding proteins such as LIN28A and -B selectively inhibit *let-7* miRNA biogenesis (Heo et al., 2008; Newman et al., 2008; Piskounova et al., 2008; Viswanathan et al., 2008), in vivo studies were conducted to delineate the role of LIN28B in *let-7*-mediated tumorigenesis. Colon cancer cells constitutively expressing LIN28B were implanted into immunocompromised mice. Biochemical evaluation of the tumors indicated increased levels of endogenous LIN28B, which strikingly negatively correlated with *let-7* levels in tumor cells relative to surrounding normal cells (King et al., 2011). This in vivo correlative study, along with additional cell-based and molecular assays, confirmed the negative feedback loop that exists between *let-7* family members and LIN28A/B (Heo et al., 2008; King et al., 2011; Madison, Liu, Zhong, et al., 2013; Newman et al., 2008; Rybak, Fuchs, Smirnova, et al., 2008). In transgenic mouse models expressing Lin28B under the *Vil1* promoter, which drives expression specifically in the intestines, tumorigenesis was directly dependent on the loss of mature *let-7* (Madison et al., 2013). High Lin28B negatively correlated with the levels of *let-7*, whereas rescuing mature *let-7* levels led to reversion of tumorigenic phenotypes (Madison et al., 2013). This study, followed by others, suggests that miRNAs are downregulated by various mechanisms in the process of tumorigenesis (Chang et al., 2008; Hermeking, 2007; Madison et al., 2013; Raver-Shapira et al., 2007; Wei et al., 2006). For example, Myc, an important oncoprotein that upregulates the *oncomiR-1* cluster (*mir-17~92*), is predominantly associated with widespread depression of miRNA expression in humans and in cells obtained from mouse models of lymphoma through direct interaction with miRNA promoters (Chang et al., 2008).

While both MYC and LIN28 are involved in downregulation of a subset of miRNAs, global miRNA downregulation was identified as a common feature in human tumors, which could not simply be explained by MYC and LIN28. Thus, experimental designs turned to evaluating major components of the miRNA processing machinery. In mouse models where the *Dicer1* locus was floxed, Dicer1 was identified as a haploinsufficient tumor suppressor (Kumar, Pester, Chen, et al., 2009). A single copy of *Dicer1* was necessary for tumorigenesis. Interestingly *DICER* is also lost in human cancers, and similar to the mouse model, only one allele is deleted (Lambertz, Nittner, Mestdagh, et al., 2010). Prior to this work multiple studies suggested that genetic mutations in components of the biogenesis machinery, such as DROSHA, DICER, and XPO5, may severely dysregulate miRNAs leading to cancer (Kim, Kim, & Kim, 2016; Rakheja, Chen, Liu, et al., 2014). However, with the use of robust murine models, the anticipated pathophysiological consequences of a disrupted biogenesis pathway on global miRNA depression resulting in tumorigenesis are now regarded as a hallmark of cancer. With the growing understanding that miRNAs are globally downregulated in cancers, innovative targeting vectors have been generated to directly ablate several evolutionarily conserved miRNAs in mouse models and to clone in reporter constructs downstream of the endogenous promoter to identify both temporal and spatial expression patterns. The targeting vectors contained floxed pre-miRNA sequences that were placed downstream of a FRT-flanked promoter-less LacZ reporter to generate LacZ-STOP (Neo)-floxed miRNA transgenic vectors. Embryos generated by crossing LacZ-Neo-flox mice with Actin-Cre animals were evaluated for LacZ expression patterns in various tissues. Data suggest that approximately one-third of miRNAs exhibit a global expression pattern, whereas about two-thirds of miRNAs demonstrate developmental stage or tissue-specific expression patterns. For example, miR-210 and miR-146a were undetected in earlier stages, but were expressed in the adult mouse, subcompartmentalized in immune cells. This investigation reconfirms the spatiotemporal nature of miRNAs, as suggested by the conventional model organisms, *C. elegans* and *D. melanogaster*. The observations made from these reporter studies can also be extrapolated to appropriately select a Cre-mouse for tissue-specific ablation/expression of specific miRNAs. This idea was verified using mice expressing the transgene and crossing them to animals that temporally or constitutively expressed Cre, and phenotypic and developmental alterations were observed for a subset of miRNAs (Park, Choi, & McManus, 2010; Park, Jeker, Carver-Moore, et al., 2012). Overall, this strategy holds promise for uncovering the roles of independent miRNA family members in various cancers, demanding further characterization.

The advantage of using a multipurpose targeting vector comes from the power to generate a conditional knock-out model of the miRNA of interest (Fig. 2). The expression of the upstream STOP cassette in the LacZ-STOP (Neo)-floxed miRNA transgenic mouse postembryonic stages, via neomycin selection allows bypassing of embryonic lethality in the absence of the miRNA, without physically excising the miRNA transgene. Nevertheless, crossing the parental transgene expressing strain with a constitutively expressed Flp strain leads to miRNA-rescue in the progeny by excision of the FRT-flanked LacZ-STOP regions upstream of the miRNA transgene. Although these murine models have been successfully generated, they remain to be characterized, which may shed additional light on the loss of miRNAs and their reexpression in various tissues and may aid in the quest to better understand the role of various miRNAs in tumorigenesis (Park et al., 2010, 2012).

To understand the role of miRNAs in promoting metastatic potential of cancer cells in vivo, in a model with an intact immune system, investigators have relied on the syngeneic mouse model. In a recent study, primary and metastatic tumors from the $Kras^{LA1/+};p53^{R172H\Delta G}$ autochthonous lung cancer model were harvested, cultured, and subcutaneously injected back into the same immunocompetent host. When xenografts of cells derived from metastasized tumors were implanted in the syngeneic mice the cells metastasized, whereas animals xenografted with cells from the primary tumor only produced localized subcutaneous tumors. Upon miRNA profiling, it was observed that the metastatic tumors expressed low levels of miR-200 family members (miR-200a, miR-200b, miR-200c, miR-141, and miR-429), relative to the localized tumors. However, when the miR-200b ~ 200a ~ 429 cluster was overexpressed in the metastatically derived cells, only primary tumors were capable of growing as compared to control cells that metastasized to the lung, heart, liver, and kidneys (Gibbons, Lin, Creighton, et al., 2009).

3. CONCLUSION

Since the discovery of *lin-14*, scientists have come a long way in delineating the roles of miRNAs in higher vertebrates, including humans. Although the biogenesis of miRNAs and their biological role in maintaining normal physiology of cells have been fairly well established, this is a rapidly growing yet constantly evolving field of research. Currently, efforts are being made to identify miRNAs that drive aberrant cellular events responsible for abnormal behavior of a cell, leading to pathogenesis of diseases such as cancer.

Initial insights into the miRNA world were furnished through the use of a simple, yet elegant model system, *C. elegans*. The miRNAs *lin-4* and *let-7* were identified through forward genetic screens, that determined these miRNAs are temporally expressed and required for normal development of *C. elegans* (Lee et al., 1993; Reinhart et al., 2000; Roush & Slack, 2008; Wightman et al., 1993). Nevertheless, with the advent of transgenesis and the use of more complex systems with closer homology to humans, such as *D. melanogaster*, *D. rerio*, and *M. musculus*, a greater understanding of miRNA biogenesis, function, and misregulation in disease has become evident. Advancements in bioinformatic, biochemical, and molecular approaches have revealed that miRNAs are aberrantly expressed in multiple cancers, and that in many instances, they function to promote and/or maintain the tumorigenic phenotype. The first miRNAs discovered to be involved in human disease, miR-15a/*16-1*, were determined to be downregulated in human CLL patients. Consequently, the discovery of NZB mouse that naturally developed CLL and displayed severe downregulation of miR-15/16 expression resulted in the recognition that mice have the potential to serve as a robust model system for miRNA functional studies in cancer (Calin et al., 2002; Raveche et al., 2007; Scaglione et al., 2007).

Soon after this novel discovery and the realization that the mouse genome is ~90% homologous to that of humans, potent methods to generate GEMM prospered. These pioneering studies in the miRNA field led to implementing intriguing strategies to generate efficient gene-targeting vectors that can not only replace an endogenous gene of interest (in this case, a miRNA gene) but also control its expression spatially and temporally. Various GEMMs have since been generated that have made immense contributions in uncovering the functions of numerous miRNAs as global and specific tumor-associated miRNAs, and have also supported the transition of miRNAs into the clinic. Models with intact immune systems are also be extensively evaluated as they more faithfully recapitulate human tumorigenesis and thus, are more accurate for studying cancer and therapeutic response, especially for agents targeted to cells of the microenvironment (reviews: Ito, Takahashi, Katano, & Ito, 2012; Kersten et al., 2017; Rupaimoole & Slack, 2017; Zhao, Huang, & Ye, 2015).

D. rerio or zebrafish is a vertebrate that has benefitted the research community through experiments that have helped to define the role of miRNAs as developmental factors, supported by biochemical experiments conducted in these visually traceable optically transparent embryos. Moreover, embryonic lethality, due to altered expression of critical heterochronic miRNAs, is easily observed as fertilization of the egg is external, whereas in mice, lethal embryos are rapidly reabsorbed, thereby obstructing evaluation of miRNAs involved in development (review: Wienholds & Plasterk, 2005). Although zebrafish have proven to be convenient model organisms for identifying miRNA involvement in development, they have also indirectly contributed to the understanding of various aspects of cancer, including the role of miRNAs in the development of cancers. While zebrafish do not develop cancer, many of the hallmarks of tumorigenesis such as proliferation, migration, differentiation, and apoptosis are conserved. As such, miRNAs ascertained through zebrafish studies have been validated and characterized in human cancer cell lines and/or in mouse models that recapitulate human cancer more closely. Mouse models, like zebrafish, have gained immense recognition in modeling various human cancers, since they belong to the class of vertebrates. However, the contributions made by invertebrates such as *C. elegans* and *D. melanogaster* have undeniably influenced this revolutionary area of research. *C. elegans* and *D. melanogaster* have been useful in dissecting the molecular machinery of miRNA biogenesis and contributed toward the identification of miRNA-mediated alterations of cellular pathways. Such developments achieved in simple model systems have since been extrapolated to more complex in vitro and in vivo systems to understand the effect of disturbed expression patterns of critical miRNAs, leading to cancer initiation, development, and progression.

Generation of transgenic mice has come a long way since using the conventional method of random integration of an exogenous DNA into the mouse genome, via nonhomologous recombination. One major drawback with this method is the lack of specificity at the site of transgene incorporation and high propensity of off-targeting effects. On the contrary, the primary advantage of generating GEMMs via the contemporary methods of gene-targeting vectors is that the endogenous gene becomes replaced by the transgene (contained in the targeting vector) via site-specific homologous recombination (Fig. 2). This ensures that the transgene is incorporated precisely at a specific genomic location. This same method also allows for evaluating tissue-specific effects of the miRNA through integration of tissue-specific promoters in the targeting vector.

Spatially and/or temporally controlling the expression of a transgenic-miRNA has led to the discovery of miRNAs with tissue-specific or stage-specific roles in the development of cancers. A powerful contribution of GEMMs is the generation of autochthonous mice that spontaneously develop cancers at the true anatomical location once the conditional-mutant allele of a known oncogene is activated. These models, as well as multiple other GEMM and xenograft models, have been utilized to uncover the contribution that miRNAs make in the presence of other genetic lesions, for example, miRNA involvement in potentiating the process of tumor generation or regressing tumors growth. Thus, these preclinical models have applications in segregating miRNAs that may have potential therapeutic applications.

The importance of the immune system on tumorigenesis can only be appreciated using GEMM or syngeneic models. Injecting tumor cells derived from autochthonous mice into another mouse of the same genetic background has established that the ability of a tumor cell to become invasive and migrate requires a complete and efficient microenvironment, including an intact and functional immune system. Therefore, the advantage of using immunocompetent mouse models such as the autochthonous and syngeneic GEMMs over the immune-deficient xenograft models is that the former possesses an intact immune system recapitulating the natural scenario of tumor initiation, development, and progression. Each of the stages of tumor development typically requires cancer cells to cross talk with surrounding stromal cells, normal cells, and the immune cells. This is the possible reason why cancer cells show metastatic potential in syngeneic models, but often not in subcutaneous, or PDX xenograft mouse models (Table 1).

The most common strategy in the field to predict the function of a miRNA in cancer development is to first utilize the straightforward and basic subcutaneous (SC) xenograft mouse model. Advantages of modeling human cancer in SC xenograft models include the ability of cells to grow inadvertently, develop into primary tumors in vivo, and the ease of tracking and measuring the tumors. Although the associated disadvantage is that the genetic, histological, and therapeutic responsiveness of the xenografted tumors are not comparable to the source human tumor. Moreover, the reduced propensity of SC xenografted cancer cell lines to metastasize in vivo owing to the lack of a conducive environment for the implanted cancer cells is perhaps another disadvantage that provides evidence that this model does not completely recapitulate an endogenous human cancer. Such models have immensely contributed toward a collection of preliminary evidence in understanding the effects of oncomiRs such as miR-21(Medina et al., 2010), or tumor-suppressive miRNAs such as *let*-7 (Trang et al., 2010) in tumorigenesis. Despite the fact that SC xenograft models demonstrate a response to therapeutic antisense oncomiRs or tumor-suppressive miRNA mimics, follow-up studies in more advanced models with intact immune systems are essential prior to clinical advancement (Humphrey & Kasinski, 2015; Kasinski et al., 2015; Xue et al., 2014).

Table 1 The Various Mouse Models That Have Been Used to Identify miRNAs That Contribute to Tumorigenesis

Mouse Model	Immune Status of Mouse Background	Generation Method	Applications
Spontaneous models	Immune competent	Spontaneously occurring tumors (Scaglione et al., 2007)	To identify miRNA genes that are lost or amplified in nature, resulting in the development of cancers
Xenograft models	Immunedeficient mice—SCID/NOD-SCID, NOG/NSG, RAG	Subcutaneous injection of human tumor cell lines (Medina et al., 2010)	To evaluate in vivo oncogenic or tumor-suppressive potential of miRNAs To determine delivery and efficacy of miRNAs that have therapeutic potential
		Orthotopic implantation of human tumor-derived cells or cell lines (Sicard et al., 2013)	The above applications. And, orthotopic xenograft mice also serve as a model for metastasis of tumor cells, therefore, they can be utilized to delineate miRNA functions in invasion and metastasis
Patient-derived xenograft (PDX) models	Immunedeficient NOD-SCID mice	Primary tumor engraftment (Hannafon, Trigoso, Calloway, et al., 2016; Sun, Wang, Zhou, et al., 2017)	To maintain tumor heterogeneity and allow personalization of treatment. Additionally, evaluate exosomal miRNAs released from tumors and evaluate miRNAs that may have therapeutic potential

Continued

Table 1 The Various Mouse Models That Have Been Used to Identify miRNAs That Contribute to Tumorigenesis—cont'd

Mouse Model	Immune Status of Mouse Background	Generation Method	Applications
Genetically engineered mouse models (GEMMs)	Immune competent	Autochthonous mouse (Kasinski et al., 2015; Xue et al., 2014)—conditionally activated mutations in transgenic oncogenes that result in endogenous tumors evolving spontaneously from normal cells in the correct anatomical location	Evaluation of spontaneous tumors to identify miRNAs that drive tumorigenesis in normal cells Evaluation of miRNAs with therapeutic potential in tumors arising in endogenous microenvironments
		Transgenic mouse—random incorporation of transgene, to overexpress or knock-down/out a miRNA, in the genome	To analyze the in vivo roles of oncogenic or tumor-suppressive miRNAs
		Conditional or inducible expression of gene-targeting vectors, via homologous recombination (Medina et al., 2010; Park et al., 2012)	To analyze the tissue-specific or cancer-stage specific roles of oncogenic or tumor-suppressive miRNAs
		Site-specific recombination (Esquela-Kerscher et al., 2008; Park et al., 2012) can also be virus mediated—retrovirus, lentivirus, adenovirus	The above, and to determine delivery and efficacy of miRNAs that have therapeutic potential
		CRISPR–Cas9-mediated knock-in or knock-out (Humphrey & Kasinski, 2015; Yang, Wang, & Jaenisch, 2014)	Has the potential to analyze the tissue-specific or cancer-stage specific roles of oncogenic or tumor-suppressive miRNAs, with greater efficacy relative to conventional site-specific recombination

Syngeneic mouse models	Immune competent	Implantation of tumors cells derived from mice from the same strain of origin (Gibbons et al., 2009)	Intact immune system and microenvironment support the development of a tumor in a *bona fide* cancerous environment. This model aids in the identification of miRNAs that potentiate metastasis of tumors
Humanized mouse models	Immunodeficient mice—NSG or NOG	Engraftment of human immune cells (peripheral blood mononuclear cell (PBMCs) or CD34 + cells) (Humanized Mice in Tumor Studies, 2017; Ito et al., 2012)	Possible immune response in patients against a treatment can be evaluated in mice Not utilized in preclinical studies of miRNA-based cancer therapies yet
		GEMMs expressing a human gene in mouse genome (Takehara, Schulz, Abe, et al., 2014)	Compensates for a gene that is missing in the mouse genome, to further study its role in oncology. MiRNAs that are not conserved can be evaluated Has potential in human-specific miRNA functional studies

On the contrary, an orthotopic model is generated when cancer cells derived from a specific anatomical location are implanted into the same location to develop primary tumors that are genetically and histologically more representative of human tumors. Orthotopic models also lack an intact immune system, similar to xenograft model systems; however, the advantage here is that the cancer cells are introduced in vivo in their original anatomical location where the tumor was harvested from. Thus, the interaction between cells of similar origin, in the correct anatomical location, may provide a moderately conducive environment for the orthotopically developed cancer to metastasize, and respond to miRNA based therapeutics. Additionally, another xenograft mouse model system, the PDX model, when treated with certain drugs have shown to favor predictability of patient response to the same. PDX models stably maintain the heterogeneity of the engrafted human tumor in the mice for multiple passages, owing to the animals'-deficient immune system. Therefore, the potential of PDX models in preclinical evaluation of therapeutics, personalized-treatment development, and useful biomarker identification is being extensively explored. The use of PDX models has recently been extended to the miRNA field. Exosomal miRNAs such as miR-21 and miR-1246 identified in PDX models of breast cancer are predicted as *bona fide* circulating miRNA biomarkers indicative of breast cancer in patients (Hannafon et al., 2016). Moreover, the therapeutic potential of replacement of miR-100 has recently been evaluated in PDX models via targeted therapy of the oncogene, fibroblast growth factor receptor 3 (FGFR3), in FGFR3-driven PDX tumors (Sun et al., 2017). Therefore, although it is becoming accepted that PDX models may be applicable to developing personalized miRNA-based cancer therapies, this model is at its infancy and requires more explicit evidence to be recognized as a faithful preclinical model system. With the recently accepted notion that accurate tumor progression, and therapeutic response requires an intact immune system, an emerging concept of using model systems to perform preclinical evaluations of miRNA-based therapies or other drugs is on the cutting edge. To this end, another mouse model that deserves a mention here is the humanized mouse model. A nude mouse is humanized by engraftment of human-donor-derived immune cells. This model system is not yet in practice in miRNA functional research in cancers per se, but it may be beneficial in preclinical evaluation of miRNA-based therapeutics, which may eventually positively impact the time span required for translating miRNA-based cancer treatments into the clinic.

Although advancements are being made to models for use in evaluating miRNA-based cancer therapeutics at the preclinical stage, many contributions in the field related to uncovering miRNAs that maintain normal archetype of a cell or miRNAs that drive the process of tumorigenesis emerged from the extensive use of GEMMs. Additionally, GEMMs have also made unparalleled contributions toward the discovery of miRNAs that confer therapeutic potential as replacement therapies or as therapeutic targets. While the technologies used to generate transgenic mice have advanced remarkably, innovative strategies to generate genetically engineered preclinical models for miRNA-based therapeutic research requires attention (Orellana et al., 2017). It is anticipated that additional use of the CRISPR–Cas9 system (Yang et al., 2014) will support miRNA research by rapidly accelerating the process of generating GEMMs by robustly overexpressing or knocking out a miRNA gene of interest. This will not only aid in the precise knock-out of a miRNA gene of interest to evaluate replacement therapies, but will also facilitate knock-in/out of discrete miRNA family members to finally dissect the functions of individual miRNAs that was not as easily achievable using conventional methods. The CRISPR–Cas9 system is superior in many ways to the now more traditional methods of generating GEMMs (reviews: Guernet & Grumolato, 2017; Humphrey & Kasinski, 2015; Kersten et al., 2017; Shalem, Sanjana, & Zhang, 2015; Wen, Yuan, Ma, Xu, & Yuan, 2016; Yang et al., 2014). Genetic manipulation of multiple genes can be simultaneously achieved in a single-embryonic cell, not requiring the laborious selection of ES cells posttransgenic vector injection. The CRISPR–Cas9 system is capable of producing miRNA knock-in/out mouse models in 4 weeks, and therefore, has immense potential in miRNA functional delineation in multiple human cancers. Efforts toward generating model systems utilizing the CRISPR–Cas9 system targeting miRNA genes in vivo in the correct anatomical location of existing autochthonous or syngeneic models are required, which can further accelerate the fate of modeling human cancer efficiently and robustly in model organisms.

REFERENCES

Abbott, A. L., Alvarez-Saavedra, E., Miska, E. A., et al. (2005). The let-7 MicroRNA family members mir-48, mir-84, and mir-241 function together to regulate developmental timing in Caenorhabditis elegans. *Developmental Cell*, *9*(3), 403–414.

Adams, B. D., Kasinski, A. L., & Slack, F. J. (2014). Aberrant regulation and function of microRNAs in cancer. *Current Biology*, *24*(16), R762–R776.

An, T., Qin, S., Xu, Y., et al. (2015). Exosomes serve as tumour markers for personalized diagnostics owing to their important role in cancer metastasis. *Journal of Extracellular Vesicles*, *4*, 27522.

Aqeilan, R. I., Calin, G. A., & Croce, C. M. (2010). miR-15a and miR-16-1 in cancer: Discovery, function and future perspectives. *Cell Death and Differentiation*, *17*(2), 215–220.

Babar, I. A., Cheng, C. J., Booth, C. J., et al. (2012). Nanoparticle-based therapy in an in vivo microRNA-155 (miR-155)-dependent mouse model of lymphoma. *Proceedings of the National Academy of Sciences of the United States of America*, *109*(26), E1695–E1704.

Balca-Silva, J., Sousa Neves, S., Goncalves, A. C., et al. (2012). Effect of miR-34b overexpression on the radiosensitivity of non-small cell lung cancer cell lines. *Anticancer Research*, *32*(5), 1603–1609.

Bandres, E., Agirre, X., Bitarte, N., et al. (2009). Epigenetic regulation of microRNA expression in colorectal cancer. *International Journal of Cancer*, *125*(11), 2737–2743.

Bartel, D. P. (2004). MicroRNAs: Genomics, biogenesis, mechanism, and function. *Cell*, *116*(2), 281–297.

Bartel, D. P. (2009). MicroRNAs: Target recognition and regulatory functions. *Cell*, *136*(2), 215–233.

Berezikov, E., Bargmann, C. I., & Plasterk, R. H. (2004). Homologous gene targeting in Caenorhabditis elegans by biolistic transformation. *Nucleic Acids Research*, *32*, 4):e40.

Bohlig, L., Friedrich, M., & Engeland, K. (2011). p53 activates the PANK1/miRNA-107 gene leading to downregulation of CDK6 and p130 cell cycle proteins. *Nucleic Acids Research*, *39*(2), 440–453.

Bonci, D., Coppola, V., Musumeci, M., et al. (2008). The miR-15a-miR-16-1 cluster controls prostate cancer by targeting multiple oncogenic activities. *Nature Medicine*, *14*(11), 1271–1277.

Bou Kheir, T., Futoma-Kazmierczak, E., Jacobsen, A., et al. (2011). miR-449 inhibits cell proliferation and is down-regulated in gastric cancer. *Molecular Cancer*, *10*, 29.

Bouabe, H., & Okkenhaug, K. (2013). Gene targeting in mice: A review. *Methods in Molecular Biology*, *1064*, 315–336.

Bracht, J., Hunter, S., Eachus, R., Weeks, P., & Pasquinelli, A. E. (2004). Trans splicing and polyadenylation of let-7 microRNA primary transcripts. *RNA*, *10*(10), 1586–1594.

Brennecke, J., Hipfner, D. R., Stark, A., Russell, R. B., & Cohen, S. M. (2003). Bantam encodes a developmentally regulated microRNA that controls cell proliferation and regulates the proapoptotic gene hid in Drosophila. *Cell*, *113*(1), 25–36.

Brueckner, B., Stresemann, C., Kuner, R., et al. (2007). The human let-7a-3 locus contains an epigenetically regulated microRNA gene with oncogenic function. *Cancer Research*, *67*(4), 1419–1423.

Brumby, A., & Richardson, H. (2005). Using Drosophila melanogaster to map human cancer pathways. *Nature Reviews. Cancer*, *5*(8), 626–639.

Burroughs, A. M., Ando, Y., de Hoon, M. J., et al. (2011). Deep-sequencing of human Argonaute-associated small RNAs provides insight into miRNA sorting and reveals Argonaute association with RNA fragments of diverse origin. *RNA Biology*, *8*(1), 158–177.

Buscaglia, L. E., & Li, Y. (2011). Apoptosis and the target genes of microRNA-21. *Chinese Journal of Cancer*, *30*(6), 371–380.

Calin, G. A., Cimmino, A., Fabbri, M., et al. (2008). MiR-15a and miR-16-1 cluster functions in human leukemia. *Proceedings of the National Academy of Sciences of the United States of America*, *105*(13), 5166–5171.

Calin, G. A., Dumitru, C. D., Shimizu, M., et al. (2002). Frequent deletions and down-regulation of micro-RNA genes miR15 and miR16 at 13q14 in chronic lymphocytic leukemia. *Proceedings of the National Academy of Sciences of the United States of America*, *99*, 15524–15529.

Calin, G. A., & Fabbri, M. (2010). Chapter 4—Epigenetics and miRNAs in human cancer. *Advances in Genetics*, *70*, 87–99.

Calin, G. A., Sevignani, C., Dumitru, C. D., et al. (2004). Human microRNA genes are frequently located at fragile sites and genomic regions involved in cancers. *Proceedings of the National Academy of Sciences of the United States of America*, *101*(9), 2999–3004.

Caygill, E. E., & Johnston, L. A. (2008). Temporal regulation of metamorphic processes in Drosophila by the let-7 and miR-125 heterochronic microRNAs. *Current Biology*, *18*(13), 943–950.

Cech, T. R., & Steitz, J. A. (2014). The noncoding RNA revolution—Trashing old rules to forge new ones. *Cell*, *157*(1), 77–94.

Cenik, E. S., & Zamore, P. D. (2011). Argonaute proteins. *Current Biology*, *21*(12), R446–449.

Chang, C. J., Chao, C. H., Xia, W., et al. (2011). p53 regulates epithelial-mesenchymal transition and stem cell properties through modulating miRNAs. *Nature Cell Biology*, *13*(3), 317–323.

Chang, K. W., Kao, S. Y., YH, W., et al. (2013). Passenger strand miRNA miR-31* regulates the phenotypes of oral cancer cells by targeting RhoA. *Oral Oncology*, *49*(1), 27–33.

Chang, T. C., Yu, D., Lee, Y. S., et al. (2008). Widespread microRNA repression by Myc contributes to tumorigenesis. *Nature Genetics*, *40*(1), 43–50.

Chawla, G., Deosthale, P., Childress, S., Wu, Y. C., & Sokol, N. S. (2016). A let-7-to-miR-125 MicroRNA switch regulates neuronal integrity and lifespan in Drosophila. *PLoS Genetics*, *12*(8), e1006247.

Chendrimada, T. P., Gregory, R. I., Kumaraswamy, E., et al. (2005). TRBP recruits the Dicer complex to Ago2 for microRNA processing and gene silencing. *Nature*, *436*(7051), 740–744.

Cheon, D. J., & Orsulic, S. (2011). Mouse models of cancer. *Annual Review of Pathology*, *6*, 95–119.

Cho, S., Jang, I., Jun, Y., et al. (2013). MiRGator v3.0: A microRNA portal for deep sequencing, expression profiling and mRNA targeting. *Nucleic Acids Research*, *41*(Database issue), D252–257.

Choi, P. S., Zakhary, L., Choi, W. Y., et al. (2008). Members of the miRNA-200 family regulate olfactory neurogenesis. *Neuron*, *57*(1), 41–55.

Cimmino, A., Calin, G. A., Fabbri, M., et al. (2005). miR-15 and miR-16 induce apoptosis by targeting BCL2. *Proceedings of the National Academy of Sciences of the United States of America*, *102*(39), 13944–13949.

Cipolla, G. A. (2014). A non-canonical landscape of the microRNA system. *Frontiers in Genetics*, *5*, 337.

Comazzetto, S., Di Giacomo, M., Rasmussen, K. D., et al. (2014). Oligoasthenoteratozoospermia and infertility in mice deficient for miR-34b/c and miR-449 loci. *PLoS Genetics*, *10*(10), e1004597.

Concepcion, C. P., Han, Y. C., Mu, P., et al. (2012). Intact p53-dependent responses in miR-34-deficient mice. *PLoS Genetics*, *8*(7), e1002797.

Correa-Medina, M., Bravo-Egana, V., Rosero, S., et al. (2009). MicroRNA miR-7 is preferentially expressed in endocrine cells of the developing and adult human pancreas. *Gene Expression Patterns*, *9*(4), 193–199.

Costinean, S., Sandhu, S. K., Pedersen, I. M., et al. (2009). Src homology 2 domain-containing inositol-5-phosphatase and CCAAT enhancer-binding protein beta are targeted by miR-155 in B cells of Emicro-MiR-155 transgenic mice. *Blood*, *114*(7), 1374–1382.

Costinean, S., Zanesi, N., Pekarsky, Y., et al. (2006). Pre-B cell proliferation and lymphoblastic leukemia/high-grade lymphoma in E(mu)-miR155 transgenic mice. *Proceedings of the National Academy of Sciences of the United States of America*, *103*(18), 7024–7029.

Coumoul, X., & Deng, C. X. (2006). RNAi in mice: A promising approach to decipher gene functions in vivo. *Biochimie, 88*(6), 637–643.

Crick, F. H. (1958). On protein synthesis. *Symposia of the Society for Experimental Biology, 12*, 138–163.

Crick, F. (1970). Central Dogma of molecular biology. *Nature, 227*(5258), 561–563.

Deng, C.-X. (2014). Conditional knockout mouse models of cancer. *Cold Spring Harbor Protocols, 2014*(12), 1217–1233.

Doench, J. G., Petersen, C. P., & Sharp, P. A. (2003). siRNAs can function as miRNAs. *Genes & Development, 17*(4), 438–442.

du Rieu, M. C., Torrisani, J., Selves, J., et al. (2010). MicroRNA-21 is induced early in pancreatic ductal adenocarcinoma precursor lesions. *Clinical Chemistry, 56*(4), 603–612.

Du, P., Wang, L., Sliz, P., & Gregory, R. I. A. (2015). Biogenesis step upstream of microprocessor controls miR-17 approximately 92 expression. *Cell, 162*(4), 885–899.

Dueck, A., Ziegler, C., Eichner, A., Berezikov, E., & Meister, G. (2012). microRNAs associated with the different human Argonaute proteins. *Nucleic Acids Research, 40*(19), 9850–9862.

Dugas, J. C., Cuellar, T. L., Scholze, A., et al. (2010). Dicer1 and miR-219 are required for normal oligodendrocyte differentiation and myelination. *Neuron, 65*(5), 597–611.

Ecsedi, M., Rausch, M., & Grosshans, H. (2015). The let-7 microRNA directs vulval development through a single target. *Developmental Cell, 32*(3), 335–344.

Esquela-Kerscher, A., & Slack, F. J. (2006). Oncomirs—MicroRNAs with a role in cancer. *Nature Reviews. Cancer, 6*(4), 259–269.

Esquela-Kerscher, A., Trang, P., Wiggins, J. F., et al. (2008). The let-7 microRNA reduces tumor growth in mouse models of lung cancer. *Cell Cycle*, 7(6), 759–764.

Evans, T. C. (Ed.), Transformation and microinjection (April 6, 2006), *WormBook*, ed. The *C. elegans* Research Community, WormBook, http://dx.doi.org/10.1895/wormbook.1.108.1. http://www.wormbook.org.

Fabbri, M. (2012). TLRs as miRNA receptors. *Cancer Research*, 72(24), 6333–6337.

Fabbri, M., Bottoni, A., Shimizu, M., et al. (2011). Association of a microRNA/TP53 feedback circuitry with pathogenesis and outcome of B-cell chronic lymphocytic leukemia. *Journal of the American Medical Association, 305*(1), 59–67.

Filipowicz, W., & Kiss, T. (1993). Structure and function of nucleolar snRNPs. *Molecular Biology Reports, 18*(2), 149–156.

Fish, J. E., Santoro, M. M., Morton, S. U., et al. (2008). miR-126 regulates angiogenic signaling and vascular integrity. *Developmental Cell, 15*(2), 272–284.

Forman, J. J., Legesse-Miller, A., & Coller, H. A. (2008). A search for conserved sequences in coding regions reveals that the let-7 microRNA targets Dicer within its coding sequence. *Proceedings of the National Academy of Sciences of the United States of America, 105*(39), 14879–14884.

Fox, G. E., & Woese, C. R. (1975). The architecture of 5S rRNA and its relation to function. *Journal of Molecular Evolution, 6*(1), 61–76.

Gartner, A., Milstein, S., Ahmed, S., Hodgkin, J., & Hengartner, M. O. (2000). A conserved checkpoint pathway mediates DNA damage-induced apoptosis and cell cycle arrest in C. elegans. *Molecular Cell, 5*(3), 435–443.

Geiger, G. A., Fu, W., & Kao, G. D. (2008). Temozolomide-mediated Radiosensitization of human glioma cells in a zebrafish embryonic system. *Cancer Research, 68*(9), 3396–3404.

Gibbons, D. L., Lin, W., Creighton, C. J., et al. (2009). Contextual extracellular cues promote tumor cell EMT and metastasis by regulating miR-200 family expression. *Genes & Development, 23*(18), 2140–2151.

Giraldez, A. J., Cinalli, R. M., Glasner, M. E., et al. (2005). MicroRNAs regulate brain morphogenesis in zebrafish. *Science, 308*(5723), 833–838.

Girard, M., Jacquemin, E., Munnich, A., Lyonnet, S., & Henrion-Caude, A. (2008). miR-122, a paradigm for the role of microRNAs in the liver. *Journal of Hepatology*, *48*(4), 648–656.
Goodwin, D., Rosenzweig, B., Zhang, J., et al. (2014). Evaluation of miR-216a and miR-217 as potential biomarkers of acute pancreatic injury in rats and mice. *Biomarkers*, *19*(6), 517–529.
Gregory, R. I., Yan, K. P., Amuthan, G., et al. (2004). The microprocessor complex mediates the genesis of microRNAs. *Nature*, *432*(7014), 235–240.
Griffiths-Jones, S., Saini, H. K., van Dongen, S., & Enright, A. J. (2008). miRBase: Tools for microRNA genomics. *Nucleic Acids Research*, *36*(Database issue), D154–158.
Grishok, A., Pasquinelli, A. E., Conte, D., et al. (2001). Genes and mechanisms related to RNA interference regulate expression of the small temporal RNAs that control C. elegans developmental timing. *Cell*, *106*(1), 23–34.
Guernet, A., & Grumolato, L. (2017). CRISPR/Cas9 editing of the genome for cancer modeling. *Methods*, *121–122*, 130–137.
Ha, M., & Kim, V. N. (2014). Regulation of microRNA biogenesis. *Nature Reviews. Molecular Cell Biology*, *15*, 509–524.
Han, J., Lee, Y., Yeom, K. H., et al. (2006). Molecular basis for the recognition of primary microRNAs by the Drosha-DGCR8 complex. *Cell*, *125*(5), 887–901.
Hanahan, D., Wagner, E. F., & Palmiter, R. D. (2007). The origins of oncomice: A history of the first transgenic mice genetically engineered to develop cancer. *Genes & Development*, *21*(18), 2258–2270.
Hannafon, B. N., Trigoso, Y. D., Calloway, C. L., et al. (2016). Plasma exosome microRNAs are indicative of breast cancer. *Breast Cancer Research*, *18*(1), 90.
Hansen, T. B., Jensen, T. I., Clausen, B. H., et al. (2013). Natural RNA circles function as efficient microRNA sponges. *Nature*, *495*(7441), 384–388.
Hatley, M. E., Patrick, D. M., Garcia, M. R., et al. (2010). Modulation of K-Ras-dependent lung tumorigenesis by MicroRNA-21. *Cancer Cell*, *18*(3), 282–293.
Hausser, J., & Zavolan, M. (2014). Identification and consequences of miRNA-target interactions—Beyond repression of gene expression. *Nature Reviews. Genetics*, *15*, 599–612.
Hayashita, Y., Osada, H., Tatematsu, Y., et al. (2005). A polycistronic microRNA cluster, miR-17-92, is overexpressed in human lung cancers and enhances cell proliferation. *Cancer Research*, *65*(21), 9628–9632.
He, L., Thomson, J. M., Hemann, M. T., et al. (2005). A microRNA polycistron as a potential human oncogene. *Nature*, *435*(7043), 828–833.
Helwak, A., Kudla, G., Dudnakova, T., & Tollervey, D. (2013). Mapping the human miRNA interactome by CLASH reveals frequent noncanonical binding. *Cell*, *153*(3), 654–665.
Henson, B. J., Bhattacharjee, S., O'Dee, D. M., Feingold, E., & Gollin, S. M. (2009). Decreased expression of miR-125b and miR-100 in oral cancer cells contributes to malignancy. *Genes, Chromosomes & Cancer*, *48*(7), 569–582.
Heo, I., Joo, C., Cho, J., Ha, M., Han, J., & Kim, V. N. (2008). Lin28 mediates the terminal uridylation of let-7 precursor MicroRNA. *Molecular Cell*, *32*(2), 276–284.
Hermeking, H. (2007). p53 enters the microRNA world. *Cancer Cell*, *12*(5), 414–418.
Hertel, J., Lindemeyer, M., Missal, K., et al. (2006). The expansion of the metazoan microRNA repertoire. *BMC Genomics*, 7(1), 25.
Hidalgo, M., Amant, F., Biankin, A. V., et al. (2014). Patient-derived xenograft models: An emerging platform for translational cancer research. *Cancer Discovery*, *4*(9), 998–1013.
Hirata, Y., Murai, N., Yanaihara, N., et al. (2014). MicroRNA-21 is a candidate driver gene for 17q23-25 amplification in ovarian clear cell carcinoma. *BMC Cancer*, *14*, 799.
Holley, R. W., Apgar, J., Everett, G. A., et al. (1965). Structure of a ribonucleic acid. *Science*, *147*(3664), 1462–1465.

Hong, L., Pan, F., Jiang, H., et al. (2016). miR-125b inhibited epithelial-mesenchymal transition of triple-negative breast cancer by targeting MAP2K7. *OncoTargets and Therapy*, *9*, 2639–2648.

Hornstein, E., & Shomron, N. (2006). Canalization of development by microRNAs. *Nature Genetics*, *38*, S20–24.

Hudish, L. I., Blasky, A. J., & Appel, B. (2013). miR-219 regulates neural precursor differentiation by direct inhibition of apical par polarity proteins. *Developmental Cell*, *27*(4), 387–398.

Humanized Mice in Tumor Studies, Charles River. (2017). from http://www.criver.com/products-services/drug-discovery/therapeuticareas/oncology/pharmacology/in-vivo-study-models/humanized-mice(July 21, 2017).

Hummel, R., Hussey, D. J., & Haier, J. (2010). MicroRNAs: Predictors and modifiers of chemo- and radiotherapy in different tumour types. *European Journal of Cancer*, *46*(2), 298–311.

Humphrey, S. E., & Kasinski, A. L. (2015). RNA-guided CRISPR-Cas technologies for genome-scale investigation of disease processes. *Journal of Hematology & Oncology*, *8*, 31.

Hutvagner, G., McLachlan, J., Pasquinelli, A. E., Balint, E., Tuschl, T., & Zamore, P. D. A. (2001). Cellular function for the RNA-interference enzyme Dicer in the maturation of the let-7 small temporal RNA. *Science*, *293*(5531), 834–838.

Ibáñez-Ventoso, C., Vora, M., & Driscoll, M. (2008). Sequence relationships among C. elegans, D. melanogaster and human microRNAs highlight the extensive conservation of microRNAs in biology. *PloS One*, *3*.

Iorio, M. V., Ferracin, M., Liu, C. G., et al. (2005). MicroRNA gene expression deregulation in human breast cancer. *Cancer Research*, *65*(16), 7065–7070.

Ito, R., Takahashi, T., Katano, I., & Ito, M. (2012). Current advances in humanized mouse models. *Cellular & Molecular Immunology*, *9*(3), 208–214.

Johnson, S. M., Grosshans, H., Shingara, J., et al. (2005). RAS is regulated by the let-7 microRNA family. *Cell*, *120*(5), 635–647.

Kalinowski, F. C., Brown, R. A., Ganda, C., et al. (2014). microRNA-7: A tumor suppressor miRNA with therapeutic potential. *The International Journal of Biochemistry & Cell Biology*, *54*, 312–317.

Kasinski, A. L., Kelnar, K., Stahlhut, C., et al. (2015). A combinatorial microRNA therapeutics approach to suppressing non-small cell lung cancer. *Oncogene*, *34*(27), 3547–3555.

Kasinski, A. L., & Slack, F. J. (2010). Potential microRNA therapies targeting Ras. NFkappaB and p53 signaling. *Current Opinion in Molecular Therapeutics*, *12*(2), 147–157.

Kasinski, A. L., & Slack, F. J. (2011). MicroRNAs en route to the clinic: Progress in validating and targeting microRNAs for cancer therapy. *Nature Reviews. Cancer*, *11*(12), 849–864.

Kasinski, A. L., & Slack, F. J. (2012). miRNA-34 prevents cancer initiation and progression in a therapeutically resistant K-ras and p53-induced mouse model of lung adenocarcinoma. *Cancer Research*, *72*(21), 5576–5587.

Kasinski, A. L., & Slack, F. J. (2013). Generation of mouse lung epithelial cells. *Bio Protocol*, *3*(15), e837.

Kato, M., Paranjape, T., Muller, R. U., et al. (2009). The mir-34 microRNA is required for the DNA damage response in vivo in C. elegans and in vitro in human breast cancer cells. *Oncogene*, *28*(25), 2419–2424.

Kato, M., & Slack, F. J. (2008). microRNAs: Small molecules with big roles—C. elegans to human cancer. *Biology of the Cell*, *100*(2), 71–81.

Kawamata, T., Yoda, M., & Tomari, Y. (2011). Multilayer checkpoints for microRNA authenticity during RISC assembly. *EMBO Reports*, *12*, 944–949.

Kefas, B., Godlewski, J., Comeau, L., et al. (2008). microRNA-7 inhibits the epidermal growth factor receptor and the Akt pathway and is down-regulated in glioblastoma. *Cancer Research*, *68*(10), 3566–3572.

Kersten, K., de Visser, K. E., van Miltenburg, M. H., & Jonkers, J. (2017). Genetically engineered mouse models in oncology research and cancer medicine. *EMBO Molecular Medicine*, *9*(2), 137–153.

Ketting, R. F., Fischer, S. E., Bernstein, E., Sijen, T., Hannon, G. J., & Plasterk, R. H. (2001). Dicer functions in RNA interference and in synthesis of small RNA involved in developmental timing in C. elegans. *Genes & Development*, *15*(20), 2654–2659.

Khuu, C., Utheim, T. P., & Sehic, A. (2016). The three paralogous MicroRNA clusters in development and disease, miR-17-92, miR-106a-363, and miR-106b-25, Scientifica (Cairo) 2016. *miR-17-92, miR-106a-363, and miR-106b-25, Scientifica (Cairo)* (pp. 10). Article ID 1379643. http://dx.doi.org/10.1155/2016/1379643.

Kim, K., Chadalapaka, G., Lee, S. O., et al. (2012). Identification of oncogenic microRNA-17-92/ZBTB4/specificity protein axis in breast cancer. *Oncogene*, *31*(8), 1034–1044.

Kim, M., Kasinski, A. L., & Slack, F. J. (2011). MicroRNA therapeutics in preclinical cancer models. *Lancet Oncology*, *12*(4), 319–321.

Kim, Y. K., Kim, B., & Kim, V. N. (2016). Re-evaluation of the roles of DROSHA, export in 5, and DICER in microRNA biogenesis. *Proceedings of the National Academy of Sciences of the United States of America*, *113*(13), E1881–1889.

Kim, D. H., Saetrom, P., Snove, O., Jr., & Rossi, J. J. (2008). MicroRNA-directed transcriptional gene silencing in mammalian cells. *Proceedings of the National Academy of Sciences of the United States of America*, *105*(42), 16230–16235.

Kim, T., Veronese, A., Pichiorri, F., et al. (2011). p53 regulates epithelial-mesenchymal transition through microRNAs targeting ZEB1 and ZEB2. *The Journal of Experimental Medicine*, *208*(5), 875–883.

King, C. E., Cuatrecasas, M., Castells, A., Sepulveda, A. R., Lee, J.-S., & Rustgi, A. K. (2011). LIN28B promotes colon cancer progression and metastasis. *Cancer Research*, *71*(12), 4260–4268.

Kirienko, N. V., Mani, K., & Fay, D. S. (2010). Cancer models in Caenorhabditis elegans. *Developmental Dynamics*, *239*(5), 1413–1448.

Klein, U., Lia, M., Crespo, M., et al. (2010). The DLEU2/miR-15a/16-1 cluster controls B cell proliferation and its deletion leads to chronic lymphocytic leukemia. *Cancer Cell*, *17*(1), 28–40.

Kloosterman, W. P., Wienholds, E., Ketting, R. F., & Plasterk, R. H. A. (2004). Substrate requirements for let-7 function in the developing zebrafish embryo. *Nucleic Acids Research*, *32*(21), 6284–6291.

Kondo, S., & Ueda, R. (2013). Highly improved gene targeting by germline-specific Cas9 expression in Drosophila. *Genetics*, *195*(3), 715–721.

Korpal, M., & Kang, Y. (2008). The emerging role of miR-200 family of microRNAs in epithelial-mesenchymal transition and cancer metastasis. *RNA Biology*, *5*(3), 115–119.

Kozomara, A., & Griffiths-Jones, S. (2014). miRBase: Annotating high confidence microRNAs using deep sequencing data. *Nucleic Acids Research*, *42*(Database issue), D68–D73.

Krutzfeldt, J., Rajewsky, N., Braich, R., et al. (2005). Silencing of microRNAs in vivo with 'antagomirs'. *Nature*, *438*(7068), 685–689.

Kumar, M. S., Erkeland, S. J., Pester, R. E., et al. (2008). Suppression of non-small cell lung tumor development by the let-7 microRNA family. *Proceedings of the National Academy of Sciences of the United States of America*, *105*(10), 3903–3908.

Kumar, M. S., Pester, R. E., Chen, C. Y., et al. (2009). Dicer1 functions as a haploinsufficient tumor suppressor. *Genes & Development*, *23*(23), 2700–2704.

Kyriakakis, E., Markaki, M., & Tavernarakis, N. (2015). Caenorhabditis elegans as a model for cancer research. *Molecular & Cellular Oncology*, *2*, e975027. http://dx.doi.org/10.4161/23723556.2014.975027.

Lacombe, J., & Zenhausern, F. (2017). Emergence of miR-34a in radiation therapy. *Critical Reviews in Oncology/Hematology*, *109*, 69–78.

Lagos-Quintana, M., Rauhut, R., Borkhardt, A., & Tuschl, T. (2003). New microRNAs from mouse and human. *RNA*, *9*(2), 175–179.

Lagos-Quintana, M., Rauhut, R., Lendeckel, W., & Tuschl, T. (2001). Identification of novel genes coding for small expressed RNAs. *Science*, *294*(5543), 853–858.

Lahdaoui, F., Delpu, Y., Vincent, A., et al. (2015). miR-219-1-3p is a negative regulator of the mucin MUC4 expression and is a tumor suppressor in pancreatic cancer. *Oncogene*, *34*(6), 780–788.

Lambertz, I., Nittner, D., Mestdagh, P., et al. (2010). Monoallelic but not biallelic loss of Dicer1 promotes tumorigenesis in vivo. *Cell Death and Differentiation*, *17*(4), 633–641.

Lau, N. C., Lim, L. P., Weinstein, E. G., & Bartel, D. P. (2001). An abundant class of tiny RNAs with probable regulatory roles in Caenorhabditis elegans. *Science*, *294*(5543), 858–862.

Lee, I., Ajay, S. S., Yook, J. I., et al. (2009). New class of microRNA targets containing simultaneous 5′-UTR and 3′-UTR interaction sites. *Genome Research*, *19*(7), 1175–1183.

Lee, R. C., & Ambros, V. (2001). An extensive class of small RNAs in Caenorhabditis elegans. *Science*, *294*(5543), 862–864.

Lee, R. C., Feinbaum, R. L., & Ambros, V. (1993). The C. elegans heterochronic gene lin-4 encodes small RNAs with antisense complementarity to lin-14. *Cell*, *75*(5), 843–854.

Lee, H., Han, S., Kwon, C. S., & Lee, D. (2016). Biogenesis and regulation of the let-7 miRNAs and their functional implications. *Protein & Cell*, 7(2), 100–113.

Lee, Y., Kim, M., Han, J., et al. (2004). MicroRNA genes are transcribed by RNA polymerase II. *The EMBO Journal*, *23*(20), 4051–4060.

Lee, M., Kim, E. J., & Jeon, M. J. (2016). MicroRNAs 125a and 125b inhibit ovarian cancer cells through post-transcriptional inactivation of EIF4EBP1. *Oncotarget*, 7(8), 8726.

Lei, H., Zou, D., Li, Z., et al. (2013). MicroRNA-219-2-3p functions as a tumor suppressor in gastric cancer and is regulated by DNA methylation. *PloS One*, *8*(4), e60369.

Lerner, M., Harada, M., Loven, J., et al. (2009). DLEU2, frequently deleted in malignancy, functions as a critical host gene of the cell cycle inhibitory microRNAs miR-15a and miR-16-1. *Experimental Cell Research*, *315*(17), 2941–2952.

Lerner, M. R., & Steitz, J. A. (1979). Antibodies to small nuclear RNAs complexed with proteins are produced by patients with systemic lupus erythematosus. *Proceedings of the National Academy of Sciences of the United States of America*, *76*(11), 5495–5499.

Lewis, B. P., Burge, C. B., & Bartel, D. P. (2005). Conserved seed pairing, often flanked by adenosines, indicates that thousands of human genes are microRNA targets. *Cell*, *120*(1), 15–20.

Li, X., & Carthew, R. W. (2005). A microRNA mediates EGF receptor signaling and promotes photoreceptor differentiation in the Drosophila eye. *Cell*, *123*(7), 1267–1277.

Li, X., Cassidy, J. J., Reinke, C. A., Fischboeck, S., & Carthew, R. W. (2009). A microRNA imparts robustness against environmental fluctuation during development. *Cell*, *137*(2), 273–282.

Liang, L., Wong, C. M., Ying, Q., et al. (2010). MicroRNA-125b suppressesed human liver cancer cell proliferation and metastasis by directly targeting oncogene LIN28B2. *Hepatology*, *52*(5), 1731–1740.

Lin, S., & Gregory, R. I. (2015). MicroRNA biogenesis pathways in cancer. *Nature Reviews. Cancer*, *15*(6), 321–333.

Lin, S. Y., Johnson, S. M., Abraham, M., et al. (2003). The C. elegans hunchback homolog, hbl-1, controls temporal patterning and is a probable microRNA target. *Developmental Cell*, *4*(5), 639–650.

Liu, J., Carmell, M. A., Rivas, F. V., et al. (2004). Argonaute2 is the catalytic engine of mammalian RNAi. *Science*, *305*(5689), 1437–1441.

Liu, S., & Leach, S. D. (2011). Zebrafish models for cancer. *Annual Review of Pathology*, *6*, 71–93.

Liu, Y., Luo, D., Zhao, H., Zhu, Z., Hu, W., & Cheng, C. H. (2013). Inheritable and precise large genomic deletions of non-coding RNA genes in zebrafish using TALENs. *PloS One*, *8*(10), e76387.

Lize, M., Herr, C., Klimke, A., Bals, R., & Dobbelstein, M. (2010). MicroRNA-449a levels increase by several orders of magnitude during mucociliary differentiation of airway epithelia. *Cell Cycle*, *9*(22), 4579–4583.

Loeb, G. B., Khan, A. A., Canner, D., et al. (2012). Transcriptome-wide miR-155 binding map reveals widespread noncanonical microRNA targeting. *Molecular Cell*, *48*(5), 760–770.

Lu, J., Getz, G., Miska, E. A., et al. (2005). MicroRNA expression profiles classify human cancers. *Nature*, *435*(7043), 834–838.

Lu, Z., Liu, M., Stribinskis, V., et al. (2008). MicroRNA-21 promotes cell transformation by targeting the programmed cell death 4 gene. *Oncogene*, *27*(31), 4373–4379.

Lujambio, A., Ropero, S., Ballestar, E., et al. (2007). Genetic unmasking of an epigenetically silenced microRNA in human cancer cells. *Cancer Research*, *67*(4), 1424–1429.

Madison, B. B., Liu, Q., Zhong, X., et al. (2013). LIN28B promotes growth and tumorigenesis of the intestinal epithelium via let-7. *Genes & Development*, *27*(20), 2233–2245.

Mao, A., Zhao, Q., Zhou, X., et al. (2016). MicroRNA-449a enhances radiosensitivity by downregulation of c-Myc in prostate cancer cells. *Scientific Reports*, *6*, 27346.

Martin, H. C., Wani, S., Steptoe, A. L., et al. (2014). Imperfect centered miRNA binding sites are common and can mediate repression of target mRNAs. *Genome Biology*, *15*(3), R51.

Medina, P. P., Nolde, M., & Slack, F. J. (2010). OncomiR addiction in an in vivo model of microRNA-21-induced pre-B-cell lymphoma. *Nature*, *467*(7311), 86–U119.

Meister, J., & Schmidt, M. H. (2010). miR-126 and miR-126*: New players in cancer. *ScientificWorldJournal*, *10*, 2090–2100.

Mello, C. C., Kramer, J. M., Stinchcomb, D., & Ambros, V. (1991). Efficient gene transfer in C. elegans: Extrachromosomal maintenance and integration of transforming sequences. *The EMBO Journal*, *10*(12), 3959–3970.

Metheetrairut, C., Adams, B. D., Nallur, S., Weidhaas, J. B., & Slack, F. J. (2017). Cel-mir-237 and its homologue, hsa-miR-125b, modulate the cellular response to ionizing radiation. *Oncogene*, *36*(4), 512–524.

Milane, L., Singh, A., Mattheolabakis, G., Suresh, M., & Amiji, M. M. (2015). Exosome mediated communication within the tumor microenvironment. *Journal of Controlled Release*, *219*, 278–294.

Misso, G., Di Martino, M. T., De Rosa, G., et al. (2014). Mir-34: A new weapon against cancer? *Molecular Therapy Nucleic Acids*, *3*, e194.

Mogilyansky, E., & Rigoutsos, I. (2013). The miR-17/92 cluster: A comprehensive update on its genomics, genetics, functions and increasingly important and numerous roles in health and disease. *Cell Death & Differentiation*, *20*(12), 1603–1614.

Moss, E. G., Lee, R. C., & Ambros, V. (1997). The cold shock domain protein LIN-28 controls developmental timing in C. elegans and is regulated by the lin-4 RNA. *Cell*, *88*(5), 637–646.

Mu, P., Han, Y. C., Betel, D., et al. (2009). Genetic dissection of the miR-17∼92 cluster of microRNAs in Myc-induced B-cell lymphomas. *Genes & Development*, *23*(24), 2806–2811.

Murphy, B. L., Obad, S., Bihannic, L., et al. (2013). Silencing of the miR-17∼92 cluster family inhibits medulloblastoma progression. *Cancer Research*, *73*(23), 7068–7078.

Nadal, E., Chen, G., Gallegos, M., et al. (2013). Epigenetic inactivation of microRNA-34b/c predicts poor disease-free survival in early-stage lung adenocarcinoma. *Clinical Cancer Research*, *19*(24), 6842–6852.

Nakanishi, K. (2016). Anatomy of RISC: How do small RNAs and chaperones activate Argonaute proteins? *Wiley Interdisciplinary Reviews: RNA*, 7, 637–660.

Narayanan, A., Hill-Teran, G., Moro, A., et al. (2016). In vivo mutagenesis of miRNA gene families using a scalable multiplexed CRISPR/Cas9 nuclease system. *Scientific Reports*, *6*, 32386.

Newman, M. A., Thomson, J. M., & Hammond, S. M. (2008). Lin-28 interaction with the let-7 precursor loop mediates regulated microRNA processing. *RNA*, *14*(8), 1539–1549.

Olive, V., Bennett, M. J., Walker, J. C., et al. (2009). miR-19 is a key oncogenic component of mir-17-92. *Genes & Development*, *23*(24), 2839–2849.

Olive, V., Jiang, I., & He, L. (2010). mir-17-92, a cluster of miRNAs in the midst of the cancer network. *The International Journal of Biochemistry & Cell Biology*, *42*(8), 1348–1354.

Olive, V., Sabio, E., Bennett, M. J., et al. (2013). A component of the mir-17-92 polycistronic oncomir promotes oncogene-dependent apoptosis. *eLife*, *2*, e00822.

Orellana, E. A., & Kasinski, A. L. (2015). MicroRNAs in cancer: A historical perspective on the path from discovery to therapy. *Cancers (Basel)*, 7(3), 1388–1405.

Orellana, E. A., Tenneti, S., Rangasamy, L., et al. (2017). FolamiRs: Ligand-targeted, vehicle-free delivery of microRNAs for the treatment of cancer. *Science Translational Medicine*, *9*, 1–10.

Ota, A., Tagawa, H., Karnan, S., et al. (2004). Identification and characterization of a novel gene, C13orf25, as a target for 13q31-q32 amplification in malignant lymphoma. *Cancer Research*, *64*(9), 3087–3095.

Park, C. Y., Choi, Y. S., & McManus, M. T. (2010). Analysis of microRNA knockouts in mice. *Human Molecular Genetics*, *19*(R2), R169–175.

Park, J. E., Heo, I., Tian, Y., et al. (2011). Dicer recognizes the 5′ end of RNA for efficient and accurate processing. *Nature*, *475*(7355), 201–205.

Park, C. Y., Jeker, L. T., Carver-Moore, K., et al. (2012). A resource for the conditional ablation of microRNAs in the mouse. *Cell Reports*, *1*(4), 385–391.

Pasquinelli, A. E., Reinhart, B. J., Slack, F., et al. (2000). Conservation of the sequence and temporal expression of let-7 heterochronic regulatory RNA. *Nature*, *408*(6808), 86–89.

Pekarsky, Y., & Croce, C. M. (2015). Role of miR-15/16 in CLL. *Cell Death and Differentiation*, *22*(1), 6–11.

Piskounova, E., Viswanathan, S. R., Janas, M., et al. (2008). Determinants of microRNA processing inhibition by the developmentally regulated RNA-binding protein Lin28. *The Journal of Biological Chemistry*, *283*(31), 21310–21314.

Praitis, V., Casey, E., Collar, D., & Austin, J. (2001). Creation of low-copy integrated transgenic lines in Caenorhabditis elegans. *Genetics*, *157*(3), 1217–1226.

Rakheja, D., Chen, K. S., Liu, Y., et al. (2014). Somatic mutations in DROSHA and DICER1 impair microRNA biogenesis through distinct mechanisms in Wilms tumours. *Nature Communications*, *2*, 4802.

Rane, J. K., Erb, H. H., Nappo, G., et al. (2016). Inhibition of the glucocorticoid receptor results in an enhanced miR-99a/100-mediated radiation response in stem-like cells from human prostate cancers. *Oncotarget*, 7(32), 51965–51980.

Raveche, E. S., Salerno, E., Scaglione, B. J., et al. (2007). Abnormal microRNA-16 locus with synteny to human 13q14 linked to CLL in NZB mice. *Blood*, *109*(12), 5079–5086.

Raver-Shapira, N., Marciano, E., Meiri, E., et al. (2007). Transcriptional activation of miR-34a contributes to p53-mediated apoptosis. *Molecular Cell*, *26*(5), 731–743.

Reinhart, B. J., Slack, F. J., Basson, M., et al. (2000). The 21-nucleotide let-7 RNA regulates developmental timing in Caenorhabditis elegans. *Nature*, *403*(6772), 901–906.

Roush, S., & Slack, F. J. (2008). The let-7 family of microRNAs. *Trends in Cell Biology*, *18*(10), 505–516.

Ruby, J. G., Jan, C. H., & Bartel, D. P. (2007). Intronic microRNA precursors that bypass Drosha processing. *Nature*, *448*(7149), 83–86.

Rupaimoole, R., & Slack, F. J. (2017). MicroRNA therapeutics: Towards a new era for the management of cancer and other diseases. *Nature Reviews. Drug Discovery*, *16*, 203–222.

Rybak, A., Fuchs, H., Smirnova, L., et al. (2008). A feedback loop comprising lin-28 and let-7 controls pre-let-7 maturation during neural stem-cell commitment. *Nature Cell Biology*, *10*(8), 987–993.

Saito, Y., Friedman, J. M., Chihara, Y., Egger, G., Chuang, J. C., & Liang, G. (2009). Epigenetic therapy upregulates the tumor suppressor microRNA-126 and its host gene EGFL7 in human cancer cells. *Biochemical and Biophysical Research Communications*, *379*(3), 726–731.

Saito, K., Ishizuka, A., Siomi, H., & Siomi, M. C. (2005). Processing of pre-microRNAs by the Dicer-1-loquacious complex in Drosophila cells. *PLoS Biology*, *3*(7), e235.

Saito, Y., Liang, G., Egger, G., et al. (2006). Specific activation of microRNA-127 with downregulation of the proto-oncogene BCL6 by chromatin-modifying drugs in human cancer cells. *Cancer Cell*, *9*(6), 435–443.

Sampson, V. B., Rong, N. H., Han, J., et al. (2007). MicroRNA let-7a down-regulates MYC and reverts MYC-induced growth in Burkitt lymphoma cells. *Cancer Research*, *67*(20), 9762–9770.

Sandhu, S. K., Fassan, M., Volinia, S., et al. (2013). B-cell malignancies in microRNA Eμ-miR-17∼92 transgenic mice. *Proceedings of the National Academy of Sciences of the United States of America*, *110*, 18208–18213.

Sato, F., Tsuchiya, S., Meltzer, S. J., & Shimizu, K. (2011). MicroRNAs and epigenetics. *The FEBS Journal*, *278*(10), 1598–1609.

Scaglione, B. J., Salerno, E., Balan, M., et al. (2007). Murine models of chronic lymphocytic leukaemia: Role of microRNA-16 in the New Zealand black mouse model. *British Journal of Haematology*, *139*(5), 645–657.

Seok, H., Ham, J., Jang, E.-S., & Chi, aSW. (2016). MicroRNA target recognition: Insights from transcriptome-wide non-canonical interactions. *Molecules and Cells*, *39*(5), 375–381.

Shalem, O., Sanjana, N. E., & Zhang, F. (2015). High-throughput functional genomics using CRISPR-Cas9. *Nature Reviews. Genetics*, *16*(5), 299–311.

Shi, J. A., DL, L., Huang, X., & Tan, W. (2014). miR-219 inhibits the proliferation, migration and invasion of medulloblastoma cells by targeting CD164. *International Journal of Molecular Medicine*, *34*(1), 237–243.

Sicard, F., Gayral, M., Lulka, H., Buscail, L., & Cordelier, P. (2013). Targeting miR-21 for the therapy of pancreatic cancer. *Molecular Therapy*, *21*(5), 986–994.

Sokol, N. S., Xu, P., Jan, Y. N., & Ambros, V. (2008). Drosophila let-7 microRNA is required for remodeling of the neuromusculature during metamorphosis. *Genes & Development*, *22*(12), 1591–1596.

Song, R., Walentek, P., Sponer, N., et al. (2014). miR-34/449 miRNAs are required for motile ciliogenesis by repressing cp110. *Nature*, *510*(7503), 115–120.

Sonoki, T., Iwanaga, E., Mitsuya, H., & Asou, N. (2005). Insertion of microRNA-125b-1, a human homologue of lin-4, into a rearranged immunoglobulin heavy chain gene locus in a patient with precursor B-cell acute lymphoblastic leukemia. *Leukemia*, *19*, 2009–2010. England.

Steinkraus, B. R., Toegel, M., & Fulga, T. A. (2016). Tiny giants of gene regulation: Experimental strategies for microRNA functional studies. *Wiley Interdisciplinary Reviews: Developmental Biology*, *5*(3), 311–362.

Sun, J., Gao, B., Zhou, M., et al. (2013). Comparative genomic analysis reveals evolutionary characteristics and patterns of microRNA clusters in vertebrates. *Gene*, *512*(2), 383–391.

Sun, S., Wang, Y., Zhou, R., et al. (2017). Targeting and regulating of an oncogene via nanovector delivery of MicroRNA using patient-derived xenografts. *Theranostics*, 7(3), 677–693.

Sun, X., Zhang, S., & Ma, X. (2015). Prognostic value of MicroRNA-125 in various human malignant neoplasms: A meta-analysis. *Clinical Laboratory*, *61*(11), 1667–1674.

Takamizawa, J., Konishi, H., Yanagisawa, K., et al. (2004). Reduced expression of the let-7 microRNAs in human lung cancers in association with shortened postoperative survival. *Cancer Research*, *64*(11), 3753–3756.

Takehara, S., Schulz, T. C., Abe, S., et al. (2014). A novel transchromosomic system: Stable maintenance of an engineered Mb-sized human genomic fragment translocated to a mouse chromosome terminal region. *Transgenic Research*, *23*(3), 441–453.

Tanaka, N., Toyooka, S., Soh, J., et al. (2012). Frequent methylation and oncogenic role of microRNA-34b/c in small-cell lung cancer. *Lung Cancer*, *76*(1), 32–38.

Trang, P., Medina, P. P., Wiggins, J. F., et al. (2010). Regression of murine lung tumors by the let-7 microRNA. *Oncogene*, *29*(11), 1580–1587.

Trang, P., Wiggins, J. F., Daige, C. L., et al. (2011). Systemic delivery of tumor suppressor microRNA mimics using a neutral lipid emulsion inhibits lung tumors in mice. *Molecular Therapy*, *19*(6), 1116–1122.

Tsujimoto, Y., Finger, L. R., Yunis, J., Nowell, P. C., & Croce, C. M. (1984). Cloning of the chromosome breakpoint of neoplastic B cells with the t(14;18) chromosome translocation. *Science*, *226*(4678), 1097–1099.

Venken, K. J., & Bellen, H. J. (2007). Transgenesis upgrades for Drosophila melanogaster. *Development*, *134*(20), 3571–3584.

Ventura, A., Young, A. G., Winslow, M. M., et al. (2008). Targeted deletion reveals essential and overlapping functions of the miR-17 through 92 family of miRNA clusters. *Cell*, *132*(5), 875–886.

Visone, R., Pallante, P., Vecchione, A., et al. (2007). Specific microRNAs are downregulated in human thyroid anaplastic carcinomas. *Oncogene*, *26*(54), 7590–7595.

Viswanathan, S. R., Daley, G. Q., & Gregory, R. I. (2008). Selective blockade of microRNA processing by Lin28. *Science*, *320*(5872), 97–100.

Vogt, M., Munding, J., Gruner, M., et al. (2011). Frequent concomitant inactivation of miR-34a and miR-34b/c by CpG methylation in colorectal, pancreatic, mammary, ovarian, urothelial, and renal cell carcinomas and soft tissue sarcomas. *Virchows Archiv*, *458*(3), 313–322.

Wang, Z., Lin, S., Li, J. J., et al. (2011). MYC protein inhibits transcription of the microRNA cluster MC-let-7a-1∼let-7d via noncanonical E-box. *The Journal of Biological Chemistry*, *286*(46), 39703–39714.

Wang, J., Yu, M., Guan, S., Zhang, G., & Cheng, Y. (2017). Prognostic significance of microRNA-100 in solid tumors: An updated meta-analysis. *OncoTargets and Therapy*, *10*, 493–502.

Wang, F., Yu, D., Liu, Z., et al. (2016). MiR-125b functions as a tumor suppressor and enhances chemosensitivity to cisplatin in osteosarcoma. *Technology in Cancer Research & Treatment*, *15*(6), NP105–NP112.

Wei, C. L., Wu, Q., Vega, V. B., et al. (2006). A global map of p53 transcription-factor binding sites in the human genome. *Cell*, *124*(1), 207–219.

Weidhaas, J. B., Eisenmann, D. M., Holub, J. M., & Nallur, S. V. A. (2006). Caenorhabditis elegans tissue model of radiation-induced reproductive cell death. *Proceedings of the National Academy of Sciences of the United States of America*, *103*(26), 9946–9951.

Wen, W. S., Yuan, Z. M., Ma, S. J., Xu, J., & Yuan, D. T. (2016). CRISPR-Cas9 systems: Versatile cancer modelling platforms and promising therapeutic strategies. *International Journal of Cancer*, *138*(6), 1328–1336.

Wienholds, E., Kloosterman, W. P., Miska, E., et al. (2005). MicroRNA expression in zebrafish embryonic development. *Science*, *309*(5732), 310–311.

Wienholds, E., & Plasterk, R. H. (2005). MicroRNA function in animal development. *FEBS Letters*, *579*(26), 5911–5922.

Wightman, B., Ha, I., & Ruvkun, G. (1993). Posttranscriptional regulation of the heterochronic gene lin-14 by lin-4 mediates temporal pattern formation in C. elegans. *Cell*, *75*(5), 855–862.

Wu, L., & Belasco, J. G. (2005). Micro-RNA regulation of the mammalian lin-28 gene during neuronal differentiation of embryonal carcinoma cells. *Molecular and Cellular Biology*, *25*(21), 9198–9208.

Wu, L., Nguyen, L. H., Zhou, K., et al. (2015). Precise let-7 expression levels balance organ regeneration against tumor suppression. *eLife*, *4*, e09431.

Xiao, C., Srinivasan, L., Calado, D. P., et al. (2008). Lymphoproliferative disease and autoimmunity in mice with increased miR-17-92 expression in lymphocytes. *Nature Immunology*, *9*(4), 405–414.

Xiao, A., Wang, Z., Hu, Y., et al. (2013). Chromosomal deletions and inversions mediated by TALENs and CRISPR/Cas in zebrafish. *Nucleic Acids Research*, *41*(14), e141.

Xiong, F., Wei, Z. Q., Zhu, Z. Y., & Sun, Y. H. (2013). Targeted expression in zebrafish primordial germ cells by Cre/loxP and Gal4/UAS systems. *Marine Biotechnology (New York, N.Y.)*, *15*(5), 526–539.

Xiong, G. B., Zhang, G. N., Xiao, Y., et al. (2015). MicroRNA-219-5p functions as a tumor suppressor partially by targeting platelet-derived growth factor receptor alpha in colorectal cancer. *Neoplasma*, *62*(6), 855–863.

Xu, P., Vernooy, S. Y., Guo, M., & Hay, B. A. (2003). The Drosophila microRNA Mir-14 suppresses cell death and is required for normal fat metabolism. *Current Biology*, *13*(9), 790–795.

Xue, W., Dahlman, J. E., Tammela, T., et al. (2014). Small RNA combination therapy for lung cancer. *Proceedings of the National Academy of Sciences of the United States of America*, *111*, E3553–3561.

Yang, T., Martin, P., Fogarty, B., et al. (2015). Exosome delivered anticancer drugs across the blood-brain barrier for brain cancer therapy in Danio rerio. *Pharmaceutical Research*, *32*(6), 2003–2014.

Yang, H., Wang, H., & Jaenisch, R. (2014). Generating genetically modified mice using CRISPR/Cas-mediated genome engineering. *Nature Protocols*, *9*(8), 1956–1968.

Yi, R., Qin, Y., Macara, I. G., & Cullen, B. R. (2003). Exportin-5 mediates the nuclear export of pre-microRNAs and short hairpin RNAs. *Genes & Development*, *17*(24), 3011–3016.

Yoon, J. H., Abdelmohsen, K., & Gorospe, M. (2014). Functional interactions among microRNAs and long noncoding RNAs. *Seminars in Cell & Developmental Biology*, *34*, 9–14.

Yu, F., Yao, H., Zhu, P., et al. (2007). let-7 regulates self renewal and tumorigenicity of breast cancer cells. *Cell*, *131*(6), 1109–1123.

Zhang, J., Na, S., Liu, C., Pan, S., Cai, J., & Qiu, J. (2016). MicroRNA-125b suppresses the epithelial-mesenchymal transition and cell invasion by targeting ITGA9 in melanoma. *Tumour Biology*, *37*(5), 5941–5949.

Zhang, X., Niu, D., Carbonell, A., et al. (2014). ARGONAUTE PIWI domain and microRNA duplex structure regulate small RNA sorting in Arabidopsis. *Nature Communications*, *5*(5468). http://dx.doi.org/10.1038/ncomms6468.

Zhang, J. G., Wang, J. J., Zhao, F., Liu, Q., Jiang, K., & Yang, G. H. (2010). MicroRNA-21 (miR-21) represses tumor suppressor PTEN and promotes growth and invasion in non-small cell lung cancer (NSCLC). *Clinica Chimica Acta*, *411*(11 – 12), 846–852.

Zhang, X., & Zeng, Y. (2010). Regulation of mammalian microRNA expression. *Journal of Cardiovascular Translational Research*, *3*(3), 197–203.

Zhang, J., Zhao, J., Jiang, W. J., Shan, X. W., Yang, X. M., & Gao, J. G. (2012). Conditional gene manipulation: Cre-ating a new biological era. *Journal of Zhejiang University. Science. B*, *13*(7), 511–524.

Zhao, X., He, X., Han, X., et al. (2010). MicroRNA-mediated control of oligodendrocyte differentiation. *Neuron*, *65*(5), 612–626.

Zhao, X., He, W., Li, J., et al. (2015). MiRNA-125b inhibits proliferation and migration by targeting SphK1 in bladder cancer. *American Journal of Translational Research*, 7(11), 2346–2354.

Zhao, S., Huang, J., & Ye, J. (2015). A fresh look at zebrafish from the perspective of cancer research. *Journal of Experimental & Clinical Cancer Research*, *34*, 80.

Zhao, J., Tao, Y., Zhou, Y., et al. (2015). MicroRNA-7: A promising new target in cancer therapy. *Cancer Cell International*, *15*, 103.

Zhu, H., Han, C., & Wu, T. (2015). MiR-17-92 cluster promotes hepatocarcinogenesis. *Carcinogenesis*, *36*(10), 1213–1222.

CHAPTER FOUR

Cancer Hallmarks and MicroRNAs: The Therapeutic Connection

Katrien Van Roosbroeck*,[1], George A. Calin*,†,[1]
*The University of Texas MD Anderson Cancer Center, Houston, TX, United States
†Center for RNA Interference and Non-Coding RNAs, The University of Texas MD Anderson Cancer Center, Houston, TX, United States
[1]Corresponding authors: e-mail address: kvan1@mdanderson.org; gcalin@mdanderson.org

Contents

1. Introduction 119
2. Regulation of Cell Proliferation and Cell Growth 120
3. Regulation of Apoptosis and Induction of Genomic Instability and Mutations 129
4. Regulation of Angiogenesis 129
5. Regulation of Invasion and Metastasis 133
6. Regulation of Immune Response 136
7. Conclusion 137
References 138

Abstract

Human cancers are characterized by a number of hallmarks, including sustained proliferative signaling, evasion of growth suppressors, activated invasion and metastasis, replicative immortality, angiogenesis, resistance to cell death, and evasion of immune destruction. As microRNAs (miRNAs) are deregulated in virtually all human cancers, they show involvement in each of the cancer hallmarks as well. In this chapter, we describe the involvement of miRNAs in cancer from a cancer hallmarks and targeted therapeutics point of view. As no miRNA-based cancer therapeutics are available to date, and the only clinical trial on miRNA-based cancer therapeutics (MRX34) was terminated prematurely due to serious adverse events, we are focusing on protein-coding miRNA targets for which targeted therapeutics in oncology are already approved by the FDA. For each of the cancer hallmarks, we selected major protein-coding players and describe the miRNAs that target them.

1. INTRODUCTION

In 2000, Hanahan and Weinberg proposed six hallmarks of cancer that encompass biological characteristics that are acquired during the

Advances in Cancer Research, Volume 135
ISSN 0065-230X
http://dx.doi.org/10.1016/bs.acr.2017.06.002

development of human tumors: sustaining proliferative signaling, evading growth suppressors, activating invasion and metastasis, enabling replicative immortality, inducing angiogenesis, and resisting cell death (Hanahan & Weinberg, 2000). In 2011, they published an update in which they added the emerging hallmarks "deregulating cellular energetics" (or reprogramming energy metabolism) and "avoiding immune destruction," as well as the enabling characteristics "genome instability and mutation" and "tumor-promoting inflammation" (Hanahan & Weinberg, 2011).

In this review, we will describe the involvement of microRNAs (miRNAs) in these cancer hallmarks from a targeted therapeutics point of view. As no miRNA-based cancer therapeutics are available to date, and the only clinical trial on miRNA-based cancer therapeutics (MRX34) was terminated prematurely due to serious adverse events (Van Roosbroeck & Calin, 2016), we decided to focus on protein-coding miRNA targets for which targeted therapeutics in oncology are already approved by the FDA (Table 1). For each of the cancer hallmarks, we selected a few major protein-coding players and describe the miRNAs that target them. As to date, there are no therapeutics that are FDA-approved and that target tumor metabolism or the immortalization process, we only discuss the remaining cancer hallmarks.

2. REGULATION OF CELL PROLIFERATION AND CELL GROWTH

A first key attribute of cancer cells is constitutive cell proliferation and, related to that, avoidance of growth arrest. Many genes and proteins, especially kinases and kinase receptors, are involved in these processes and continue to attract a lot of attention as potential therapeutic targets. Cyclin-dependent kinases (CDKs) are a family of protein kinases that are important for progression through the cell cycle and, consequently, proliferation. The pathways they are involved in are often deregulated in human cancers, leading to increased proliferation. Especially CDK4 and CDK6, which are activated by cyclin D and which are important for progression through the G1 phase of the cell cycle, are popular anticancer targets, with the specific CDK4/6 inhibitors ribociclib and palbociclib recently being approved by the FDA for treatment of several human cancers (Table 1) (Sanchez-Martinez, Gelbert, Lallena, & de Dios, 2015; Whittaker, Mallinger, Workman, & Clarke, 2017). *CDK4* and *CDK6* are directly

Table 1 List of Targeted Drugs Approved by the FDA for Treatment in Oncology (Between January 2000 and May 2017)

Name of Drug	Company	Target(s)	Approved for the Treatment of	Approval Date
Ribociclib	Novartis	CDK4/6	Breast cancer	March 2017
Palbociclib	Pfizer	CDK4/6	ER-positive, HER2-negative breast cancer	February 2015
Necitumumab	Eli Lilly	EGFR	Metastatic squamous non-small cell lung cancer	November 2015
Osimertinib	AstraZeneca	EGFR	EGFR T790M mutation positive non-small cell lung cancer	November 2015
Vandetanib	AstraZeneca	EGFR, VEGFR, RET, BRK, TIE2, EPHR, SCR	Thyroid cancer	April 2011
Lapatinib	GlaxoSmithKline	EGFR	Breast cancer	March 2017
Panitumumab	Amgen	EGFR	Colorectal cancer	September 2006
Cetuximab	Bristol-Myers Squibb	EGFR, HER1, c-ErbB-1	EGFR-expressing, metastatic colorectal cancer	February 2004
Erlotinib	Genentech, OSI Pharmaceuticals	HER1/EGFR	Advanced refractory metastatic non-small cell lung cancer	November 2004
Gefitinib	AstraZeneca	EGFR	Second-line treatment of non-small cell lung cancer	May 2003
Cobimetinib	Genentech	MEK1/2	BRAF V600E or V600K melanoma	November 2015

Continued

Table 1 List of Targeted Drugs Approved by the FDA for Treatment in Oncology (Between January 2000 and May 2017)—cont'd

Name of Drug	Company	Target(s)	Approved for the Treatment of	Approval Date
Trametinib	GlaxoSmithKline	MEK1/2	Unresectable or metastatic melanoma with BRAF V600E or V600K mutations	May 2013
Venetoclax	AbbVie	BCL2	Chronic lymphocytic leukemia with 17p deletion	April 2016
Niraparib	Tesaro	PARP1/2	Recurrent epithelial ovarian cancer, fallopian tube cancer, or primary peritoneal cancer	March 2017
Olaparib	AstraZeneca	PARP1/2	Previously treated BRCA mutated advanced ovarian cancer	December 2014
Rucaparib	Clovis Oncology	PARP	Advanced ovarian cancer in women with deleterious germline or somatic BRCA mutation	December 2016
Bevacizumab	Genentech	VEGFA	Metastatic carcinoma of the colon or rectum	February 2004
			Renal cell carcinoma	July 2009
Cabozantinib	Exelixis	Pan-tyrosine kinase inhibitor (RET, MET, VEGFR1/2/3, KIT, TRKB, FLT-3, AXL, TIE-2, ROS1, TYRO3, MER)	Metastatic medullary thyroid cancer	November 2012
			Advanced renal cell carcinoma	April 2016

Lenvatinib	Eisai	VEGFR1/2/3, FGFR1/2/3/4, PDGFRA, KIT, RET	Thyroid cancer	February 2015
			Advanced renal cell carcinoma	May 2016
Ramucirumab	Eli Lilly	VEGFR2	Gastric cancer	April 2014
Regorafenib	Bayer HealthCare Pharmaceuticals	Multikinase inhibitor (VEGFR2, TIE2)	Previously treated metastatic colorectal cancer	September 2012
			Gastrointestinal stromal tumor	February 2013
Axitinib	Pfizer	VEGFR1/2/3	Advanced renal cell carcinoma	January 2012
Pazopanib	GlaxoSmithKline	VEGFR1/2/3	Renal cell carcinoma	October 2009
			Soft tissue sarcoma	April 2012
Sunitinib	Pfizer	Multikinase inhibitor (VEGFR, PDGFR, KIT, FLT3, RET, CSF-1R)	Kidney cancer and gastrointestinal stromal tumors	January 2006
			Pancreatic neuroendocrine tumors	May 2011
Sorafenib	Bayer/Onyx	Multikinase inhibitor (CRAF, BRAF, KIT, FLT-3, VEGFR2/3, PDGFRB)	Renal cell carcinoma	December 2005
Pembrolizumab	Merck	PD-1	Unresectable or metastatic melanoma	September 2014
			PD-L1 positive advanced non-small cell lung cancer	October 2015
			Head and neck squamous cell cancer	August 2016

Continued

Table 1 List of Targeted Drugs Approved by the FDA for Treatment in Oncology (Between January 2000 and May 2017)—cont'd

Name of Drug	Company	Target(s)	Approved for the Treatment of	Approval Date
Nivolumab	Bristol-Myers Squibb	PD-1	Unresectable or metastatic melanoma	December 2014
			Metastatic squamous non-small cell lung cancer	March 2015
			Advanced renal cell carcinoma	November 2015
			Classical Hodgkin lymphoma	May 2016
			Recurrent or metastatic squamous cell carcinoma of the head and neck	November 2016
Avelumab	Merck	PD-L1	Merkel cell carcinoma	March 2017
Atezolizumab	Genentech	PD-L1	Urothelial carcinoma and metastatic non-small cell lung cancer	May 2016
Ipilimumab	Bristol-Myers Squibb	CTLA-4	Metastatic melanoma	March 2011
Ibrutinib	Pharmacyclics	BTK	Mantle cell lymphoma	November 2013
			Chronic lymphocytic leukemia and Waldenstrom macroglobulinemia	February 2014
Olaratumab	Eli Lilly	PDGFRA	Soft tissue sarcoma	October 2016
Nilotinib	Novartis	BCR-ABL, KIT, PDGFR	Chronic myelogenous leukemia	October 2007

Dasatinib	Bristol-Myers Squibb	Multityrosine kinase inhibitor (BCR-ABL, SRC, LCK, YES, FYN, c-KIT, EPHA2, and PDGFRB)	Imatinib-resistant chronic myeloid leukemia	June 2006
Imatinib	Novartis	Multityrosine kinase inhibitor (BCR-ABL, PDGFR, KIT)	Chronic myeloid leukemia	May 2001
			Gastrointestinal stromal tumors	February 2002
Ceritinib	Novartis	ALK	ALK+ metastatic non-small cell lung cancer	April 2014
Alectinib	Roche	ALK, RET	ALK+ metastatic non-small cell lung cancer	December 2015
Trastuzumab	Genentech	HER2	Gastric cancer	October 2010
			HER2+ metastatic breast cancer	February 2013
Pertuzumab	Genentech	HER2/neu	First-line treatment of HER2+ metastatic breast cancer	June 2012
Vemurafenib	Roche	BRAF	BRAF+ melanoma	August 2011
Dabrafenib	GlaxoSmithKline	BRAF	Unresectable or metastatic melanoma with BRAF V600E mutation	May 2013
Belinostat	Spectrum Pharmaceuticals	HDAC	Relapsed or refractory peripheral T-cell lymphoma	July 2014
Panobinostat	Novartis	HDAC	Multiple myeloma	February 2015

Continued

Table 1 List of Targeted Drugs Approved by the FDA for Treatment in Oncology (Between January 2000 and May 2017)—cont'd

Name of Drug	Company	Target(s)	Approved for the Treatment of	Approval Date
Idelalisib	Gilead	PI3K-delta	Relapsed chronic lymphocytic leukemia, follicular B-cell non-Hodgkin lymphoma and small lymphocytic lymphoma	July 2014
Gemtuzumab	Wyeth	CD33	CD33 positive acute myeloid leukemia	May 2000
Ofatumumab	GlaxoSmithKline	CD20	Chronic lymphocytic leukemia	October 2009
Obinutuzumab	Genentech	CD20	Previously untreated chronic lymphocytic leukemia	October 2013

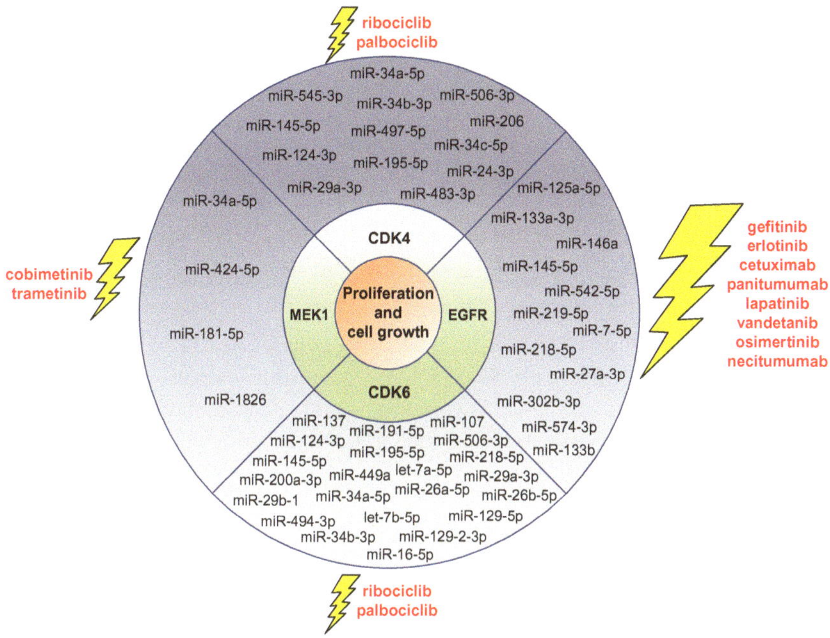

Fig. 1 miRNAs involved in the regulation of proliferation and growth of cancer cells. The *blue circle* contains miRNAs that are validated to target the protein-coding genes in the *green circle*. In *red* are the drugs that specifically target these protein-coding genes and that are approved by the FDA for use in oncology.

regulated by a variety of miRNAs, many of which target both CDKs (Fig. 1). Examples are miR-29a-3p (Schmitt et al., 2012; Zhao et al., 2015), miR-34a-5p (He et al., 2007; Hermeking, 2010; Sun et al., 2008), miR-34b-3p (Hermeking, 2010; Lujambio et al., 2008), miR-124-3p (Chen, He, et al., 2013; Pierson, Hostager, Fan, & Vibhakar, 2008), miR-145-5p (Chang et al., 2015; Chen et al., 2010), miR-195-5p (Deng et al., 2013; Furuta et al., 2013; Lin et al., 2012; Xu et al., 2009), and miR-506-3p (Liu et al., 2014). In addition, each CDK has unique miRNAs that, so far, have only been found to target either CDK4 or CDK6. For example, CDK4 is directly targeted by miR-24-3p (Lal et al., 2009), miR-34c-5p (Hermeking, 2010), miR-206 (Georgantas et al., 2014), miR-483-3p (Bertero et al., 2013), miR-497-5p (Furuta et al., 2013) and miR-545-3p (Du et al., 2014), and CDK6 is directly regulated by let-7a-5p (Zhang, Huang, et al., 2014), let-7b-5p (Johnson et al., 2007), miR-16-5p (Liu et al., 2008), miR-26a-5p (Zhu, Lu, et al., 2012), miR-26b-5p (Zhu, Lu, et al., 2012), miR-29b-1 (Ouyang et al., 2016), miR-107 (Bohlig, Friedrich, & Engeland, 2011), miR-129-2-3p (Ouyang et al., 2016), miR-129-5p (Wu et al., 2010), miR-137 (Silber et al., 2008), miR-191-5p (Lena et al., 2012), miR-200a-3p (Xiao et al., 2013), miR-218-5p (Venkataraman et al., 2013), miR-449a (Yang et al., 2009), and miR-494-3p (Olaru et al., 2011). However, it is possible that some of the unique miRNAs also target the other CDKs, but that the miRNA–protein-coding gene interaction has not been evaluated to date.

The epidermal growth factor receptor (EGFR) is a receptor tyrosine kinase that is activated by binding of its ligand, the epidermal growth factor (EGF), resulting in receptor dimerization and autophosphorylation, and activation of signaling pathways promoting proliferation. In cancer, deregulation occurs through overexpression, amplification, and mutation of *EGFR*, making it an interesting therapeutic target. Several inhibitors targeting EGFR are FDA approved for the treatment of many different cancers and include necitumumab, osimertinib, vandetanib, lapatinib, panitumumab, cetuximab, erlotinib, and gefitinib (Table 1). miRNAs that directly target *EGFR* and that have consequence for cell proliferation and cell growth are miR-7-5p (Kefas et al., 2008; Suto et al., 2015; Webster et al., 2009), miR-27a-3p (Wang, Cheng, Miao, Mei, & Wu, 2013), miR-125a-5p (Ufkin et al., 2014), miR-133a-3p (Cui, Zhang, Shan, Zhou, & Zhou, 2013; Tao et al., 2012; Wang, Hsiao, et al., 2014), miR-133b (Liu, Shao, et al., 2012; Tao et al., 2012), miR-145-5p (Cho, Chow, & Au, 2011), miR-146a (Ali et al., 2014; Chen, Umelo, et al., 2013; Xu et al., 2012), miR-218-5p (Mathew et al., 2014), miR-219-5p (Rao et al., 2013), miR-302b-3p (Wang, Yao, et al., 2013), miR-542-5p (Yamaguchi et al., 2012), and miR-574-3p (Chiyomaru et al., 2013) (Fig. 1).

The mitogen-activated protein kinases (MAPK)/extracellular signal-related kinases MEK1 (MAP2K1) and MEK2 (MAP2K2) are another pair of protein-coding genes that have FDA-approved targeted drugs and which are involved in cell proliferation. Cobimetinib and trametinib are both approved by the FDA for the treatment of unresectable or metastatic melanoma with a BRAF V600E or V600K mutation (Table 1). To the best of our knowledge, no miRNA directly interacting with *MEK2* has been described so far, but a few microRNAs directly targeting *MEK1* and influencing cell proliferation have been reported. These include miR-34a-5p (Ichimura, Ruike, Terasawa, Shimizu, & Tsujimoto, 2010), miR-181-5p (He et al., 2013), miR-424-5p (Nakashima et al., 2010), and miR-1826 (Hirata, Hinoda, Ueno, Nakajima, et al., 2012; Hirata, Hinoda, Ueno, Shahryari, et al., 2012) (Fig. 1).

3. REGULATION OF APOPTOSIS AND INDUCTION OF GENOMIC INSTABILITY AND MUTATIONS

Another cancer hallmark is the evasion of programmed cell death or apoptosis. Many pro- and antiapoptotic factors are known to be deregulated in cancer, of which the B-cell lymphoma 2 (Bcl-2) family, consisting of at least 25 pro- and antiapoptotic members, is the most prominent. Venetoclax was the first apoptosis-targeting therapeutic for cancer (Croce & Reed, 2016) approved by the FDA in 2016 for the treatment of patients with chronic lymphocytic leukemia (CLL) that harbor deletion of the short arm of chromosome 17 (del17p) (Table 1). It targets the antiapoptotic protein BCL2, which has been proven to be directly targeted by more than 50 miRNAs, many of which are cancer-related and documented to be involved in apoptosis (Table 2). Among these are the 13q14 clustered miR-15a and miR-16 that are targeting both antiapoptotic oncogenes *BCL2* and *MCL1*, and which were the first ever noncoding RNAs demonstrated to be altered in human disease, more specifically CLL (Calin et al., 2008, 2002).

Poly(ADP-ribose) polymerases (PARP) and topoisomerases are involved in a number of cellular processes, including DNA repair, genomic instability, and apoptosis. The activity of PARPs is indispensable for the base excision repair of single-stranded DNA breaks. Therefore, inhibition of PARP will result in attenuated DNA damage repair and apoptosis. In addition, PARP inhibition boosts the cytotoxicity of DNA-damaging agents and is, consequently, able to reverse radio- and chemoresistance of tumor cells. PARP inhibitors that have been approved by the FDA for oncology target PARP1/2 and include niraparib (for the treatment of recurrent epithelial ovarian cancer, fallopian tube cancer, or primary peritoneal cancer) and olaparib and rucaparib (for the treatment of BRCA-mutated ovarian cancer) (Table 1). So far, no cancer-associated miRNA-*PARP2* interaction has been identified, but *PARP1* was found to be targeted by miR-124 (Chen, Chou, et al., 2015), miR-216b (Liu, Niu, Lin, & Tian, 2017), miR-223-3p (Streppel et al., 2013), miR-335 (Luo et al., 2017), and miR-708-5p (Guo et al., 2013).

4. REGULATION OF ANGIOGENESIS

Another key factor in the survival of cancer cells, is the development of blood vessels (vasculo- and angiogenesis), on which they rely for the

Table 2 Cancer-Related miRNAs Validated to Directly Target *BCL2* and Have a Demonstrated Role in Apoptosis

miRNA	Type of cancer	References
hsa-miR-7-5p	Non-small cell lung cancer	Xiong et al. (2011)
hsa-miR-15a-5p	Chronic lymphocytic leukemia, breast cancer, hypopharyngeal squamous cell carcinoma	Cimmino et al. (2005), Cittelly et al. (2010), and Lu et al. (2016)
hsa-miR-15b-5p	Gastric cancer	Sun, Meng, et al. (2013) and Xia et al. (2008)
hsa-miR-16-5p	Gastric cancer, chronic lymphocytic leukemia, hepatocellular carcinoma, breast cancer, malignant pleural mesothelioma, glioma	Cimmino et al. (2005), Cittelly et al. (2010), Mobarra et al. (2015), Reid et al. (2013), Sun, Meng, et al. (2013), Tsang, and Kwok (2010), Xia et al. (2008), and Yang, Lu, et al. (2014)
hsa-miR-17-5p	Acute lymphoblastic leukemia	Scherr et al. (2014)
hsa-miR-18a-5p	Acute lymphoblastic leukemia	Scherr et al. (2014)
hsa-miR-24-2-5p	Breast cancer	Srivastava et al. (2011)
hsa-miR-29a-3p	Hepatocellular carcinoma	Xiong et al. (2010)
hsa-miR-29b-3p	Hepatocellular carcinoma	Xiong et al. (2010)
hsa-miR-29c-3p	Hepatocellular carcinoma	Xiong et al. (2010)
hsa-miR-30b-5p	Colorectal cancer	Liao et al. (2014)
hsa-miR-34a-3p	Meningioma	Werner et al. (2017)
hsa-miR-34a-5p	Gastric cancer, hepatocellular carcinoma, non-small cell lung cancer, breast cancer	Bommer et al. (2007), Ji et al. (2008), Li et al. (2013), and Yang, Li, et al. (2014)

hsa-miR-34b-5p	Gastric cancer	Ji et al. (2008)
hsa-miR-34c-5p	Gastric cancer	Ji et al. (2008)
hsa-miR-125b-5p	Hepatocellular carcinoma, nasopharyngeal carcinoma	Yuan et al. (2017) and Zhao et al. (2012)
hsa-miR-135a-5p	Lung cancer	Zhao et al. (2014)
hsa-miR-136-5p	Glioma	Yang et al. (2012)
hsa-miR-143-3p	Gastric cancer, cervical cancer, colorectal cancer, prostate cancer, osteosarcoma	Borralho et al. (2009), Li, Wu, et al. (2016), Liu, Wang, Liu, and Chen (2015), Liu, Yu, et al. (2012), Ma, Luo, and Qiu (2017), and Zhuang et al. (2015)
hsa-miR-148a-3p	Colorectal cancer, breast cancer, pancreatic cancer	Li et al. (2017), Zhang et al. (2011), and Zhang, Li, et al. (2014)
hsa-miR-153-3p	Glioblastoma	Xu, Liao, and Wong (2010)
hsa-miR-181a-5p	Chronic myeloid leukemia, chronic lymphocytic leukemia, acute myeloid leukemia, breast cancer	Bai, Cao, Deng, Zhou, and Wang (2012), Li, Hui, and Xu (2012), Zhu, Zhu, Fang, et al. (2012), and Zhu, Wu, et al. (2013)
hsa-miR-181b-5p	Chronic lymphocytic leukemia, gastric cancer, lung cancer, pancreatic ductal adenocarcinoma	Cai et al. (2013), Zhu, Shan, Wang, Shu, and Liu (2010), and Zhu, Zhu, Fang, et al. (2012)
hsa-miR-182-5p	Uveal melanoma	Yan et al. (2012)
hsa-miR-195-5p	Colorectal cancer, tongue squamous cell carcinoma, breast cancer	Jia, Wei, Gong, Gan, and Yu (2013), Liu, Chen, Xu, Li, and Du (2010), and Zhu, Ye, et al. (2015)
hsa-miR-200b-3p	Gastric cancer, lung cancer	Zhu, Xu, et al. (2012)

Continued

Table 2 Cancer-Related miRNAs Validated to Directly Target *BCL2* and Have a Demonstrated Role in Apoptosis—cont'd

miRNA	Type of cancer	References
hsa-miR-200c-3p	Gastric cancer, lung cancer	Zhu, Xu, et al. (2012)
hsa-miR-204-5p	Neuroblastoma, gastric cancer, prostate cancer	Lin et al. (2016), Ryan et al. (2012), and Sacconi et al. (2012)
hsa-miR-205-5p	Prostate cancer, laryngeal squamous cell carcinoma, adrenocortical carcinoma	Tian et al. (2014), Verdoodt et al. (2013), and Wu et al. (2015)
hsa-miR-206	Non-small cell lung cancer	Sun et al. (2015)
hsa-miR-429	Gastric cancer, lung cancer, glioblastoma, esophageal cancer	Wang, Li, Zang, et al. (2013), Zhu, Xu, et al. (2012), and Zhu, Zhang, et al. (2015)
hsa-miR-449a	Osteosarcoma, gastric cancer, myeloid leukemia	Chen, Zhou, et al. (2015), De Weer et al. (2011), Hu, Fang, Cao, Qin, and Chen (2014), and Wei, Song, Zhang, and Hu (2013)
hsa-miR-497-5p	Gastric cancer, lung cancer, glioma, breast cancer	Wei et al. (2015), Zhu, Zhu, Lu, et al. (2012), and Zhu et al. (2017)
hsa-miR-503-5p	Non-small cell lung cancer, gastric cancer	Qiu et al. (2013) and Wang, Ge, et al. (2014)
hsa-miR-1915-3p	Colorectal cancer	Xu et al. (2013)

supply of nutrients and oxygen and the removal of metabolic wastes and carbon dioxide. Key proteins in the process of angiogenesis are vascular endothelial growth factors (VEGF) and their receptors (VEGFR). There are five VEGF ligands (VEGFA through VEGFE) and three VEGF receptors: VEGFR1 (FLT1), VEGFR2 (KDR), and VEGFR3 (FLT4). FDA-approved drugs for cancer treatment are available that specifically target the VEGFA (VEGF) ligand (bevacizumab) and all three VEGFRs (cabozantinib, lenvatinib, ramucirumab, regorafenib, axitinib, pazopanib, sunitinib, vandetanib, and sorafenib) (Table 1). The majority of the drugs that target VEGFRs are kinase inhibitors that, in addition to VEGFRs, also target a plethora of other tyrosine kinases. miRNAs that directly regulate the *VEGFRs* are: (i) miR-10 (Hassel et al., 2012) and the miR-200 family (b and c) (Choi, Yoon, Jeong, Yoon, & Baek, 2011; Hur et al., 2013; Panda et al., 2012; Roybal et al., 2011) for *FLT1*, (ii) miR-15b-5p (Chan, Yue, Wong, & Wong, 2013), miR-16-5p (Chamorro-Jorganes et al., 2011), miR-17-3p (Yin, Wang, Guo, Zhang, & Lu, 2013), miR-195 (Sun et al., 2016), the miR-200 family (b and c) (Chan, Roy, Khanna, & Sen, 2012; Choi et al., 2011; Shi et al., 2013), and miR-370 (Wang & Chen, 2017) for *KDR*, and (iii) miR-1236-3p (Jones et al., 2012) for *FLT4*. To date, more than 40 miRNAs have been validated, which directly target *VEGF*. The cancer-related miRNAs found to directly bind to VEGF are listed in Table 3.

5. REGULATION OF INVASION AND METASTASIS

As cancers become more aggressive, tumor cells gain the ability to invade local tissues and migrate to distal sites, causing metastasis, which is an important cause of cancer morbidity and mortality. Although no specific protein-coding genes involved in invasion and metastasis have FDA-approved targeted drugs, many of the genes and their interacting miRNAs described in the other sections of this chapter also showed involvement in migration and invasion. For example, miR-133a-3p suppresses cell invasion and cell growth in non-small cell lung cancer through inhibition of *EGFR* (Wang, Hsiao, et al., 2014); miR-206 directly targets *BCL2* and decreases cell growth, migration and invasion, and promotes apoptosis in non-small cell lung cancer (Sun et al., 2015); miR-223 directly targets *PARP1*, increases migration and invasion, and modulates chemosensitivity in Barrett's esophagus-associated esophageal adenocarcinoma (Streppel et al., 2013).

Table 3 Cancer-Related miRNAs Validated to Directly Target *VEGF*

miRNA	Type of cancer	References
hsa-miR-15a-5p	Multiple myeloma	Li, Zhang, et al. (2016) and Sun, She, et al. (2013)
hsa-miR-16-5p	Multiple myeloma, anaplastic large cell lymphoma	Dejean et al. (2011), Li, Zhang, et al. (2016), and Sun, She, et al. (2013)
hsa-miR-29a-3p	Gastric cancer	Chen, Xiao, et al. (2014)
hsa-miR-29b-3p	Breast cancer, osteosarcoma, prostate cancer	Chou et al. (2013), Szczyrba et al. (2013), and Zhang, Zhang, Liu, and Zhou (2014)
hsa-miR-101-3p	Cholangiocarcinoma	Zhang, Han, Zhu, Song, and Wu (2013)
hsa-miR-126-3p	Breast cancer, lung cancer, gastric cancer, colorectal cancer, non-small cell lung cancer, oral cancer	Chen, Li, et al. (2014), Liu, Peng, Zheng, Wang, and Qin (2009), Sasahira et al. (2012), Zhang, Wang, et al. (2013), Zhu et al. (2011), and Zhu, Li, et al. (2012)
hsa-miR-145-5p	Breast cancer, osteosarcoma	Fan, Wu, Xing, Wei, and Shao (2012) and Zou et al. (2012)
hsa-miR-185-5p	Breast cancer, clear cell renal cell carcinoma	Ma et al. (2015) and Wang, Tian, et al. (2014)
hsa-miR-195-5p	Hepatocellular carcinoma, bone cancer	Almeida et al. (2016), Wang, Zhang, Tong, Ma, and Qiu (2015), and Wang, Zhao, et al. (2013)
hasa-miR-199a-3p	Hepatocellular carcinoma	Ghosh et al. (2017)
hsa-miR-200b-3p	Lung cancer	Choi et al. (2011)
hsa-miR-200c-3p	Leiomyoma, endometrial cancer	Chuang, Panda, Luo, and Chegini (2012) and Panda et al. (2012)

hsa-miR-203a-3p	Cervical cancer	Zhu, Er, et al. (2013)
hsa-miR-205-5p	Glioblastoma, breast cancer, osteosarcoma, thyroid cancer, ovarian cancer	Li et al. (2015), Salajegheh et al. (2015), Wang, Shan, et al. (2015), Wu, Zhu, and Mo (2009) and Yue et al. (2012)
hsa-miR-361-5p	Cutaneous squamous cell carcinoma	Kanitz et al. (2012)
hsa-miR-503-5p	Hepatocellular carcinoma	Zhou et al. (2013)
hsa-miR-718	Ovarian cancer	Leng, Zha, and Tang (2014)

6. REGULATION OF IMMUNE RESPONSE

The regulation of the immune response in cancer is dual. On one hand, tumor cells are able to evade immune-surveillance and subsequent immune destruction; on the other hand, the tumor microenvironment is infiltrated by immune cells that, although they migrated to the tumor site to enable tumor eradication, have the paradoxical effect of enhancing tumor formation and progression by supplying the tumor microenvironment with factors and molecules (such as growth factors, survival factors, and proangiogenic factors) necessary to form and maintain the tumor, making it more aggressive (Hanahan & Weinberg, 2011). Based on recent successes in the clinical arena in treating advanced cancers, including melanoma, renal and lung, regulation of the immune response as a potential viable, and effective cancer therapy is currently an area of significant interest, and many so-called "immune checkpoint inhibitors," inhibitors of molecules expressed on immune cells that need to be (in)activated to initiate an immune response, were recently approved by the FDA for the treatment of cancer (Table 1). The programmed death 1 receptor (PD-1)/programmed death ligand 1 (PD-L1) pathway provides such a coinhibitory checkpoint. When the PD-1 receptor, expressed on T-cells and upregulated on tumor-infiltrating lymphocytes, interacts with PD-L1 ligand, expressed on tumor cells, an inhibitory signal is initiated, which prevents T-cell receptor signaling and subsequent eradication of the tumor cell by the immune cell (Keir, Butte, Freeman, & Sharpe, 2008). PD-1 is targeted by the monoclonal antibodies pembrolizumab and nivolumab, while PD-L1 is targeted by the monoclonal antibodies avelumab and atezolizumab (Table 1). To date, only miR-138-5p has been demonstrated to directly target *PD-1* in cancer (Smolle, Calin, Pichler, & Calin, 2017; Wei et al., 2016). In addition, in hepatitis B virus (HBV) infection, the leading cause of hepatocellular carcinoma, miR-4717 interacts with the 3′UTR of *PD-1* in an allele-specific manner, as the binding site contains a single-nucleotide polymorphism (SNP) that is associated with HBV infection (Zhang et al., 2015). In contrast, *PD-L1* is regulated by several miRNAs, including miR-34a-5p (Cortez et al., 2016; Wang, Li, et al., 2015), miR-138-5p (Zhao et al., 2016), miR-200 (Chen, Gibbons, et al., 2014), miR-424(322) (Xu et al., 2016), miR-513a-5p (Wu, Chen, Zhang, & Xing, 2012), and miR-570-3p (Guo et al., 2015; Wang, Li, Mao, et al., 2013; Wang et al., 2012) (Fig. 2). Interestingly, also the miR-570-3p binding site in *PD-L1* contains

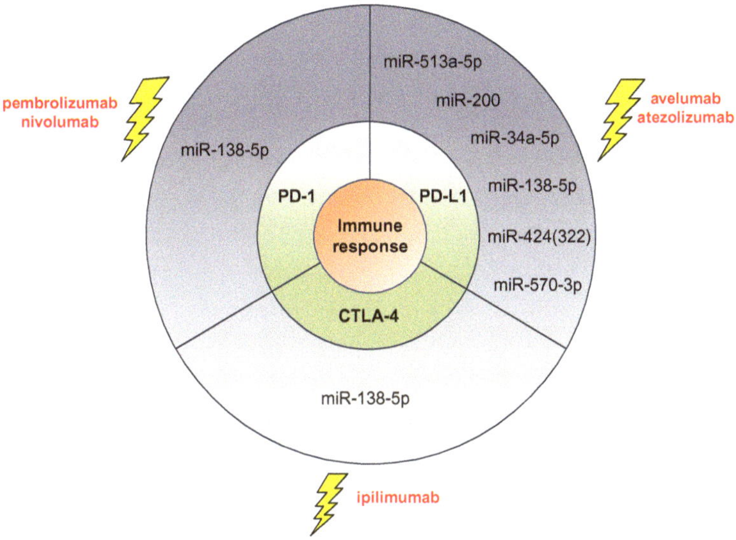

Fig. 2 miRNAs involved in the regulation of the immune response in cancer. The *blue circle* contains miRNAs that are validated to target the protein-coding genes in the *green circle*. In *red* are the drugs that specifically target these protein-coding genes and that are approved by the FDA for use in oncology.

a SNP which results in allele-specific binding and overexpression of the PD-L1 protein due to the disruption of the binding site (Wang, Li, Mao, et al., 2013; Wang et al., 2012).

Cytotoxic T-lymphocyte-associated antigen 4 (CTLA-4), expressed on the surface of T-cells, is another immune checkpoint molecule, which negatively affects T-cells signaling through binding with CD80 and CD86 on antigen-presenting cells. It is specifically targeted by ipilimumab, a monoclonal anti-CTLA-4 antibody, which is approved for the treatment of metastatic melanoma (Table 1). So far, *CTLA-4* has only been described to be directly negatively regulated by miR-138-5p (Wei et al., 2016) (Fig. 2).

7. CONCLUSION

Targeted therapeutics are already available for many proteins involved in most of the cancer hallmarks, and efforts are continued to expand the current availability of targeted drugs. For example, an inhibitor of glutaminase (GLS), an enzyme controlling glutaminolysis which plays a key role in cancer cell metabolism (Xiang et al., 2015), is being tested in clinical trials (CB-839, clinicaltrials.gov), and also the "enabling replicative immortality" hallmark will soon be tackled, as an inhibitor of telomerase, an enzyme responsible for maintenance of telomeres (through which cells acquire immortalization), is currently being evaluated in clinical trials as well (imetelstat, clinicaltrials.gov).

Virtually all protein-coding genes for which targeted therapeutics are available, are under regulation by miRNAs, which by themselves also form interesting druggable targets. However, with the recent withdrawal of MRX34, a miR-34 mimic that was being tested as the first miRNA-based therapy in cancer, due to adverse events, it might still require further research and testing before miRNA therapeutics will enter the clinic and provide a therapeutic impact.

REFERENCES

Ali, S., Ahmad, A., Aboukameel, A., Ahmed, A., Bao, B., Banerjee, S., et al. (2014). Deregulation of miR-146a expression in a mouse model of pancreatic cancer affecting EGFR signaling. *Cancer Letters*, *351*, 134–142.

Almeida, M. I., Silva, A. M., Vasconcelos, D. M., Almeida, C. R., Caires, H., Pinto, M. T., et al. (2016). miR-195 in human primary mesenchymal stromal/stem cells regulates proliferation, osteogenesis and paracrine effect on angiogenesis. *Oncotarget*, *7*, 7–22.

Bai, H., Cao, Z., Deng, C., Zhou, L., & Wang, C. (2012). miR-181a sensitizes resistant leukaemia HL-60/Ara-C cells to Ara-C by inducing apoptosis. *Journal of Cancer Research and Clinical Oncology*, *138*, 595–602.

Bertero, T., Gastaldi, C., Bourget-Ponzio, I., Mari, B., Meneguzzi, G., Barbry, P., et al. (2013). CDC25A targeting by miR-483-3p decreases CCND-CDK4/6 assembly and contributes to cell cycle arrest. *Cell Death and Differentiation*, *20*, 800–811.

Bohlig, L., Friedrich, M., & Engeland, K. (2011). p53 activates the PANK1/miRNA-107 gene leading to downregulation of CDK6 and p130 cell cycle proteins. *Nucleic Acids Research*, *39*, 440–453.

Bommer, G. T., Gerin, I., Feng, Y., Kaczorowski, A. J., Kuick, R., Love, R. E., et al. (2007). p53-mediated activation of miRNA34 candidate tumor-suppressor genes. *Current Biology*, *17*, 1298–1307.

Borralho, P. M., Kren, B. T., Castro, R. E., da Silva, I. B., Steer, C. J., & Rodrigues, C. M. (2009). MicroRNA-143 reduces viability and increases sensitivity to 5-fluorouracil in HCT116 human colorectal cancer cells. *The FEBS Journal*, *276*, 6689–6700.

Cai, B., An, Y., Lv, N., Chen, J., Tu, M., Sun, J., et al. (2013). miRNA-181b increases the sensitivity of pancreatic ductal adenocarcinoma cells to gemcitabine in vitro and in nude mice by targeting BCL-2. *Oncology Reports*, *29*, 1769–1776.

Calin, G. A., Cimmino, A., Fabbri, M., Ferracin, M., Wojcik, S. E., Shimizu, M., et al. (2008). MiR-15a and miR-16-1 cluster functions in human leukemia. *Proceedings of the National Academy of Sciences of the United States of America*, *105*, 5166–5171.

Calin, G. A., Dumitru, C. D., Shimizu, M., Bichi, R., Zupo, S., Noch, E., et al. (2002). Frequent deletions and down-regulation of micro-RNA genes miR15 and miR16 at 13q14 in chronic lymphocytic leukemia. *Proceedings of the National Academy of Sciences of the United States of America*, *99*, 15524–15529.

Chamorro-Jorganes, A., Araldi, E., Penalva, L. O., Sandhu, D., Fernandez-Hernando, C., & Suarez, Y. (2011). MicroRNA-16 and microRNA-424 regulate cell-autonomous angiogenic functions in endothelial cells via targeting vascular endothelial growth factor receptor-2 and fibroblast growth factor receptor-1. *Arteriosclerosis, Thrombosis, and Vascular Biology*, *31*, 2595–2606.

Chan, Y. C., Roy, S., Khanna, S., & Sen, C. K. (2012). Downregulation of endothelial microRNA-200b supports cutaneous wound angiogenesis by desilencing GATA binding protein 2 and vascular endothelial growth factor receptor 2. *Arteriosclerosis, Thrombosis, and Vascular Biology*, *32*, 1372–1382.

Chan, L. S., Yue, P. Y., Wong, Y. Y., & Wong, R. N. (2013). MicroRNA-15b contributes to ginsenoside-Rg1-induced angiogenesis through increased expression of VEGFR-2. *Biochemical Pharmacology*, *86*, 392–400.

Chang, S., Gao, L., Yang, Y., Tong, D., Guo, B., Liu, L., et al. (2015). miR-145 mediates the antiproliferative and gene regulatory effects of vitamin D3 by directly targeting E2F3 in gastric cancer cells. *Oncotarget*, *6*, 7675–7685.

Chen, S. M., Chou, W. C., Hu, L. Y., Hsiung, C. N., Chu, H. W., Huang, Y. L., et al. (2015). The effect of MicroRNA-124 overexpression on anti-tumor drug sensitivity. *PLoS One*, *10*, e0128472.

Chen, L., Gibbons, D. L., Goswami, S., Cortez, M. A., Ahn, Y. H., Byers, L. A., et al. (2014). Metastasis is regulated via microRNA-200/ZEB1 axis control of tumour cell PD-L1 expression and intratumoral immunosuppression. *Nature Communications*, *5*, 5241.

Chen, X., He, D., Dong, X. D., Dong, F., Wang, J., Wang, L., et al. (2013). MicroRNA-124a is epigenetically regulated and acts as a tumor suppressor by controlling multiple targets in uveal melanoma. *Investigative Ophthalmology & Visual Science*, *54*, 2248–2256.

Chen, H., Li, L., Wang, S., Lei, Y., Ge, Q., Lv, N., et al. (2014). Reduced miR-126 expression facilitates angiogenesis of gastric cancer through its regulation on VEGF-A. *Oncotarget*, *5*, 11873–11885.

Chen, G., Umelo, I. A., Lv, S., Teugels, E., Fostier, K., Kronenberger, P., et al. (2013). miR-146a inhibits cell growth, cell migration and induces apoptosis in non-small cell lung cancer cells. *PLoS One*, *8*, e60317.

Chen, L., Xiao, H., Wang, Z. H., Huang, Y., Liu, Z. P., Ren, H., et al. (2014). miR-29a suppresses growth and invasion of gastric cancer cells in vitro by targeting VEGF-A. *BMB Reports*, *47*, 39–44.

Chen, Z., Zeng, H., Guo, Y., Liu, P., Pan, H., Deng, A., et al. (2010). miRNA-145 inhibits non-small cell lung cancer cell proliferation by targeting c-Myc. *Journal of Experimental & Clinical Cancer Research*, *29*, 151.

Chen, J., Zhou, J., Chen, X., Yang, B., Wang, D., Yang, P., et al. (2015). miRNA-449a is downregulated in osteosarcoma and promotes cell apoptosis by targeting BCL2. *Tumour Biology*, *36*, 8221–8229.

Chiyomaru, T., Yamamura, S., Fukuhara, S., Hidaka, H., Majid, S., Saini, S., et al. (2013). Genistein up-regulates tumor suppressor microRNA-574-3p in prostate cancer. *PLoS One*, *8*, e58929.

Cho, W. C., Chow, A. S., & Au, J. S. (2011). MiR-145 inhibits cell proliferation of human lung adenocarcinoma by targeting EGFR and NUDT1. *RNA Biology*, *8*, 125–131.

Choi, Y. C., Yoon, S., Jeong, Y., Yoon, J., & Baek, K. (2011). Regulation of vascular endothelial growth factor signaling by miR-200b. *Molecules and Cells*, *32*, 77–82.

Chou, J., Lin, J. H., Brenot, A., Kim, J. W., Provot, S., & Werb, Z. (2013). GATA3 suppresses metastasis and modulates the tumour microenvironment by regulating microRNA-29b expression. *Nature Cell Biology*, *15*, 201–213.

Chuang, T. D., Panda, H., Luo, X., & Chegini, N. (2012). miR-200c is aberrantly expressed in leiomyomas in an ethnic-dependent manner and targets ZEBs, VEGFA, TIMP2, and FBLN5. *Endocrine-Related Cancer*, *19*, 541–556.

Cimmino, A., Calin, G. A., Fabbri, M., Iorio, M. V., Ferracin, M., Shimizu, M., et al. (2005). miR-15 and miR-16 induce apoptosis by targeting BCL2. *Proceedings of the National Academy of Sciences of the United States of America*, *102*, 13944–13949.

Cittelly, D. M., Das, P. M., Salvo, V. A., Fonseca, J. P., Burow, M. E., & Jones, F. E. (2010). Oncogenic HER2{Delta}16 suppresses miR-15a/16 and deregulates BCL-2 to promote endocrine resistance of breast tumors. *Carcinogenesis*, *31*, 2049–2057.

Cortez, M. A., Ivan, C., Valdecanas, D., Wang, X., Peltier, H. J., Ye, Y., et al. (2016). PDL1 regulation by p53 via miR-34. *Journal of the National Cancer Institute*, *108*, djv303.

Croce, C. M., & Reed, J. C. (2016). Finally, an apoptosis-targeting therapeutic for cancer. *Cancer Research*, *76*, 5914–5920.

Cui, W., Zhang, S., Shan, C., Zhou, L., & Zhou, Z. (2013). microRNA-133a regulates the cell cycle and proliferation of breast cancer cells by targeting epidermal growth factor receptor through the EGFR/Akt signaling pathway. *The FEBS Journal*, *280*, 3962–3974.

De Weer, A., Van der Meulen, J., Rondou, P., Taghon, T., Konrad, T. A., De Preter, K., et al. (2011). EVI1-mediated down regulation of MIR449A is essential for the survival of EVI1 positive leukaemic cells. *British Journal of Haematology*, *154*, 337–348.

Dejean, E., Renalier, M. H., Foisseau, M., Agirre, X., Joseph, N., de Paiva, G. R., et al. (2011). Hypoxia-microRNA-16 downregulation induces VEGF expression in anaplastic lymphoma kinase (ALK)-positive anaplastic large-cell lymphomas. *Leukemia*, *25*, 1882–1890.

Deng, H., Guo, Y., Song, H., Xiao, B., Sun, W., Liu, Z., et al. (2013). MicroRNA-195 and microRNA-378 mediate tumor growth suppression by epigenetical regulation in gastric cancer. *Gene*, *518*, 351–359.

Du, B., Wang, Z., Zhang, X., Feng, S., Wang, G., He, J., et al. (2014). MicroRNA-545 suppresses cell proliferation by targeting cyclin D1 and CDK4 in lung cancer cells. *PLoS One*, *9*, e88022.

Fan, L., Wu, Q., Xing, X., Wei, Y., & Shao, Z. (2012). MicroRNA-145 targets vascular endothelial growth factor and inhibits invasion and metastasis of osteosarcoma cells. *Acta Biochimica et Biophysica Sinica*, *44*, 407–414.

Furuta, M., Kozaki, K., Tanimoto, K., Tanaka, S., Arii, S., Shimamura, T., et al. (2013). The tumor-suppressive miR-497-195 cluster targets multiple cell-cycle regulators in hepatocellular carcinoma. *PLoS One*, *8*, e60155.

Georgantas, R. W., 3rd, Streicher, K., Luo, X., Greenlees, L., Zhu, W., Liu, Z., et al. (2014). MicroRNA-206 induces G1 arrest in melanoma by inhibition of CDK4 and cyclin D. *Pigment Cell & Melanoma Research*, *27*, 275–286.

Ghosh, A., Dasgupta, D., Ghosh, A., Roychoudhury, S., Kumar, D., Gorain, M., et al. (2017). MiRNA199a-3p suppresses tumor growth, migration, invasion and angiogenesis in hepatocellular carcinoma by targeting VEGFA, VEGFR1, VEGFR2, HGF and MMP2. *Cell Death & Disease*, *8*, e2706.

Guo, P., Lan, J., Ge, J., Nie, Q., Mao, Q., & Qiu, Y. (2013). miR-708 acts as a tumor suppressor in human glioblastoma cells. *Oncology Reports*, *30*, 870–876.

Guo, W., Tan, W., Liu, S., Huang, X., Lin, J., Liang, R., et al. (2015). MiR-570 inhibited the cell proliferation and invasion through directly targeting B7-H1 in hepatocellular carcinoma. *Tumour Biology*, *36*, 9049–9057.

Hanahan, D., & Weinberg, R. A. (2000). The hallmarks of cancer. *Cell*, *100*, 57–70.

Hanahan, D., & Weinberg, R. A. (2011). Hallmarks of cancer: The next generation. *Cell*, *144*, 646–674.

Hassel, D., Cheng, P., White, M. P., Ivey, K. N., Kroll, J., Augustin, H. G., et al. (2012). MicroRNA-10 regulates the angiogenic behavior of zebrafish and human endothelial cells by promoting vascular endothelial growth factor signaling. *Circulation Research*, *111*, 1421–1433.

He, L., He, X., Lim, L. P., de Stanchina, E., Xuan, Z., Liang, Y., et al. (2007). A microRNA component of the p53 tumour suppressor network. *Nature*, *447*, 1130–1134.

He, Q., Zhou, X., Li, S., Jin, Y., Chen, Z., Chen, D., et al. (2013). MicroRNA-181a suppresses salivary adenoid cystic carcinoma metastasis by targeting MAPK-Snai2 pathway. *Biochimica et Biophysica Acta*, *1830*, 5258–5266.

Hermeking, H. (2010). The miR-34 family in cancer and apoptosis. *Cell Death and Differentiation*, *17*, 193–199.

Hirata, H., Hinoda, Y., Ueno, K., Nakajima, K., Ishii, N., & Dahiya, R. (2012). MicroRNA-1826 directly targets beta-catenin (CTNNB1) and MEK1 (MAP2K1) in VHL-inactivated renal cancer. *Carcinogenesis*, *33*, 501–508.

Hirata, H., Hinoda, Y., Ueno, K., Shahryari, V., Tabatabai, Z. L., & Dahiya, R. (2012). MicroRNA-1826 targets VEGFC, beta-catenin (CTNNB1) and MEK1 (MAP2K1) in human bladder cancer. *Carcinogenesis*, *33*, 41–48.

Hu, J., Fang, Y., Cao, Y., Qin, R., & Chen, Q. (2014). miR-449a regulates proliferation and chemosensitivity to cisplatin by targeting cyclin D1 and BCL2 in SGC7901 cells. *Digestive Diseases and Sciences*, *59*, 336–345.

Hur, K., Toiyama, Y., Takahashi, M., Balaguer, F., Nagasaka, T., Koike, J., et al. (2013). MicroRNA-200c modulates epithelial-to-mesenchymal transition (EMT) in human colorectal cancer metastasis. *Gut*, *62*, 1315–1326.

Ichimura, A., Ruike, Y., Terasawa, K., Shimizu, K., & Tsujimoto, G. (2010). MicroRNA-34a inhibits cell proliferation by repressing mitogen-activated protein kinase kinase 1 during megakaryocytic differentiation of K562 cells. *Molecular Pharmacology*, *77*, 1016–1024.

Ji, Q., Hao, X., Meng, Y., Zhang, M., Desano, J., Fan, D., et al. (2008). Restoration of tumor suppressor miR-34 inhibits human p53-mutant gastric cancer tumorspheres. *BMC Cancer*, *8*, 266.

Jia, L. F., Wei, S. B., Gong, K., Gan, Y. H., & Yu, G. Y. (2013). Prognostic implications of micoRNA miR-195 expression in human tongue squamous cell carcinoma. *PLoS One*, *8*, e56634.

Johnson, C. D., Esquela-Kerscher, A., Stefani, G., Byrom, M., Kelnar, K., Ovcharenko, D., et al. (2007). The let-7 microRNA represses cell proliferation pathways in human cells. *Cancer Research*, *67*, 7713–7722.

Jones, D., Li, Y., He, Y., Xu, Z., Chen, H., & Min, W. (2012). Mirtron microRNA-1236 inhibits VEGFR-3 signaling during inflammatory lymphangiogenesis. *Arteriosclerosis, Thrombosis, and Vascular Biology*, *32*, 633–642.

Kanitz, A., Imig, J., Dziunycz, P. J., Primorac, A., Galgano, A., Hofbauer, G. F., et al. (2012). The expression levels of microRNA-361-5p and its target VEGFA are inversely correlated in human cutaneous squamous cell carcinoma. *PLoS One*, *7*, e49568.

Kefas, B., Godlewski, J., Comeau, L., Li, Y., Abounader, R., Hawkinson, M., et al. (2008). microRNA-7 inhibits the epidermal growth factor receptor and the Akt pathway and is down-regulated in glioblastoma. *Cancer Research*, *68*, 3566–3572.

Keir, M. E., Butte, M. J., Freeman, G. J., & Sharpe, A. H. (2008). PD-1 and its ligands in tolerance and immunity. *Annual Review of Immunology*, *26*, 677–704.

Lal, A., Navarro, F., Maher, C. A., Maliszewski, L. E., Yan, N., O'Day, E., et al. (2009). miR-24 inhibits cell proliferation by targeting E2F2, MYC, and other cell-cycle genes via binding to "seedless" 3'UTR microRNA recognition elements. *Molecular Cell*, *35*, 610–625.

Lena, A. M., Mancini, M., Rivetti di Val Cervo, P., Saintigny, G., Mahe, C., Melino, G., et al. (2012). MicroRNA-191 triggers keratinocytes senescence by SATB1 and CDK6 downregulation. *Biochemical and Biophysical Research Communications*, *423*, 509–514.

Leng, R., Zha, L., & Tang, L. (2014). MiR-718 represses VEGF and inhibits ovarian cancer cell progression. *FEBS Letters*, *588*, 2078–2086.

Li, H., Hui, L., & Xu, W. (2012). miR-181a sensitizes a multidrug-resistant leukemia cell line K562/A02 to daunorubicin by targeting BCL-2. *Acta Biochimica et Biophysica Sinica*, *44*, 269–277.

Li, J., Li, L., Li, Z., Gong, G., Chen, P., Liu, H., et al. (2015). The role of miR-205 in the VEGF-mediated promotion of human ovarian cancer cell invasion. *Gynecologic Oncology*, *137*, 125–133.

Li, Q., Ren, P., Shi, P., Chen, Y., Xiang, F., Zhang, L., et al. (2017). MicroRNA-148a promotes apoptosis and suppresses growth of breast cancer cells by targeting B-cell lymphoma 2. *Anticancer Drugs*, *28*(6), 588–595.

Li, W. H., Wu, H. J., Li, Y. X., Pan, H. G., Meng, T., & Wang, X. (2016). MicroRNA-143 promotes apoptosis of osteosarcoma cells by caspase-3 activation via targeting Bcl-2. *Biomedicine & Pharmacotherapy*, *80*, 8–15.

Li, L., Yuan, L., Luo, J., Gao, J., Guo, J., & Xie, X. (2013). MiR-34a inhibits proliferation and migration of breast cancer through down-regulation of Bcl-2 and SIRT1. *Clinical and Experimental Medicine*, *13*, 109–117.

Li, Y., Zhang, B., Li, W., Wang, L., Yan, Z., Li, H., et al. (2016). MiR-15a/16 regulates the growth of myeloma cells, angiogenesis and antitumor immunity by inhibiting Bcl-2, VEGF-A and IL-17 expression in multiple myeloma. *Leukemia Research*, *49*, 73–79.

Liao, W. T., Ye, Y. P., Zhang, N. J., Li, T. T., Wang, S. Y., Cui, Y. M., et al. (2014). MicroRNA-30b functions as a tumour suppressor in human colorectal cancer by targeting KRAS, PIK3CD and BCL2. *The Journal of Pathology*, *232*, 415–427.

Lin, Y. C., Lin, J. F., Tsai, T. F., Chou, K. Y., Chen, H. E., & Hwang, T. I. (2016). Tumor suppressor miRNA-204-5p promotes apoptosis by targeting BCL2 in prostate cancer cells. *Asian Journal of Surgery*, (in press, epub ahead of print).

Lin, Y., Wu, J., Chen, H., Mao, Y., Liu, Y., Mao, Q., et al. (2012). Cyclin-dependent kinase 4 is a novel target in micoRNA-195-mediated cell cycle arrest in bladder cancer cells. *FEBS Letters*, *586*, 442–447.

Liu, L., Chen, L., Xu, Y., Li, R., & Du, X. (2010). microRNA-195 promotes apoptosis and suppresses tumorigenicity of human colorectal cancer cells. *Biochemical and Biophysical Research Communications*, *400*, 236–240.

Liu, Q., Fu, H., Sun, F., Zhang, H., Tie, Y., Zhu, J., et al. (2008). miR-16 family induces cell cycle arrest by regulating multiple cell cycle genes. *Nucleic Acids Research*, *36*, 5391–5404.

Liu, Y., Niu, Z., Lin, X., & Tian, Y. (2017). MiR-216b increases cisplatin sensitivity in ovarian cancer cells by targeting PARP1. *Cancer Gene Therapy*, *24*(5), 208–214.

Liu, B., Peng, X. C., Zheng, X. L., Wang, J., & Qin, Y. W. (2009). MiR-126 restoration down-regulate VEGF and inhibit the growth of lung cancer cell lines in vitro and in vivo. *Lung Cancer*, *66*, 169–175.

Liu, L., Shao, X., Gao, W., Zhang, Z., Liu, P., Wang, R., et al. (2012). MicroRNA-133b inhibits the growth of non-small-cell lung cancer by targeting the epidermal growth factor receptor. *The FEBS Journal*, *279*, 3800–3812.

Liu, G., Sun, Y., Ji, P., Li, X., Cogdell, D., Yang, D., et al. (2014). MiR-506 suppresses proliferation and induces senescence by directly targeting the CDK4/6-FOXM1 axis in ovarian cancer. *The Journal of Pathology*, *233*, 308–318.

Liu, H., Wang, H., Liu, H., & Chen, Y. (2015). Effect of miR-143 on the apoptosis of osteosarcoma cells. *International Journal of Clinical and Experimental Pathology*, *8*, 14241–14246.

Liu, L., Yu, X., Guo, X., Tian, Z., Su, M., Long, Y., et al. (2012). miR-143 is downregulated in cervical cancer and promotes apoptosis and inhibits tumor formation by targeting Bcl-2. *Molecular Medicine Reports*, *5*, 753–760.

Lu, W., Feng, L., Zhang, Y., Ma, Y., Li, P., Wang, Y., et al. (2016). miR-15a induces cell apoptosis by targeting BCL2L2 and BCL2 in HPV-positive hypopharyngeal squamous cell carcinoma. *Oncology Reports*, *36*, 2169–2176.

Lujambio, A., Calin, G. A., Villanueva, A., Ropero, S., Sanchez-Cespedes, M., Blanco, D., et al. (2008). A microRNA DNA methylation signature for human cancer metastasis. *Proceedings of the National Academy of Sciences of the United States of America*, *105*, 13556–13561.

Luo, Y., Tong, L., Meng, H., Zhu, W., Guo, L., Wei, T., et al. (2017). MiR-335 regulates the chemo-radio resistance of small cell lung cancer cells by targeting PARP-1. *Gene, 600*, 9–15.

Ma, Z., Luo, Y., & Qiu, M. (2017). miR-143 induces the apoptosis of prostate cancer LNCap cells by suppressing Bcl-2 expression. *Medical Science Monitor, 23*, 359–365.

Ma, X., Shen, D., Li, H., Zhang, Y., Lv, X., Huang, Q., et al. (2015). MicroRNA-185 inhibits cell proliferation and induces cell apoptosis by targeting VEGFA directly in von Hippel–Lindau-inactivated clear cell renal cell carcinoma. *Urologic Oncology, 33*, 169.e1–169.e11.

Mathew, L. K., Skuli, N., Mucaj, V., Lee, S. S., Zinn, P. O., Sathyan, P., et al. (2014). miR-218 opposes a critical RTK-HIF pathway in mesenchymal glioblastoma. *Proceedings of the National Academy of Sciences of the United States of America, 111*, 291–296.

Mobarra, N., Shafiee, A., Rad, S. M., Tasharrofi, N., Soufi-Zomorod, M., Hafizi, M., et al. (2015). Overexpression of microRNA-16 declines cellular growth, proliferation and induces apoptosis in human breast cancer cells. *In Vitro Cellular & Developmental Biology. Animal, 51*, 604–611.

Nakashima, T., Jinnin, M., Etoh, T., Fukushima, S., Masuguchi, S., Maruo, K., et al. (2010). Down-regulation of mir-424 contributes to the abnormal angiogenesis via MEK1 and cyclin E1 in senile hemangioma: Its implications to therapy. *PloS One, 5*, e14334.

Olaru, A. V., Ghiaur, G., Yamanaka, S., Luvsanjav, D., An, F., Popescu, I., et al. (2011). MicroRNA down-regulated in human cholangiocarcinoma control cell cycle through multiple targets involved in the G1/S checkpoint. *Hepatology, 54*, 2089–2098.

Ouyang, Q., Chen, G., Zhou, J., Li, L., Dong, Z., Yang, R., et al. (2016). Neurotensin signaling stimulates glioblastoma cell proliferation by upregulating c-Myc and inhibiting miR-29b-1 and miR-129-3p. *Neuro-Oncology, 18*, 216–226.

Panda, H., Pelakh, L., Chuang, T. D., Luo, X., Bukulmez, O., & Chegini, N. (2012). Endometrial miR-200c is altered during transformation into cancerous states and targets the expression of ZEBs, VEGFA, FLT1, IKKbeta, KLF9, and FBLN5. *Reproductive Sciences, 19*, 786–796.

Pierson, J., Hostager, B., Fan, R., & Vibhakar, R. (2008). Regulation of cyclin dependent kinase 6 by microRNA 124 in medulloblastoma. *Journal of Neuro-Oncology, 90*, 1–7.

Qiu, T., Zhou, L., Wang, T., Xu, J., Wang, J., Chen, W., et al. (2013). miR-503 regulates the resistance of non-small cell lung cancer cells to cisplatin by targeting Bcl-2. *International Journal of Molecular Medicine, 32*, 593–598.

Rao, S. A., Arimappamagan, A., Pandey, P., Santosh, V., Hegde, A. S., Chandramouli, B. A., et al. (2013). miR-219-5p inhibits receptor tyrosine kinase pathway by targeting EGFR in glioblastoma. *PLoS One, 8*, e63164.

Reid, G., Pel, M. E., Kirschner, M. B., Cheng, Y. Y., Mugridge, N., Weiss, J., et al. (2013). Restoring expression of miR-16: A novel approach to therapy for malignant pleural mesothelioma. *Annals of Oncology, 24*, 3128–3135.

Roybal, J. D., Zang, Y., Ahn, Y. H., Yang, Y., Gibbons, D. L., Baird, B. N., et al. (2011). miR-200 inhibits lung adenocarcinoma cell invasion and metastasis by targeting Flt1/VEGFR1. *Molecular Cancer Research, 9*, 25–35.

Ryan, J., Tivnan, A., Fay, J., Bryan, K., Meehan, M., Creevey, L., et al. (2012). MicroRNA-204 increases sensitivity of neuroblastoma cells to cisplatin and is associated with a favourable clinical outcome. *British Journal of Cancer, 107*, 967–976.

Sacconi, A., Biagioni, F., Canu, V., Mori, F., Di Benedetto, A., Lorenzon, L., et al. (2012). miR-204 targets Bcl-2 expression and enhances responsiveness of gastric cancer. *Cell Death & Disease, 3*, e423.

Salajegheh, A., Vosgha, H., Md Rahman, A., Amin, M., Smith, R. A., & Lam, A. K. (2015). Modulatory role of miR-205 in angiogenesis and progression of thyroid cancer. *Journal of Molecular Endocrinology, 55*, 183–196.

Sanchez-Martinez, C., Gelbert, L. M., Lallena, M. J., & de Dios, A. (2015). Cyclin dependent kinase (CDK) inhibitors as anticancer drugs. *Bioorganic & Medicinal Chemistry Letters*, *25*, 3420–3435.

Sasahira, T., Kurihara, M., Bhawal, U. K., Ueda, N., Shimomoto, T., Yamamoto, K., et al. (2012). Downregulation of miR-126 induces angiogenesis and lymphangiogenesis by activation of VEGF-A in oral cancer. *British Journal of Cancer*, *107*, 700–706.

Scherr, M., Elder, A., Battmer, K., Barzan, D., Bomken, S., Ricke-Hoch, M., et al. (2014). Differential expression of miR-17∼92 identifies BCL2 as a therapeutic target in BCR-ABL-positive B-lineage acute lymphoblastic leukemia. *Leukemia*, *28*, 554–565.

Schmitt, M. J., Philippidou, D., Reinsbach, S. E., Margue, C., Wienecke-Baldacchino, A., Nashan, D., et al. (2012). Interferon-gamma-induced activation of signal transducer and activator of transcription 1 (STAT1) up-regulates the tumor suppressing microRNA-29 family in melanoma cells. *Cell Communication and Signaling: CCS*, *10*, 41.

Shi, L., Zhang, S., Wu, H., Zhang, L., Dai, X., Hu, J., et al. (2013). MiR-200c increases the radiosensitivity of non-small-cell lung cancer cell line A549 by targeting VEGF-VEGFR2 pathway. *PLoS One*, *8*, e78344.

Silber, J., Lim, D. A., Petritsch, C., Persson, A. I., Maunakea, A. K., Yu, M., et al. (2008). miR-124 and miR-137 inhibit proliferation of glioblastoma multiforme cells and induce differentiation of brain tumor stem cells. *BMC Medicine*, *6*, 14.

Smolle, M. A., Calin, H. N., Pichler, M., & Calin, G. A. (2017). Noncoding RNAs and immune checkpoints-clinical implications as cancer therapeutics. *The FEBS Journal*, *284*, 1952–1966.

Srivastava, N., Manvati, S., Srivastava, A., Pal, R., Kalaiarasan, P., Chattopadhyay, S., et al. (2011). miR-24-2 controls H2AFX expression regardless of gene copy number alteration and induces apoptosis by targeting antiapoptotic gene BCL-2: A potential for therapeutic intervention. *Breast Cancer Research*, 13, R39.

Streppel, M. M., Pai, S., Campbell, N. R., Hu, C., Yabuuchi, S., Canto, M. I., et al. (2013). MicroRNA 223 is upregulated in the multistep progression of Barrett's esophagus and modulates sensitivity to chemotherapy by targeting PARP1. *Clinical Cancer Research*, *19*, 4067–4078.

Sun, F., Fu, H., Liu, Q., Tie, Y., Zhu, J., Xing, R., et al. (2008). Downregulation of CCND1 and CDK6 by miR-34a induces cell cycle arrest. *FEBS Letters*, *582*, 1564–1568.

Sun, C., Liu, Z., Li, S., Yang, C., Xue, R., Xi, Y., et al. (2015). Down-regulation of c-Met and Bcl2 by microRNA-206, activates apoptosis, and inhibits tumor cell proliferation, migration and colony formation. *Oncotarget*, *6*, 25533–25574.

Sun, H., Meng, X., Han, J., Zhang, Z., Wang, B., Bai, X., et al. (2013). Anti-cancer activity of DHA on gastric cancer—An in vitro and in vivo study. *Tumour Biology*, *34*, 3791–3800.

Sun, C. Y., She, X. M., Qin, Y., Chu, Z. B., Chen, L., Ai, L. S., et al. (2013). miR-15a and miR-16 affect the angiogenesis of multiple myeloma by targeting VEGF. *Carcinogenesis*, *34*, 426–435.

Sun, P., Wang, L., Lu, Y., Liu, Y., Li, L., Yin, L., et al. (2016). MicroRNA-195 targets VEGFR2 and has a tumor suppressive role in ACHN cells via PI3K/Akt and Raf/MEK/ERK signaling pathways. *International Journal of Oncology*, *49*, 1155–1163.

Suto, T., Yokobori, T., Yajima, R., Morita, H., Fujii, T., Yamaguchi, S., et al. (2015). MicroRNA-7 expression in colorectal cancer is associated with poor prognosis and regulates cetuximab sensitivity via EGFR regulation. *Carcinogenesis*, *36*, 338–345.

Szczyrba, J., Nolte, E., Hart, M., Doll, C., Wach, S., Taubert, H., et al. (2013). Identification of ZNF217, hnRNP-K, VEGF-A and IPO7 as targets for microRNAs that are downregulated in prostate carcinoma. *International Journal of Cancer*, *132*, 775–784.

Tao, J., Wu, D., Xu, B., Qian, W., Li, P., Lu, Q., et al. (2012). microRNA-133 inhibits cell proliferation, migration and invasion in prostate cancer cells by targeting the epidermal growth factor receptor. *Oncology Reports*, *27*, 1967–1975.

Tian, L., Zhang, J., Ge, J., Xiao, H., Lu, J., Fu, S., et al. (2014). MicroRNA-205 suppresses proliferation and promotes apoptosis in laryngeal squamous cell carcinoma. *Medical Oncology*, *31*, 785.

Tsang, W. P., & Kwok, T. T. (2010). Epigallocatechin gallate up-regulation of miR-16 and induction of apoptosis in human cancer cells. *The Journal of Nutritional Biochemistry*, *21*, 140–146.

Ufkin, M. L., Peterson, S., Yang, X., Driscoll, H., Duarte, C., & Sathyanarayana, P. (2014). miR-125a regulates cell cycle, proliferation, and apoptosis by targeting the ErbB pathway in acute myeloid leukemia. *Leukemia Research*, *38*, 402–410.

Van Roosbroeck, K., & Calin, G. A. (2016). MicroRNAs in chronic lymphocytic leukemia: miRacle or miRage for prognosis and targeted therapies? *Seminars in Oncology*, *43*, 209–214.

Venkataraman, S., Birks, D. K., Balakrishnan, I., Alimova, I., Harris, P. S., Patel, P. R., et al. (2013). MicroRNA 218 acts as a tumor suppressor by targeting multiple cancer phenotype-associated genes in medulloblastoma. *The Journal of Biological Chemistry*, *288*, 1918–1928.

Verdoodt, B., Neid, M., Vogt, M., Kuhn, V., Liffers, S. T., Palisaar, R. J., et al. (2013). MicroRNA-205, a novel regulator of the anti-apoptotic protein Bcl2, is downregulated in prostate cancer. *International Journal of Oncology*, *43*, 307–314.

Wang, X. H., & Chen, L. (2017). MicroRNA-370 suppresses the retinal capillary endothelial cell growth by targeting KDR gene. *Bratislavské Lekárske Listy*, *118*, 202–207.

Wang, W., Cheng, B., Miao, L., Mei, Y., & Wu, M. (2013). Mutant p53-R273H gains new function in sustained activation of EGFR signaling via suppressing miR-27a expression. *Cell Death & Disease*, *4*, e574.

Wang, T., Ge, G., Ding, Y., Zhou, X., Huang, Z., Zhu, W., et al. (2014). MiR-503 regulates cisplatin resistance of human gastric cancer cell lines by targeting IGF1R and BCL2. *Chinese Medical Journal*, *127*, 2357–2362.

Wang, L. K., Hsiao, T. H., Hong, T. M., Chen, H. Y., Kao, S. H., Wang, W. L., et al. (2014). MicroRNA-133a suppresses multiple oncogenic membrane receptors and cell invasion in non-small cell lung carcinoma. *PLoS One*, *9*, e96765.

Wang, X., Li, J., Dong, K., Lin, F., Long, M., Ouyang, Y., et al. (2015). Tumor suppressor miR-34a targets PD-L1 and functions as a potential immunotherapeutic target in acute myeloid leukemia. *Cellular Signalling*, *27*, 443–452.

Wang, W., Li, F., Mao, Y., Zhou, H., Sun, J., Li, R., et al. (2013). A miR-570 binding site polymorphism in the B7-H1 gene is associated with the risk of gastric adenocarcinoma. *Human Genetics*, *132*, 641–648.

Wang, Y., Li, M., Zang, W., Ma, Y., Wang, N., Li, P., et al. (2013). MiR-429 up-regulation induces apoptosis and suppresses invasion by targeting Bcl-2 and SP-1 in esophageal carcinoma. *Cellular Oncology (Dordrecht)*, *36*, 385–394.

Wang, L., Shan, M., Liu, Y., Yang, F., Qi, H., Zhou, L., et al. (2015). miR-205 suppresses the proliferative and migratory capacity of human osteosarcoma mg-63 cells by targeting VEGFA. *OncoTargets and Therapy*, *8*, 2635–2642.

Wang, W., Sun, J., Li, F., Li, R., Gu, Y., Liu, C., et al. (2012). A frequent somatic mutation in CD274 3′-UTR leads to protein over-expression in gastric cancer by disrupting miR-570 binding. *Human Mutation*, *33*, 480–484.

Wang, R., Tian, S., Wang, H. B., Chu, D. P., Cao, J. L., Xia, H. F., et al. (2014). MiR-185 is involved in human breast carcinogenesis by targeting Vegfa. *FEBS Letters*, *588*, 4438–4447.

Wang, L., Yao, J., Shi, X., Hu, L., Li, Z., Song, T., et al. (2013). MicroRNA-302b suppresses cell proliferation by targeting EGFR in human hepatocellular carcinoma SMMC-7721 cells. *BMC Cancer*, *13*, 448.

Wang, M., Zhang, J., Tong, L., Ma, X., & Qiu, X. (2015). MiR-195 is a key negative regulator of hepatocellular carcinoma metastasis by targeting FGF2 and VEGFA. *International Journal of Clinical and Experimental Pathology*, *8*, 14110–14120.

Wang, R., Zhao, N., Li, S., Fang, J. H., Chen, M. X., Yang, J., et al. (2013). MicroRNA-195 suppresses angiogenesis and metastasis of hepatocellular carcinoma by inhibiting the expression of VEGF, VAV2, and CDC42. *Hepatology*, *58*, 642–653.

Webster, R. J., Giles, K. M., Price, K. J., Zhang, P. M., Mattick, J. S., & Leedman, P. J. (2009). Regulation of epidermal growth factor receptor signaling in human cancer cells by microRNA-7. *The Journal of Biological Chemistry*, *284*, 5731–5741.

Wei, C., Luo, Q., Sun, X., Li, D., Song, H., Li, X., et al. (2015). MicroRNA-497 induces cell apoptosis by negatively regulating Bcl-2 protein expression at the posttranscriptional level in human breast cancer. *International Journal of Clinical and Experimental Pathology*, *8*, 7729–7739.

Wei, J., Nduom, E. K., Kong, L. Y., Hashimoto, Y., Xu, S., Gabrusiewicz, K., et al. (2016). MiR-138 exerts anti-glioma efficacy by targeting immune checkpoints. *Neuro-Oncology*, *18*, 639–648.

Wei, B., Song, Y., Zhang, Y., & Hu, M. (2013). microRNA-449a functions as a tumor-suppressor in gastric adenocarcinoma by targeting Bcl-2. *Oncology Letters*, *6*, 1713–1718.

Werner, T. V., Hart, M., Nickels, R., Kim, Y. J., Menger, M. D., Bohle, R. M., et al. (2017). MiR-34a-3p alters proliferation and apoptosis of meningioma cells in vitro and is directly targeting SMAD4, FRAT1 and BCL2. *Aging (Albany, NY)*, *9*, 932–954.

Whittaker, S. R., Mallinger, A., Workman, P., & Clarke, P. A. (2017). Inhibitors of cyclin-dependent kinases as cancer therapeutics. *Pharmacology & Therapeutics*, *173*, 83–105.

Wu, L., Chen, Z., Zhang, J., & Xing, Y. (2012). Effect of miR-513a-5p on etoposide-stimulating B7-H1 expression in retinoblastoma cells. *Journal of Huazhong University of Science and Technology. Medical Sciences*, *32*, 601–606.

Wu, J., Qian, J., Li, C., Kwok, L., Cheng, F., Liu, P., et al. (2010). miR-129 regulates cell proliferation by downregulating Cdk6 expression. *Cell Cycle*, *9*, 1809–1818.

Wu, Y., Wang, W., Hu, W., Xu, W., Xiao, G., Nie, Q., et al. (2015). MicroRNA-205 suppresses the growth of adrenocortical carcinoma SW-13 cells via targeting Bcl-2. *Oncology Reports*, *34*, 3104–3110.

Wu, H., Zhu, S., & Mo, Y. Y. (2009). Suppression of cell growth and invasion by miR-205 in breast cancer. *Cell Research*, *19*, 439–448.

Xia, L., Zhang, D., Du, R., Pan, Y., Zhao, L., Sun, S., et al. (2008). miR-15b and miR-16 modulate multidrug resistance by targeting BCL2 in human gastric cancer cells. *International Journal of Cancer*, *123*, 372–379.

Xiang, Y., Stine, Z. E., Xia, J., Lu, Y., O'Connor, R. S., Altman, B. J., et al. (2015). Targeted inhibition of tumor-specific glutaminase diminishes cell-autonomous tumorigenesis. *The Journal of Clinical Investigation*, *125*, 2293–2306.

Xiao, F., Zhang, W., Zhou, L., Xie, H., Xing, C., Ding, S., et al. (2013). microRNA-200a is an independent prognostic factor of hepatocellular carcinoma and induces cell cycle arrest by targeting CDK6. *Oncology Reports*, *30*, 2203–2210.

Xiong, Y., Fang, J. H., Yun, J. P., Yang, J., Zhang, Y., Jia, W. H., et al. (2010). Effects of microRNA-29 on apoptosis, tumorigenicity, and prognosis of hepatocellular carcinoma. *Hepatology*, *51*, 836–845.

Xiong, S., Zheng, Y., Jiang, P., Liu, R., Liu, X., & Chu, Y. (2011). MicroRNA-7 inhibits the growth of human non-small cell lung cancer A549 cells through targeting BCL-2. *International Journal of Biological Sciences*, *7*, 805–814.

Xu, K., Liang, X., Cui, D., Wu, Y., Shi, W., & Liu, J. (2013). miR-1915 inhibits Bcl-2 to modulate multidrug resistance by increasing drug-sensitivity in human colorectal carcinoma cells. *Molecular Carcinogenesis*, *52*, 70–78.

Xu, J., Liao, X., & Wong, C. (2010). Downregulations of B-cell lymphoma 2 and myeloid cell leukemia sequence 1 by microRNA 153 induce apoptosis in a glioblastoma cell line DBTRG-05MG. *International Journal of Cancer, 126*, 1029–1035.

Xu, S., Tao, Z., Hai, B., Liang, H., Shi, Y., Wang, T., et al. (2016). miR-424(322) reverses chemoresistance via T-cell immune response activation by blocking the PD-L1 immune checkpoint. *Nature Communications*, 7, 11406.

Xu, B., Wang, N., Wang, X., Tong, N., Shao, N., Tao, J., et al. (2012). MiR-146a suppresses tumor growth and progression by targeting EGFR pathway and in a p-ERK-dependent manner in castration-resistant prostate cancer. *Prostate, 72*, 1171–1178.

Xu, T., Zhu, Y., Xiong, Y., Ge, Y. Y., Yun, J. P., & Zhuang, S. M. (2009). MicroRNA-195 suppresses tumorigenicity and regulates G1/S transition of human hepatocellular carcinoma cells. *Hepatology, 50*, 113–121.

Yamaguchi, G., Takanashi, M., Tanaka, M., Fujita, K., Ohira, T., Kuroda, M., et al. (2012). Isolation of miRNAs that target EGFR mRNA in human lung cancer. *Biochemical and Biophysical Research Communications, 420*, 411–416.

Yan, D., Dong, X. D., Chen, X., Yao, S., Wang, L., Wang, J., et al. (2012). Role of microRNA-182 in posterior uveal melanoma: Regulation of tumor development through MITF, BCL2 and cyclin D2. *PLoS One*, 7, e40967.

Yang, X., Feng, M., Jiang, X., Wu, Z., Li, Z., Aau, M., et al. (2009). miR-449a and miR-449b are direct transcriptional targets of E2F1 and negatively regulate pRb-E2F1 activity through a feedback loop by targeting CDK6 and CDC25A. *Genes & Development, 23*, 2388–2393.

Yang, F., Li, Q. J., Gong, Z. B., Zhou, L., You, N., Wang, S., et al. (2014). MicroRNA-34a targets Bcl-2 and sensitizes human hepatocellular carcinoma cells to sorafenib treatment. *Technology in Cancer Research & Treatment, 13*, 77–86.

Yang, T. Q., Lu, X. J., Wu, T. F., Ding, D. D., Zhao, Z. H., Chen, G. L., et al. (2014). MicroRNA-16 inhibits glioma cell growth and invasion through suppression of BCL2 and the nuclear factor-kappaB1/MMP9 signaling pathway. *Cancer Science, 105*, 265–271.

Yang, Y., Wu, J., Guan, H., Cai, J., Fang, L., Li, J., et al. (2012). MiR-136 promotes apoptosis of glioma cells by targeting AEG-1 and Bcl-2. *FEBS Letters, 586*, 3608–3612.

Yin, R., Wang, R., Guo, L., Zhang, W., & Lu, Y. (2013). MiR-17-3p inhibits angiogenesis by downregulating flk-1 in the cell growth signal pathway. *Journal of Vascular Research, 50*, 157–166.

Yuan, T. Z., Zhang, H. H., Lin, X. L., Yu, J. X., Yang, Q. X., Liang, Y., et al. (2017). microRNA-125b reverses the multidrug resistance of nasopharyngeal carcinoma cells via targeting of Bcl-2. *Molecular Medicine Reports, 15*, 2223–2228.

Yue, X., Wang, P., Xu, J., Zhu, Y., Sun, G., Pang, Q., et al. (2012). MicroRNA-205 functions as a tumor suppressor in human glioblastoma cells by targeting VEGF-A. *Oncology Reports, 27*, 1200–1206.

Zhang, J., Han, C., Zhu, H., Song, K., & Wu, T. (2013). miR-101 inhibits cholangiocarcinoma angiogenesis through targeting vascular endothelial growth factor (VEGF). *The American Journal of Pathology, 182*, 1629–1639.

Zhang, Z., Huang, L., Yu, Z., Chen, X., Yang, D., Zhan, P., et al. (2014). Let-7a functions as a tumor suppressor in Ewing's sarcoma cell lines partly by targeting cyclin-dependent kinase 6. *DNA and Cell Biology, 33*, 136–147.

Zhang, H., Li, Y., Huang, Q., Ren, X., Hu, H., Sheng, H., et al. (2011). MiR-148a promotes apoptosis by targeting Bcl-2 in colorectal cancer. *Cell Death and Differentiation, 18*, 1702–1710.

Zhang, G., Li, N., Li, Z., Zhu, Q., Li, F., Yang, C., et al. (2015). microRNA-4717 differentially interacts with its polymorphic target in the PD1 3′ untranslated region:

A mechanism for regulating PD-1 expression and function in HBV-associated liver diseases. *Oncotarget*, *6*, 18933–18944.

Zhang, R., Li, M., Zang, W., Chen, X., Wang, Y., Li, P., et al. (2014). MiR-148a regulates the growth and apoptosis in pancreatic cancer by targeting CCKBR and Bcl-2. *Tumour Biology*, *35*, 837–844.

Zhang, Y., Wang, X., Xu, B., Wang, B., Wang, Z., Liang, Y., et al. (2013). Epigenetic silencing of miR-126 contributes to tumor invasion and angiogenesis in colorectal cancer. *Oncology Reports*, *30*, 1976–1984.

Zhang, K., Zhang, C., Liu, L., & Zhou, J. (2014). A key role of microRNA-29b in suppression of osteosarcoma cell proliferation and migration via modulation of VEGF. *International Journal of Clinical and Experimental Pathology*, 7, 5701–5708.

Zhao, J., Li, X., Zou, M., He, J., Han, Y., Wu, D., et al. (2014). miR-135a inhibition protects A549 cells from LPS-induced apoptosis by targeting Bcl-2. *Biochemical and Biophysical Research Communications*, *452*, 951–957.

Zhao, Z., Wang, L., Song, W., Cui, H., Chen, G., Qiao, F., et al. (2015). Reduced miR-29a-3p expression is linked to the cell proliferation and cell migration in gastric cancer. *World Journal of Surgical Oncology*, *13*, 101.

Zhao, L., Yu, H., Yi, S., Peng, X., Su, P., Xiao, Z., et al. (2016). The tumor suppressor miR-138-5p targets PD-L1 in colorectal cancer. *Oncotarget*, 7, 45370–45384.

Zhao, A., Zeng, Q., Xie, X., Zhou, J., Yue, W., Li, Y., et al. (2012). MicroRNA-125b induces cancer cell apoptosis through suppression of Bcl-2 expression. *Journal of Genetics and Genomics*, *39*, 29–35.

Zhou, B., Ma, R., Si, W., Li, S., Xu, Y., Tu, X., et al. (2013). MicroRNA-503 targets FGF2 and VEGFA and inhibits tumor angiogenesis and growth. *Cancer Letters*, *333*, 159–169.

Zhu, X., Er, K., Mao, C., Yan, Q., Xu, H., Zhang, Y., et al. (2013). miR-203 suppresses tumor growth and angiogenesis by targeting VEGFA in cervical cancer. *Cellular Physiology and Biochemistry*, *32*, 64–73.

Zhu, X., Li, H., Long, L., Hui, L., Chen, H., Wang, X., et al. (2012). miR-126 enhances the sensitivity of non-small cell lung cancer cells to anticancer agents by targeting vascular endothelial growth factor A. *Acta Biochimica et Biophysica Sinica*, *44*, 519–526.

Zhu, Y., Lu, Y., Zhang, Q., Liu, J. J., Li, T. J., Yang, J. R., et al. (2012). MicroRNA-26a/b and their host genes cooperate to inhibit the G1/S transition by activating the pRb protein. *Nucleic Acids Research*, *40*, 4615–4625.

Zhu, W., Shan, X., Wang, T., Shu, Y., & Liu, P. (2010). miR-181b modulates multidrug resistance by targeting BCL2 in human cancer cell lines. *International Journal of Cancer*, *127*, 2520–2529.

Zhu, D., Tu, M., Zeng, B., Cai, L., Zheng, W., Su, Z., et al. (2017). Up-regulation of miR-497 confers resistance to temozolomide in human glioma cells by targeting mTOR/Bcl-2. *Cancer Medicine*, *6*, 452–462.

Zhu, Y., Wu, J., Li, S., Ma, R., Cao, H., Ji, M., et al. (2013). The function role of miR-181a in chemosensitivity to adriamycin by targeting Bcl-2 in low-invasive breast cancer cells. *Cellular Physiology and Biochemistry*, *32*, 1225–1237.

Zhu, W., Xu, H., Zhu, D., Zhi, H., Wang, T., Wang, J., et al. (2012). miR-200bc/429 cluster modulates multidrug resistance of human cancer cell lines by targeting BCL2 and XIAP. *Cancer Chemotherapy and Pharmacology*, *69*, 723–731.

Zhu, J., Ye, Q., Chang, L., Xiong, W., He, Q., & Li, W. (2015). Upregulation of miR-195 enhances the radiosensitivity of breast cancer cells through the inhibition of BCL-2. *International Journal of Clinical and Experimental Medicine*, *8*, 9142–9148.

Zhu, N., Zhang, D., Xie, H., Zhou, Z., Chen, H., Hu, T., et al. (2011). Endothelial-specific intron-derived miR-126 is down-regulated in human breast cancer and targets both VEGFA and PIK3R2. *Molecular and Cellular Biochemistry*, *351*, 157–164.

Zhu, P., Zhang, J., Zhu, J., Shi, J., Zhu, Q., & Gao, Y. (2015). MiR-429 induces gastric carcinoma cell apoptosis through Bcl-2. *Cellular Physiology and Biochemistry*, *37*, 1572–1580.

Zhu, D. X., Zhu, W., Fang, C., Fan, L., Zou, Z. J., Wang, Y. H., et al. (2012). miR-181a/b significantly enhances drug sensitivity in chronic lymphocytic leukemia cells via targeting multiple anti-apoptosis genes. *Carcinogenesis*, *33*, 1294–1301.

Zhu, W., Zhu, D., Lu, S., Wang, T., Wang, J., Jiang, B., et al. (2012). miR-497 modulates multidrug resistance of human cancer cell lines by targeting BCL2. *Medical Oncology*, *29*, 384–391.

Zhuang, M., Shi, Q., Zhang, X., Ding, Y., Shan, L., Shan, X., et al. (2015). Involvement of miR-143 in cisplatin resistance of gastric cancer cells via targeting IGF1R and BCL2. *Tumour Biology*, *36*, 2737–2745.

Zou, C., Xu, Q., Mao, F., Li, D., Bian, C., Liu, L. Z., et al. (2012). MiR-145 inhibits tumor angiogenesis and growth by N-RAS and VEGF. *Cell Cycle*, *11*, 2137–2145.

CHAPTER FIVE

microRNAs in Cancer Susceptibility

Bríd M. Ryan[1]
Laboratory of Human Carcinogenesis, Center for Cancer Research, National Cancer Institute, Bethesda, MD, United States
[1]Corresponding author: e-mail address: brid_ryan@nih.gov

Contents

1. Introduction 152
2. Functional Consequences of miRNA SNPs 152
  2.1 SNPs in miRNA Genes 153
  2.2 SNPs in miRNA-Binding Sites 156
  2.3 SNPs in the Processing Machinery 157
3. Emerging and Unresolved Concepts in miR-SNP Biology 158
  3.1 Viral miRNA SNPs 158
  3.2 IsomiRs 158
  3.3 Methylation 159
  3.4 Alternative Polyadenylation 160
  3.5 SNPs in Coding Regions and 5′UTR 161
4. Conclusions 163
References 163

Abstract

microRNAs (miRNAs) are a small RNA species without protein-coding potential. However, they are key modulators of protein translation. Many studies have linked miRNAs with cancer initiation, progression, diagnosis, and prognosis, and recent studies have also linked them with cancer etiology and susceptibility, especially through single-nucleotide polymorphisms (SNPs). This review discusses some of the recent advances in miRNA-SNP literature—including SNPs in miRNA genes, miRNA target sites, and the processing machinery. In addition, we highlight some emerging areas of interest, including isomiRs and non-3′UTR focused miRNA-binding mechanisms that could provide further novel insight into the relationship between miR-SNPs and cancer. Finally, we note that additional epidemiological and experimental research is needed to close the gap in our understanding of the genotype–phenotype relationship between miRNA-SNPs and cancer.

Advances in Cancer Research, Volume 135
ISSN 0065-230X
http://dx.doi.org/10.1016/bs.acr.2017.06.004

1. INTRODUCTION

As discussed in this series, microRNAs (miRNAs) play a rheostat-like role in regulating protein translation. Many types of cancer cells have coopted this relationship and use it for driving cancer initiation and progression (Di Leva, Garofalo, & Croce, 2014). The first evidence that miRNAs were connected to cancer came from the laboratory of Carlo Croce in 2002 when he, George Calin, and their colleagues, identified *mir-15* and *mir-16-1* on Chr13q14, a region that is frequently deleted in chronic lymphocytic leukemia (CLL) (Calin et al., 2002). Since then, studies have shown that expression of miRNAs is deregulated in almost every human cancer (Calin & Croce, 2006; Landi et al., 2010; Lu et al., 2005; Mathe et al., 2009; Schembri et al., 2009; Volinia et al., 2006; Yanaihara et al., 2006). In addition, approximately 50% of miRNA genes exist in fragile sites or areas of the genome that are associated with cancer (Calin et al., 2002, 2004; Sevignani et al., 2007). To date, miRNAs have been linked to the etiology, progression, and prognosis of cancer (Di Leva et al., 2014).

Single-nucleotide polymorphisms (SNPs) are the most common form of genetic variation in the human genome. The first SNP was identified in 1978 in the β-globin gene cluster (Kan & Dozy, 1978). Most SNPs are silent but some occur in regulatory regions, affect gene function and phenotypes, and in some cases, are also associated with disease susceptibility (Shastry, 2002). Indeed, the use of SNPs as susceptibility biomarkers and as predictors of outcome has been proposed (McCarthy et al., 2008). This review will focus on the relationship between microRNAs with cancer etiology and susceptibility.

2. FUNCTIONAL CONSEQUENCES OF miRNA SNPs

SNPs occur in many parts of the miRNA network, including miRNA genes themselves, miRNA-binding sites, and in miRNA processing machinery genes (Antoniou et al., 2014; Georges, Coppieters, & Charlier, 2007; Sun et al., 2009). While SNPs in miRNA genes can modify function by affecting transcription of the primary transcript and pri-miRNA and pre-miRNA processing, miR-SNPs can also contribute to the evolution of miRNAs that have new or novel functions (Sun et al., 2009). SNPs in a miRNA-binding site can create—as well as destroy—a binding site, while SNPs in the processing machinery can disrupt the proper maturation, export, and docking of miRNAs to their target sites (Ryan, Robles, &

Harris, 2010). The first demonstration of a miRNA-modulating SNP was described in 2005; Abelson and colleagues described a SNP in the 3′UTR of the *SLITRK1* gene, a binding site for miR-189—this SNP was associated with Tourette's syndrome (Abelson et al., 2005). Since then, numerous studies have demonstrated SNPs within the miRNA network that modulate SNP function, of which many have also been linked with disease (Chen et al., 2013; Chin et al., 2008; Christensen et al., 2009; Clague et al., 2010; Jazdzewski et al., 2009, 2008; Johnson et al., 2005; Pharoah et al., 2011; Ruzzo et al., 2011; Ryan, Robles, & Harris, 2012; Smits et al., 2011; Wang, Wang, Gao, Lv, & Fang, 2014; Yang et al., 2008; Ye et al., 2008; Zanetti et al., 2012; Zhang, Jin, et al., 2012).

In general, several elements are required for a miRNA-related SNP to be functionally and pathologically relevant: The SNP should modulate the interaction between a miRNA and mRNA, both the miRNA and the target mRNA should be expressed in the relevant tissue, and a relationship between the SNP and the cancer needs to be demonstrated. Instances where these elements have converged are described later.

2.1 SNPs in miRNA Genes

Initial estimates suggested that SNPs in miRNA genes were rare (Chen & Rajewsky, 2006; Saunders, Liang, & Li, 2007), primarily due to evolutionary selection pressures (Chen & Rajewsky, 2006; Iwai & Naraba, 2005). Recently, more detailed analyses of the genome have identified additional miR-SNPs. For example, version 138 of dbSNP detailed 1 SNP in miR-21, however, the recent version—148—documents 7 in the pre-miRNA and more than 20 in the pri-miRNA. Epidemiological studies have yet to catch up with this expanding view of miRNA-related genetic variation and determine which, if any, are associated with cancer susceptibility and/or altered functions. A recent study developed the tool SNiPer, a database of seed region miRNA SNPs across a range of species, including humans (Zorc et al., 2012). Interestingly, they found that miR-SNPs are frequently found in chromosome fragile sites, cancer susceptibility loci, and quantitative trait loci. Given that the seed region of a miRNA is the site that most strictly coordinates miRNA–mRNA binding, it has been proposed that seed miRNA SNPs would be the most deleterious to miRNA function and/or most strongly associated with phenotypes. This theory is supported by evidence that a SNP in the seed region of miR-96 is associated with hearing loss in both mice and humans (Lewis et al., 2009; Mencía et al., 2009).

The first description of a miRNA SNP was by Croce and colleagues. Specifically, they discovered a SNP within *pri-miR-16-1* in a kindred with familial CLL (Calin et al., 2005). This functional SNP leads to decreased levels of miR-16-1, most likely due to disruption of pri–pre processing. Interestingly, the same germline mutation is found in New Zealand black mice that naturally develop CLL-like disease (Raveche et al., 2007).

The cancer-associated function of a miRNA—as in whether it functions as an oncogene or a tumor suppressor—depends upon the tissue in which it is expressed (Iorio & Croce, 2012). This contextual relationship is also observed in the relationship between miRNA-SNPs and cancer. The literature has numerous examples where alleles are associated with increased risk of one cancer type and reduced risk of another. One example is rs7372209 in *miR-26a-1*, which is associated with reduced risk in bladder cancer and increased risk of premalignant oral lesions (Clague et al., 2010; Yang et al., 2008). Another example is rs6505162 in *miR-423*—it is associated with increased risk of bladder and ovarian cancer (Kontorovich, Levy, Korostishevsky, Nir, & Friedman, 2010; Yang et al., 2008), but decreased risk of esophageal cancer (Ye et al., 2008). This phenomenon is likely due to the pleomorphic relationship between miRNAs and mRNAs and tissue-specific expression profiles.

rs2910164 in *miR-146a* is one of the better-studied miR-SNPs. It is located in the 3′ passenger (3p) strand of the hairpin and leads to mispairing, altered processing, and altered expression (Jazdzewski et al., 2008; Shen et al., 2008; Xu et al., 2008). It was originally associated with risk of papillary thyroid carcinoma (Jazdzewski et al., 2009, 2008), but has since been linked with risk of many other cancers (Pastrello, Polesel, Della Puppa, Viel, & Maestro, 2010; Ren, Zhou, Cui, & Hou, 2016; Shen et al., 2008; Yin et al., 2017). rs2910164 is unusual for several reasons. First, the minor allele frequency is high and switches between populations. For example, G is the minor allele in Asian populations and the major allele in European populations. In addition, in some cases the heterozygote genotype (GC) is associated with risk of cancer, rather than either the wild-type (GG) or homozygous variant (CC) genotypes. This phenomenon, called over-dominance (Parsons & Bodmer, 1961), is rare in genetic association studies and reflects a phenotype where the heterozygote phenotype differs from the sum of the phenotypes of homozygous phenotypes. In most circumstances, each miRNA hairpin gives rise to two mature miRNA strands, regardless of whether the carrier is wild-type or homozygous variant.

However, a heterozygote will produce three potential mature miRNAs. Collectively, these three miRNAs drive a distinct biology and contribute to cancer risk, as demonstrated by Jazdzewski and colleagues (Jazdzewski et al., 2009, 2008).

rs11614913 in *miR-196a-2* is another well-studied miRNA-SNP (Guo, Jin, Zhang, & Chen, 2012). Studies have linked the C or T allele with risk of breast, lung, colorectal, and gastric cancer (Chen et al., 2013; Duan, Pak, & Jin, 2007; Guo et al., 2012; Hoffman et al., 2009; Ren et al., 2016; Tian et al., 2009; Xu et al., 2016; Yin et al., 2015; Yoon et al., 2012; Zhang, Su, Yu, & Qian, 2012). This is also an interesting SNP: It is in the 3p strand mature sequence of *miR-196a-2* yet it affects expression of the 5p mature strand and not the 3p strand. Using expression vectors containing either the C or T allele of rs11614913, Hoffman et al. observed that the T allele was linked with less efficient processing of the pre-miR 5p strand to its mature form. In addition, introducing the C or T allele caused differential target repression (Hoffman et al., 2009; Huang et al., 2013). Moreover, the cell cycle response to mutagen challenge was significantly enhanced in cells treated with miR-196a-2-T compared with cells treated with the C allele vector (Hoffman et al., 2009). Interestingly, the SNP was also associated with waist-to-hip ratio (Ghanbari et al., 2015) raising the possibility that this relationship could in part mediate the link between this SNP with cancer.

A very interesting recent study also linked this rs11614913 with recurrence of hepatocellular carcinoma following liver transplantation. Remarkably, patients who received donor livers with the CC genotype of rs11614913 had significantly higher recurrence rates of hepatocellular carcinoma (HCC) (41.7% vs 15.3%) and lower cumulative tumor-free survival compared with individuals that received liver transplants from individuals with the TT genotype. As the C allele is associated with higher levels of miR-196a-2, the authors reasoned that this miRNA played a role in liver cancer and created a tumor-favorable microenvironment. Indeed, orthotopic mouse models revealed that overexpression of miR-196a increased tumor size. Importantly, this SNP can also improve risk models that predict tumor recurrence (Xu et al., 2016). rs11614913 is another example of a SNP where the cancer-associated effect of the T or C allele depends on the context of the tumor. In addition, as a deeper appreciation for genetic variation in miRNAs has evolved, *miR-196a-2* has emerged as a SNP-rich miRNA. Moreover, there is considerable population heterogeneity in the prevalence of the T and C alleles—the GG genotype is rare in African populations. Given the evidence to date linking this SNP with a range of phenotypes, these newly identified SNPs should be investigated further.

rs4938723, which lies in the promotor of *miR-34b/c*, has also been linked with cancer risk in several epidemiological studies (Hashemi et al., 2017; Tong et al., 2016; Xu et al., 2011; Zhang et al., 2014)—many studies have linked it with susceptibility to hepatocellular carcinoma and potential interaction with hepatitis B mutations (Han et al., 2013; Xu et al., 2011). Its location in the promoter of this mirtron makes it a likely candidate as a functional polymorphism, and while some evidence suggests that it is functional, an in-depth analysis of the SNPs consequences is still lacking (Tong et al., 2016; Xie et al., 2013).

2.2 SNPs in miRNA-Binding Sites

SNPs in miRNA-binding sites can also disrupt mRNA interactions. One of the earlier examples of this phenomenon was the identification of a novel SNP (rs61764370) in the 3′UTR of *KRAS*. The SNP, which disrupts binding by Let-7, was associated with increased risk of lung cancer (Chin et al., 2008). In addition, while the SNP was not associated with lung cancer survival (Nelson et al., 2010), a more recent report suggested that it might be associated with treatment response (Ganzinelli et al., 2015). rs61764370 was also linked with response to therapy in colorectal cancer (Kjersem et al., 2012; Saridaki et al., 2014; Sclafani et al., 2015; Sebio et al., 2013; Zhang, Labonte, & Lenz, 2011); however, there is heterogeneity in the literature regarding the exact nature of the response, which may be modified by treatment type and population origin (Kjersem et al., 2012; Ryan et al., 2012; Sha et al., 2014). This SNP has also been associated with ovarian (Pharoah et al., 2011), oral (Christensen et al., 2009), and colorectal cancer outcomes (Ruzzo et al., 2011; Ryan et al., 2012; Smits et al., 2011). Interestingly, recent work suggests that in some circumstances, this SNP may mediate its phenotype by fine-tuning the expression of Let-7, rather than translational modulation of the KRAS protein (Crowley, Arena, Lamba, Di Nicolantonio, & Bardelli, 2014). This SNP was previously thought to be associated with risk of ovarian cancer, and as such, be a useful tool for high-risk screening and stratification (Ratner et al., 2010). However, recent work suggests that this SNP may not be linked with risk of ovarian cancer (Pharoah et al., 2011).

There are many other examples of 3′UTR SNPs associated with risk of cancer, many of which have been shown to be functional and too many to cover in this review (Dong et al., 2015; Hughes, Ruth, Rebbeck, & Giri, 2013; Mishra et al., 2007; Robles & Ryan, 2016; Robles et al., 2014; Ryan et al., 2010, 2012; Zanetti et al., 2012). rs1126579—in the 3′UTR of *CXCR2*—modulates the relationship between IL-8 and lung cancer risk. It is a functional allele that alters CXCR2 expression in lung cancer (Ryan et al., 2015). Moreover, a SNP in *KRT81* has also been linked with lung cancer, perhaps with a sex-specific effect (Campayo et al., 2011; Lee et al., 2015; Robles & Ryan, 2016). rs686 in the 3′UTR of dopamine receptor D1 was associated with lung cancer risk and modulated the association between secondhand smoke exposure and risk, observations that were seen in European Americans and African Americans (the minor allele frequency is low in Asian populations) (Robles et al., 2014). Interestingly, this SNP was also associated with total circulating lymphocytes and with CD3+ and CD4+ T cells (but not CD8+ T cells), raising the possibility that the SNP could modulate immune function and the relationship between chronic inflammation with lung cancer risk (Barker et al., 2007).

2.3 SNPs in the Processing Machinery

miRNA biogenesis involves five main steps: transcription, pri-to-pre cleavage, nuclear export, premature cleavage, and RISC incorporation. Genetic variation in genes involved in any of these steps could also affect miRNA function and cancer susceptibility. Indeed, some data already suggest that SNPs in the miRNA biogenesis and processing genes are linked to the etiology of cancer (Clague et al., 2010; Yang et al., 2008; Ye et al., 2008). However, most of these studies are epidemiological and lack experimental insight into the SNP–phenotype relationship.

One of the main genes responsible for pri- to pre-miRNA cleavage is *DROSHA* (Wu, Ye, Ramirez, & Manjunath, 2009; Zeng, Yi, & Cullen, 2005). Several studies have now shown that genetic polymorphisms in *DROSHA* are linked to cancer, including breast (Qian et al., 2016). However, null associations have also been reported (Horikawa et al., 2008; Permuth-Wey et al., 2011). miRNAs are exported out of the nucleus by XPO5 and Ran—SNPs in both of these genes have been associated with cancer susceptibility (Ye et al., 2008). Interestingly, an XPO5 SNP is correlated with prognosis and chemotherapy response in advanced nonsmall cell lung cancer (Geng et al., 2016; Guo et al., 2013) and multiple myeloma (de Larrea et al., 2012). Pre-to-mature miRNA processing is mediated by a partnership between DICER1 and TRBP—limited data suggest *DICER1*

SNPs are related to cancer risk (Clague et al., 2010; Ye et al., 2008). Finally, the movement of miRNAs to their targets of interest is directed by the RISC complex, of which AGO2 and GEMIN4 are two key components. SNPs in the *GEMIN4* gene were significantly associated with renal cell carcinoma (Horikawa et al., 2008) and colorectal cancer risk (Mullany, Herrick, Wolff, Buas, & Slattery, 2016) among others (Yang et al., 2008, 2014; Ye et al., 2008).

3. EMERGING AND UNRESOLVED CONCEPTS IN miR-SNP BIOLOGY

3.1 Viral miRNA SNPs

There is a strong etiological link between viral exposures and cancer risk, including HPV (human papilloma virus), HBV (hepatitis B virus), and EBV (Epstein–Barr virus) (de Martel et al., 2012). Indeed, recent work suggests that SNPs in viral miRNAs could also be linked with host viral replication and, thereby, cancer etiology. For example, a SNP in the EBV miRNA, *mir-BART22*, increases processing of pre-to-mature miRNA, giving rise to increased levels of the mature miRNA, and consequently, enhanced downregulation of its target, latent membrane protein 2A. The latter is a potent immunogenic viral antigen, thus, when downregulated, EBV is less likely to be recognized by the host's immune system. This, therefore, is a novel mechanism by which miR-SNPs can contribute to carcinogenesis (Marston et al., 2013). miR-145 regulates the HPV life cycle both by directly targeting viral sequences and by acting on cellular factors, such as KLF-4 (Gunasekharan & Laimins, 2013)—further interaction between HPV and miR-145 is also described (Shi et al., 2012). A study has described a relationship between a miR-145 SNP, rs353291, and lung cancer risk (Chacon-Cortes et al., 2015)—an interesting follow-up study is whether this SNP might also modulate susceptibility to HPV infection/persistence and/or cervical cancer.

It is not clear how far-reaching the role of viral miR-SNPs could be in relation to viral persistence and cancer. However, a recent study described the existence of many SNPs across viral genomes, raising the possibility that viruses have evolved novel mechanisms to escape host detection and in doing so, persist within the host and possibly contribute to cancer (Marston et al., 2013).

3.2 IsomiRs

IsomiRs occur as sequence variations of canonical mature miRNAs. They were first annotated in 2008, having initially been thought of as sequencing errors (Morin et al., 2008). Indeed, their discovery is largely due to the acceleration of next-generation sequencing technology. IsomiRs are formed via posttranscriptional editing of canonical miRNA sequences and can include 3′ or 5′ trimming, 3′ or 5′ additions (usually adenylation or uridylation), and internal nucleotide (nt) substitution. Importantly, novel non-DNA or nontemplated nucleotide additions can be made to miRNA sequences, in themselves a novel type of polymorphism. Given the stringency with how miRNAs function, even one nucleotide addition to the 5′end of a miRNA could change the target pool for a given miRNA, while 3′end changes can alter both target binding and stability. In many cases, mature miRNAs can have hundreds of sequence variants, and in some cases, the canonical or miRBase annotated miRNA is not the most abundant. Recent work has shown that isomiRs are generally formed via RNA editing with ADARs, variable cleavage by DICER and DROSHA or by the activity of exoribonucleases or nucleotidyl transferases. Little is known about how they function, but evidence suggests that isomiRs can be loaded onto the RISC complex and have distinct functions (Cloonan et al., 2011; Kozlowska, Krzyzosiak, & Koscianska, 2013; Llorens et al., 2013; Neilsen, Goodall, & Bracken, 2012; Wu et al., 2007), mostly related to target selection and stability (Tan et al., 2014). Research also suggested that they may be linked with phenotypes such as stem cell differentiation (Fernandez-Valverde, Taft, & Mattick, 2010; Morin et al., 2008) and cancer (Li et al., 2012; Loher, Londin, & Rigoutsos, 2014; Telonis, Loher, Jing, Londin, & Rigoutsos, 2015).

Several tools have been developed to help the downstream analysis of isomiRs, including DeAnnIso, SeqBuster, IsomiRex, miRanalyzer, isomiRID, IsomiRage, and CPSS among others. The field is still attempting to understand the landscape of isomiRs in cancer. Future work will be needed to understand if (a) germline miR-SNPs can modify isomiR profiles, (b) whether inherent and naturally occurring (nongermline) isomiR polymorphisms affect miRNA biology, and (c) whether these SNPs are linked with etiology. Further, whether SNPs in the machinery that edit miRNA can alter isomiR profiles is also unknown.

3.3 Methylation

miRNA expression can also be affected by epigenetic regulation (Chhabra, 2015; Loginov, Rykov, Fridman, & Braga, 2015; Qin, Zhang, Clarke, Weiland, & Sauter, 2014; Richardson & Patel, 2014; Schliesser et al.,

2016; Sijen, 2015; Tellez et al., 2016; Vrba, Munoz-Rodriguez, Stampfer, & Futscher, 2013; Wang & Chim, 2015; Wong, Yu, & Chim, 2011) and many, including the aforementioned cancer-associated miR-34b/c, are located in CpG islands. miR-SNPs located in these CpG islands could influence miRNA expression and contribute to cancer susceptibility. Interestingly, differences in miRNA genes in tumor-susceptible mice compared with tumor-resistant mice tend to occur in their promoters (Sevignani et al., 2007). Hypermethylation of a CpG island upstream of miR-196a-2 is associated with reduced breast cancer risk (Hoffman et al., 2009), while a promoter Let-7 SNP is linked with risk of hepatocellular carcinoma (Xie et al., 2013).

Interestingly, rs4938723, located in the promoter CpG island of the miR-34b/miR-34c mirtron, is associated with risk of hepatocellular carcinoma, prostate cancer, and renal cell carcinoma (Hashemi et al., 2017; Xie et al., 2013; Xu et al., 2011). The miR-34 family members are direct transcriptional targets of the tumor suppressor p53, and loss of miR-34 function can impair p53-mediated cell cycle arrest and apoptosis. As mentioned, rs4938723 is located within the promoter of *miR-34b/c* but it is also located within a CpG island. Recent in silico analysis found that the SNP might affect binding of the transcription factor GATA and modulate expression (Liang, Liu, Zhao, Yu, & Li, 2013). Evidence of a gene–environment interaction was also found (Xu et al., 2011). These studies indicate that the potentially functional SNP, rs4938723, in the promoter region of *miR-34b/c* could contribute to the susceptibility of HCC in a mechanism that involves differential methylation.

3.4 Alternative Polyadenylation

Many genes undergo alternative polyadenylation, a process that generates 3′UTR sequences with distinct lengths on mRNAs—it has been recognized as a major mechanism of gene regulation (Tian & Manley, 2017). Recent studies suggest that proliferating cells have shorter 3′UTRs, and by extension, fewer miRNA-binding sites (Sandberg, Neilson, Sarma, Sharp, & Burge, 2008). This mechanism is functional. For example, activated T cells have shorter 3′UTRs compared with nonactivated T cells (Sandberg et al., 2008). Mayr and Bartel also reported a global shortening of 3′UTR in cancer cells (Mayr & Bartel, 2009). These shorter isoforms can have greater stability, in part through loss of miRNA-binding sites and can give rise to 10-fold more protein. This further raises the concept of whether SNPs can be expressed, or equally expressed in tumor and normal tissue. For example, a SNP may be present in DNA, and the transcript can be expressed in a specific tissue, but unless the longer form of the 3′UTR is expressed, the effect of the SNP on miRNA binding will not be discernable or possible. Given that various environmental exposures can affect the dynamics of alternative polyadenylation (Akman, Can, & Erson-Bensan, 2012; Liu et al., 2013), this observation could in some instances contribute to the differences observed in the relationship between miR-SNPs and cancer.

Interestingly, SNPs are rare in polyadenylation sites—hexamers of which the canonical site is AAUAAAA—especially in genes where just one polyadenylation site exists. Recent data suggest that such SNPs are under negative selection (Kainov, Aushev, Naumenko, Tchevkina, & Bazykin, 2016) and that selection mainly permits PAS mutations at redundant and/or weakly functional PAS. Nevertheless, a fraction of the SNPs at PAS hexamers likely affect gene functions; in particular, some of the observed SNPs are associated with disease, including cancer (Kainov et al., 2016).

3.5 SNPs in Coding Regions and 5′UTR

Although much less explored or incorporated into prediction algorithms, miRNAs can also bind to target sites in the 5′UTR and open reading frames (Lytle, Yario, & Steitz, 2007; Tay, Zhang, Thomson, Lim, & Rigoutsos, 2008). Interestingly, these binding sites can occasionally be interrupted by introns and therefore require splicing to bind with their complementary miRNAs (Ehrenreich & Purugganan, 2008). In many instances, 5′UTR interactions between a miRNA and mRNA lead to transcript stability and an enhancement of translation, thus the opposite of 3′UTR binding. For example, miR-10a binds to the 5′UTR of ribosomal proteins, including RPS16, PRS6, and RPL9 (Orom, Nielsen, & Lund, 2008), and increases protein translation.

Another example is the interaction between miR-122 and the hepatitis C virus (HCV). Liver-specific miR-122 binds to the 5′UTR of HCV RNA to promote its translation and replication (Jopling, 2008). Recent in-depth studies have shown that miR-122 is also a relatively SNP-dense region. Whether and how these SNPs might interact with this function are not fully known. However, rs4309483 was shown to modulate expression of miR-122 and was protective against chronic infection of HBV, but associated with increased risk of hepatocellular carcinoma in HBV carriers (Liu et al., 2014).

In plants, most miRNAs bind to conserved coding regions (Rhoades et al., 2002). While this contrasts with what is observed in mammals, studies have shown that miRNAs also bind to the coding region of genes in humans. In one of the first examples, miR-148 was shown to regulate DNMT3B protein expression through a conserved site in the protein-coding region (Duursma, Kedde, Schrier, le Sage, & Agami, 2008), while Let-7 binds to the coding region of DICER1 (Forman, Legesse-Miller, & Coller, 2008). Other interesting work shows that miR-145 binds to the coding region of Nanog, OCT4, and SOX2 and in doing so, modulates pluripotency (Tay et al., 2008). In this study, the authors showed that altering the miRNA target site modulated the binding and abrogated the interaction, suggesting that SNPs in these regions could have a similar effect. Indeed, somatic mutations in these noncoding regions, while in some ways are considered silent mutations, could also affect miRNA binding and therefore represent nonsilent mutations. This area is not well explored. However, one study has found that somatic mutations in the 3′UTR of genes could alter miRNA targeting via computational methods. They found that somatic mutations can create or disrupt putative miRNA target sites in the 3′UTRs of many genes, including *MITF*, *EPHA3*, *TAL1*, *SCG3*, and *GSDMA* (Ziebarth, Bhattacharya, & Cui, 2012). One example of a functional relationship was the identification of a somatic mutation in the 3′-UTR of *TNFAIP2* identified from sequencing the genome of a patient with acute myeloid leukemia where the authors found that the mutation creates a new miRNA target site in *TNFAIP2*, a known target of the *PML–RARα* oncogene, reducing its expression (Ramsingh et al., 2010).

The "SomamiR" database catalogs the impact of somatic mutations on miRNA dysregulation in cancer (Bhattacharya, Ziebarth, & Cui, 2013). However, this approach only focusses on the 3′UTR of genes. SomamiR 2.0 is an updated database of cancer somatic mutations in miRNAs and their target sites that potentially alter the interactions between miRNAs and competing endogenous RNAs including mRNAs, circular RNAs, and long noncoding RNAs (Bhattacharya & Cui, 2016), but again, does not cover coding regions. There is a suggestion that this could be important. For example, recent studies reported recurrent mutations in miR-143 in patients with CLL, suggesting it may alter miRNA–mRNA binding and is a target of both hypermutation and selection (Piraino & Furney, 2017; Puente et al., 2015).

Collectively, this work suggests that the search for miRNA-related susceptibility loci should be expanded to include both the 5′UTR and coding regions and should explore the impact of both germline and somatic mutations.

4. CONCLUSIONS

Since first described, SNPs in the miRNA network have been linked with cancer risk, prognosis, and response to therapy. As with many of the genome wide association studies, one hope is that these signatures can be used to facilitate risk stratification and the identification of high-risk patients. Risk stratification tries to identify patients at high risk of disease so that they can be prioritized for screening programs and patient management. For SNPs to be included in such a strategy, the effect size of the risk relationship needs to be high enough that it leads to effective and actionable stratification of patients—for miRNA-related SNPs, it remains to be seen if they can be useful individually. However, given the propensity for miR-SNPs to be functional, it is possible that polyrisk scores—or the use of multiple miRNA SNPs—will be more informative.

Collectively, the evidence suggests that SNPs in the miRNA network are both functional and related with cancer susceptibility. The emerging concepts outlined earlier have not been comprehensively investigated with respect to either potential functional effects and/or relationships with cancer. There are many tools now available that help researchers identify miR-SNPs, bioinformatically predict function and prioritize for research studies. Broadly speaking, additional experimental data and approaches are needed to more fully understand the functional links between miR-SNPs and the range of phenotypes that they are associated with. Moreover, given recent evidence that germline SNPs can influence somatic mutations (Carter et al., 2017), the role of miR-SNPs in this regard is also worth exploring.

REFERENCES

Abelson, J. F., Kwan, K. Y., O'Roak, B. J., Baek, D. Y., Stillman, A. A., Morgan, T. M., et al. (2005). Sequence variants in SLITRK1 are associated with Tourette's syndrome. *Science*, *310*, 317–320.

Akman, B. H., Can, T., & Erson-Bensan, A. E. (2012). Estrogen-induced upregulation and 3′-UTR shortening of CDC6. *Nucleic Acids Research*, *40*, 10679–10688.

Antoniou, A. C., Casadei, S., Heikkinen, T., Barrowdale, D., Pylkas, K., Roberts, J., et al. (2014). Breast-cancer risk in families with mutations in PALB2. *The New England Journal of Medicine*, *371*, 497–506.

Barker, N., van Es, J. H., Kuipers, J., Kujala, P., van den Born, M., Cozijnsen, M., et al. (2007). Identification of stem cells in small intestine and colon by marker gene Lgr5. *Nature*, *449*, 1003–1007.

Bhattacharya, A., & Cui, Y. (2016). SomamiR 2.0: A database of cancer somatic mutations altering microRNA-ceRNA interactions. *Nucleic Acids Research*, *44*, D1005–D1010.

Bhattacharya, A., Ziebarth, J. D., & Cui, Y. (2013). SomamiR: A database for somatic mutations impacting microRNA function in cancer. *Nucleic Acids Research*, *41*, D977–D982.
Calin, G. A., & Croce, C. M. (2006). MicroRNA signatures in human cancers. *Nature Reviews. Cancer*, *6*, 857–866.
Calin, G. A., Dumitru, C. D., Shimizu, M., Bichi, R., Zupo, S., Noch, E., et al. (2002). Frequent deletions and down-regulation of micro-RNA genes miR15 and miR16 at 13q14 in chronic lymphocytic leukemia. *Proceedings of the National Academy of Sciences of the United States of America*, *99*, 15524–15529.
Calin, G. A., Ferracin, M., Cimmino, A., Di Leva, G., Shimizu, M., Wojcik, S. E., et al. (2005). A MicroRNA signature associated with prognosis and progression in chronic lymphocytic leukemia. *The New England Journal of Medicine*, *353*, 1793–1801.
Calin, G. A., Sevignani, C., Dumitru, C. D., Hyslop, T., Noch, E., Yendamuri, S., et al. (2004). Human microRNA genes are frequently located at fragile sites and genomic regions involved in cancer. *Proceedings of the National Academy of Sciences of the United States of America*, *101*, 2999–3004.
Campayo, M., Navarro, A., Viñolas, N., Tejero, R., Muñoz, C., Diaz, T., et al. (2011). A dual role for KRT81: A miR-SNP associated with recurrence in non-small-cell lung cancer and a novel marker of squamous cell lung carcinoma. *PLoS One*, *6*, e22509.
Carter, H., Marty, R., Hofree, M., Gross, A. M., Jensen, J., Fisch, K. M., et al. (2017). Interaction landscape of inherited polymorphisms with somatic events in cancer. *Cancer Discovery*, 7(4), 410–423.
Chacon-Cortes, D., Smith, R. A., Haupt, L. M., Lea, R. A., Youl, P. H., & Griffiths, L. R. (2015). Genetic association analysis of miRNA SNPs implicates MIR145 in breast cancer susceptibility. *BMC Medical Genetics*, *16*, 107.
Chen, K., & Rajewsky, N. (2006). Natural selection on human microRNA binding sites inferred from SNP data. *Nature Genetics*, *38*, 1452–1456.
Chen, Z., Xu, L., Ye, X., Shen, S., Li, Z., Niu, X., et al. (2013). Polymorphisms of microRNA sequences or binding sites and lung cancer: A meta-analysis and systematic review. *PLoS One*, *8*, e61008.
Chhabra, R. (2015). miRNA and methylation: A multifaceted liaison. *Chembiochem*, *16*, 195–203.
Chin, L. J., Ratner, E., Leng, S., Zhai, R., Nallur, S., Babar, I., et al. (2008). A SNP in a let-7 microRNA complementary site in the KRAS 3′ untranslated region increases non-small cell lung cancer risk. *Cancer Research*, *68*, 8535–8540.
Christensen, B. C., Moyer, B. J., Avissar, M., Ouellet, L. G., Plaza, S. L., McClean, M. D., et al. (2009). A let-7 microRNA-binding site polymorphism in the KRAS 3′ UTR is associated with reduced survival in oral cancer. *Carcinogenesis*, *30*, 1003–1007.
Clague, J., Lippman, S. M., Yang, H., Hildebrandt, M. A., Ye, Y., Lee, J. J., et al. (2010). Genetic variation in microRNA genes and risk of oral premalignant lesions. *Molecular Carcinogenesis*, *49*, 183–189.
Cloonan, N., Wani, S., Xu, Q., Gu, J., Lea, K., Heater, S., et al. (2011). MicroRNAs and their isomiRs function cooperatively to target common biological pathways. *Genome Biology*, *12*, R126.
Crowley, E. H., Arena, S., Lamba, S., Di Nicolantonio, F., & Bardelli, A. (2014). Targeted knock-in of the polymorphism rs61764370 does not affect KRAS expression but reduces let-7 levels. *Human Mutation*, *35*, 208–214.
de Larrea, C. F., Navarro, A., Tejero, R., Tovar, N., Diaz, T., Cibeira, M. T., et al. (2012). Impact of MiRSNPs on survival and progression in patients with multiple myeloma undergoing autologous stem cell transplantation. *Clinical Cancer Research*, *18*, 3697–3704.
de Martel, C., Ferlay, J., Franceschi, S., Vignat, J., Bray, F., Forman, D., et al. (2012). Global burden of cancers attributable to infections in 2008: A review and synthetic analysis. *The Lancet Oncology*, *13*, 607–615.

Di Leva, G., Garofalo, M., and Croce, C. M. (2014). *Annual Review of Pathology*. MicroRNAs in cancer 9**b** 287–314.

Dong, B., Shi, Z., Wang, J., Wu, J., Yang, Z., & Fang, K. (2015). IL-6 inhibits the targeted modulation of PDCD4 by miR-21 in prostate cancer. *PLoS One*, *10*, e0134366.

Duan, R., Pak, C., & Jin, P. (2007). Single nucleotide polymorphism associated with mature miR-125a alters the processing of pri-miRNA. *Human Molecular Genetics*, *16*, 1124–1131.

Duursma, A. M., Kedde, M., Schrier, M., le Sage, C., & Agami, R. (2008). miR-148 targets human DNMT3b protein coding the region. *RNA*, *14*, 872–877.

Ehrenreich, I. M., & Purugganan, M. D. (2008). Sequence variation of microRNAs and their binding sites in Arabidopsis. *Plant Physiology*, *146*, 1974–1982.

Fernandez-Valverde, S. L., Taft, R. J., & Mattick, J. S. (2010). Dynamic isomiR regulation in Drosophila development. *RNA*, *16*, 1881–1888.

Forman, J. J., Legesse-Miller, A., & Coller, H. A. (2008). A search for conserved sequences in coding regions reveals that the let-7 microRNA targets Dicer within its coding sequence. *Proceedings of the National Academy of Sciences of the United States of America*, *105*, 14879–14884.

Ganzinelli, M., Rulli, E., Caiola, E., Garassino, M. C., Broggini, M., Copreni, E., et al. (2015). Role of KRAS-LCS6 polymorphism in advanced NSCLC patients treated with erlotinib or docetaxel in second line treatment (TAILOR). *Scientific Reports*, *5*, 16331.

Geng, J. Q., Wang, X. C., Li, L. F., Zhao, J., Wu, S., Yu, G. P., et al. (2016). MicroRNA-related single-nucleotide polymorphism of XPO5 is strongly correlated with the prognosis and chemotherapy response in advanced non-small-cell lung cancer patients. *Tumour Biology*, *37*, 2257–2265.

Georges, M., Coppieters, W., & Charlier, C. (2007). Polymorphic miRNA-mediated gene regulation: Contribution to phenotypic variation and disease. *Current Opinion in Genetics & Development*, *17*, 166–176.

Ghanbari, M., Sedaghat, S., de Looper, H. W., Hofman, A., Erkeland, S. J., Franco, O. H., et al. (2015). The association of common polymorphisms in miR-196a2 with waist to hip ratio and miR-1908 with serum lipid and glucose. *Obesity (Silver Spring)*, *23*, 495–503.

Gunasekharan, V., & Laimins, L. A. (2013). Human papillomaviruses modulate microRNA 145 expression to directly control genome amplification. *Journal of Virology*, *87*, 6037–6043.

Guo, J., Jin, M., Zhang, M., & Chen, K. (2012). A genetic variant in miR-196a2 increased digestive system cancer risks: A meta-analysis of 15 case-control studies. *PLoS One*, *7*, e30585.

Guo, Z., Wang, H., Li, Y., Li, B., Li, C., & Ding, C. (2013). A microRNA-related single nucleotide polymorphism of the XPO5 gene is associated with survival of small cell lung cancer patients. *Biomedical Reports*, *1*, 545–548.

Han, Y., Pu, R., Han, X., Zhao, J., Zhang, Y., Zhang, Q., et al. (2013). Associations of pri-miR-34b/c and pre-miR-196a2 polymorphisms and their multiplicative interactions with hepatitis B virus mutations with hepatocellular carcinoma risk. *PLoS One*, *8*, e58564.

Hashemi, M., Danesh, H., Bizhani, F., Narouie, B., Sotoudeh, M., Nouralizadeh, A., et al. (2017). Pri-miR-34b/c rs4938723 polymorphism increased the risk of prostate cancer. *Cancer Biomarkers*, *18*, 155–159.

Hoffman, A. E., Zheng, T., Yi, C., Leaderer, D., Weidhaas, J., Slack, F., et al. (2009). microRNA miR-196a-2 and breast cancer: A genetic and epigenetic association study and functional analysis. *Cancer Research*, *69*, 5970–5977.

Horikawa, Y., Wood, C. G., Yang, H., Zhao, H., Ye, Y., Gu, J., et al. (2008). Single nucleotide polymorphisms of microRNA machinery genes modify the risk of renal cell carcinoma. *Clinical Cancer Research*, *14*, 7956–7962.

Huang, Y., Yi, X., Jian, Z., Wei, C., Li, S., Cai, C., et al. (2013). A single-nucleotide polymorphism of miR-196a-2 and vitiligo: An association study and functional analysis in a Han Chinese population. *Pigment Cell & Melanoma Research*, *26*, 338–347.

Hughes, L., Ruth, K., Rebbeck, T. R., & Giri, V. N. (2013). Genetic variation in IL-16 miRNA target site and time to prostate cancer diagnosis in African-American men. *Prostate Cancer and Prostatic Diseases*, *16*, 308–314.

Iorio, M. V., & Croce, C. M. (2012). Causes and consequences of microRNA dysregulation. *Cancer Journal*, *18*, 215–222.

Iwai, N., & Naraba, H. (2005). Polymorphisms in human pre-miRNAs. *Biochemical and Biophysical Research Communications*, *331*, 1439–1444.

Jazdzewski, K., Liyanarachchi, S., Swierniak, M., Pachucki, J., Ringel, M. D., Jarzab, B., et al. (2009). Polymorphic mature microRNAs from passenger strand of pre-miR-146a contribute to thyroid cancer. *Proceedings of the National Academy of Sciences of the United States of America*, *106*, 1502–1505.

Jazdzewski, K., Murray, E. L., Franssila, K., Jarzab, B., Schoenberg, D. R., & de la Chapelle, A. (2008). Common SNP in pre-miR-146a decreases mature miR expression and predisposes to papillary thyroid carcinoma. *Proceedings of the National Academy of Sciences of the United States of America*, *105*, 7269–7274.

Johnson, S. M., Grosshans, H., Shingara, J., Byrom, M., Jarvis, R., Cheng, A., et al. (2005). RAS is regulated by the let-7 microRNA family. *Cell*, *120*, 635–647.

Jopling, C. L. (2008). Regulation of hepatitis C virus by microRNA-122. *Biochemical Society Transactions*, *36*, 1220–1223.

Kainov, Y. A., Aushev, V. N., Naumenko, S. A., Tchevkina, E. M., & Bazykin, G. A. (2016). Complex selection on human polyadenylation signals revealed by polymorphism and divergence data. *Genome Biology and Evolution*, *8*, 1971–1979.

Kan, Y. W., & Dozy, A. M. (1978). Polymorphism of DNA sequence adjacent to human beta-globin structural gene: Relationship to sickle mutation. *Proceedings of the National Academy of Sciences of the United States of America*, *75*, 5631–5635.

Kjersem, J. B., Ikdahl, T., Guren, T., Skovlund, E., Sorbye, H., Hamfjord, J., et al. (2012). Let-7 miRNA-binding site polymorphism in the KRAS 3'UTR; colorectal cancer screening population prevalence and influence on clinical outcome in patients with metastatic colorectal cancer treated with 5-fluorouracil and oxaliplatin +/− cetuximab. *BMC Cancer*, *12*, 534.

Kontorovich, T., Levy, A., Korostishevsky, M., Nir, U., & Friedman, E. (2010). Single nucleotide polymorphisms in miRNA binding sites and miRNA genes as breast/ovarian cancer risk modifiers in Jewish high-risk women. *International Journal of Cancer*, *127*, 589–597.

Kozlowska, E., Krzyzosiak, W. J., & Koscianska, E. (2013). Regulation of huntingtin gene expression by miRNA-137, -214, -148a, and their respective isomiRs. *International Journal of Molecular Sciences*, *14*, 16999–17016.

Landi, M. T., Zhao, Y., Rotunno, M., Koshiol, J., Liu, H., Bergen, A. W., et al. (2010). MicroRNA expression differentiates histology and predicts survival of lung cancer. *Clinical Cancer Research*, *16*, 430–441.

Lee, S. Y., Choi, J. E., Jeon, H. S., Hong, M. J., Choi, Y. Y., Kang, H. G., et al. (2015). A genetic variation in microRNA target site of KRT81 gene is associated with survival in early-stage non-small-cell lung cancer. *Annals of Oncology*, *26*, 1142–1148.

Lewis, M. A., Quint, E., Glazier, A. M., Fuchs, H., De Angelis, M. H., Langford, C., et al. (2009). An ENU-induced mutation of miR-96 associated with progressive hearing loss in mice. *Nature Genetics*, *41*, 614–618.

Li, S. C., Liao, Y. L., Ho, M. R., Tsai, K. W., Lai, C. H., & Lin, W. C. (2012). miRNA arm selection and isomiR distribution in gastric cancer. *BMC Genomics*, *13*(Suppl. 1), S13.

Liang, T. J., Liu, H. J., Zhao, X. Q., Yu, C. H., & Li, C. S. (2013). Lack of association of MiR-34b/c polymorphism (rs4938723) with hepatocellular carcinoma: A meta-analysis. *PLoS One*, *8*, e68588.

Liu, Y., Hu, W., Murakawa, Y., Yin, J., Wang, G., Landthaler, M., et al. (2013). Cold-induced RNA-binding proteins regulate circadian gene expression by controlling alternative polyadenylation. *Scientific Reports*, *3*, 2054.

Liu, Y., Xie, K., Wen, J., Deng, M., Li, J., & Hu, Z. (2014). A genetic variant in microRNA-122 regulatory region confers risk for chronic hepatitis B virus infection and hepatocellular carcinoma in Han Chinese. *Journal of Medical Virology*, *86*, 1669–1674.

Llorens, F., Bañez-Coronel, M., Pantano, L., del Río, J. A., Ferrer, I., Estivill, X., et al. (2013). A highly expressed miR-101 isomiR is a functional silencing small RNA. *BMC Genomics*, *14*, 104.

Loginov, V. I., Rykov, S. V., Fridman, M. V., & Braga, E. A. (2015). Methylation of miRNA genes and oncogenesis. *Biochemistry (Mosc)*, *80*, 145–162.

Loher, P., Londin, E. R., & Rigoutsos, I. (2014). IsomiR expression profiles in human lymphoblastoid cell lines exhibit population and gender dependencies. *Oncotarget*, *5*, 8790–8802.

Lu, J., Getz, G., Miska, E. A., Alvarez-Saavedra, E., Lamb, J., Peck, D., et al. (2005). MicroRNA expression profiles classify human cancers. *Nature*, *435*, 834–838.

Lytle, J. R., Yario, T. A., & Steitz, J. A. (2007). Target mRNAs are repressed as efficiently by microRNA-binding sites in the 5′ UTR as in the 3′ UTR. *Proceedings of the National Academy of Sciences of the United States of America*, *104*, 9667–9672.

Marston, D. A., McElhinney, L. M., Ellis, R. J., Horton, D. L., Wise, E. L., Leech, S. L., et al. (2013). Next generation sequencing of viral RNA genomes. *BMC Genomics*, *14*, 444.

Mathe, E. A., Nguyen, G. H., Bowman, E. D., Zhao, Y., Budhu, A., Schetter, A. J., et al. (2009). MicroRNA expression in squamous cell carcinoma and adenocarcinoma of the esophagus: Associations with survival. *Clinical Cancer Research*, *15*, 6192–6200.

Mayr, C., & Bartel, D. P. (2009). Widespread shortening of 3'UTRs by alternative cleavage and polyadenylation activates oncogenes in cancer cells. *Cell*, *138*, 673–684.

McCarthy, M. I., Abecasis, G. R., Cardon, L. R., Goldstein, D. B., Little, J., Ioannidis, J. P., et al. (2008). Genome-wide association studies for complex traits: Consensus, uncertainty and challenges. *Nature Reviews. Genetics*, *9*, 356–369.

Mencía, A., Modamio-Høybjør, S., Redshaw, N., Morín, M., Mayo-Merino, F., Olavarrieta, L., et al. (2009). Mutations in the seed region of human miR-96 are responsible for nonsyndromic progressive hearing loss. *Nature Genetics*, *41*, 609–613.

Mishra, P. J., Humeniuk, R., Mishra, P. J., Longo-Sorbello, G. S., Banerjee, D., & Bertino, J. R. (2007). A miR-24 microRNA binding-site polymorphism in dihydrofolate reductase gene leads to methotrexate resistance. *Proceedings of the National Academy of Sciences of the United States of America*, *104*, 13513–13518.

Morin, R. D., O'Connor, M. D., Griffith, M., Kuchenbauer, F., Delaney, A., Prabhu, A. L., et al. (2008). Application of massively parallel sequencing to microRNA profiling and discovery in human embryonic stem cells. *Genome Research*, *18*, 610–621.

Mullany, L. E., Herrick, J. S., Wolff, R. K., Buas, M. F., & Slattery, M. L. (2016). Impact of polymorphisms in microRNA biogenesis genes on colon cancer risk and microRNA expression levels: A population-based, case-control study. *BMC Medical Genomics*, *9*, 21.

Neilsen, C. T., Goodall, G. J., & Bracken, C. P. (2012). IsomiRs—The overlooked repertoire in the dynamic microRNAome. *Trends in Genetics*, *28*, 544–549.

Nelson, H. H., Christensen, B. C., Plaza, S. L., Wiencke, J. K., Marsit, C. J., & Kelsey, K. T. (2010). KRAS mutation, KRAS-LCS6 polymorphism, and non-small cell lung cancer. *Lung Cancer*, *69*, 51–53.

Orom, U. A., Nielsen, F. C., & Lund, A. H. (2008). MicroRNA-10a binds the 5'UTR of ribosomal protein mRNAs and enhances their translation. *Molecular Cell*, *30*, 460–471.

Parsons, P. A., & Bodmer, W. F. (1961). The evolution of overdominance: Natural selection and heterozygote advantage. *Nature*, *190*, 7–12.

Pastrello, C., Polesel, J., Della Puppa, L., Viel, A., & Maestro, R. (2010). Association between hsa-mir-146a genotype and tumor age-of-onset in BRCA1/BRCA2-negative familial breast and ovarian cancer patients. *Carcinogenesis*, *31*, 2124–2126.

Permuth-Wey, J., Chen, Z., Tsai, Y. Y., Lin, H. Y., Chen, Y. A., Barnholtz-Sloan, J., et al. (2011). MicroRNA processing and binding site polymorphisms are not replicated in the Ovarian Cancer Association Consortium. *Cancer Epidemiology, Biomarkers & Prevention*, *20*, 1793–1797.

Pharoah, P. D., Palmieri, R. T., Ramus, S. J., Gayther, S. A., Andrulis, I. L., Anton-Culver, H., et al. (2011). The role of KRAS rs61764370 in invasive epithelial ovarian cancer: Implications for clinical testing. *Clinical Cancer Research*, *17*, 3742–3750.

Piraino, S. W., & Furney, S. J. (2017). Identification of coding and non-coding mutational hotspots in cancer genomes. *BMC Genomics*, *18*, 17.

Puente, X. S., Beà, S., Valdés-Mas, R., Villamor, N., Gutiérrez-Abril, J., Martín-Subero, J. I., et al. (2015). Non-coding recurrent mutations in chronic lymphocytic leukaemia. *Nature*, *526*, 519–524.

Qian, F., Feng, Y., Zheng, Y., Ogundiran, T. O., Ojengbede, O., Zheng, W., et al. (2016). Genetic variants in microRNA and microRNA biogenesis pathway genes and breast cancer risk among women of African ancestry. *Human Genetics*, *135*, 1145–1159.

Qin, W., Zhang, K., Clarke, K., Weiland, T., & Sauter, E. R. (2014). Methylation and miRNA effects of resveratrol on mammary tumors vs. normal tissue. *Nutrition and Cancer*, *66*, 270–277.

Ramsingh, G., Koboldt, D. C., Trissal, M., Chiappinelli, K. B., Wylie, T., Koul, S., et al. (2010). Complete characterization of the microRNAome in a patient with acute myeloid leukemia. *Blood*, *116*, 5316–5326.

Ratner, E., Lu, L., Boeke, M., Barnett, R., Nallur, S., Chin, L. J., et al. (2010). A KRAS-variant in ovarian cancer acts as a genetic marker of cancer risk. *Cancer Research*, *70*, 6509–6515.

Raveche, E. S., Salerno, E., Scaglione, B. J., Manohar, V., Abbasi, F., Lin, Y. C., et al. (2007). Abnormal microRNA-16 locus with synteny to human 13q14 linked to CLL in NZB mice. *Blood*, *109*, 5079–5086.

Ren, Y. G., Zhou, X. M., Cui, Z. G., & Hou, G. (2016). Effects of common polymorphisms in miR-146a and miR-196a2 on lung cancer susceptibility: A meta-analysis. *Journal of Thoracic Disease*, *8*, 1297–1305.

Rhoades, M. W., Reinhart, B. J., Lim, L. P., Burge, C. B., Bartel, B., & Bartel, D. P. (2002). Prediction of plant microRNA targets. *Cell*, *110*, 513–520.

Richardson, B. C., & Patel, D. R. (2014). Epigenetics in 2013. DNA methylation and miRNA: Key roles in systemic autoimmunity. *Nature Reviews. Rheumatology*, *10*, 72–74.

Robles, A. I., & Ryan, B. M. (2016). KRT81 miR-SNP rs3660 is associated with risk and survival of NSCLC. *Annals of Oncology*, *27*, 360–361.

Robles, A. I., Yang, P., Jen, J., McClary, A. C., Calhoun, K., Bowman, E. D., et al. (2014). A DRD1 polymorphism predisposes to lung cancer among those exposed to secondhand smoke during childhood. *Cancer Prevention Research (Philadelphia, Pa.)*, *7*, 1210–1218.

Ruzzo, A., Canestrari, E., Galluccio, N., Santini, D., Vincenzi, B., Tonini, G., et al. (2011). Role of KRAS let-7 LCS6 SNP in metastatic colorectal cancer patients. *Annals of Oncology*, *22*, 234–235.

Ryan, B. M., Robles, A. I., & Harris, C. C. (2010). Genetic variation in microRNA networks: The implications for cancer research. *Nature Reviews. Cancer*, *10*, 389–402.

Ryan, B. M., Robles, A. I., & Harris, C. C. (2012). KRAS-LCS6 genotype as a prognostic marker in early-stage CRC—Letter. *Clinical Cancer Research*, *18*, 3487–3488 [author reply 3489].

Ryan, B. M., Robles, A. I., McClary, A. C., Haznadar, M., Bowman, E. D., Pine, S. R., et al. (2015). Identification of a functional SNP in the 3'UTR of CXCR2 that is associated with reduced risk of lung cancer. *Cancer Research*, *75*, 566–575.

Sandberg, R., Neilson, J. R., Sarma, A., Sharp, P. A., & Burge, C. B. (2008). Proliferating cells express mRNAs with shortened 3′ untranslated regions and fewer microRNA target sites. *Science*, *320*, 1643–1647.

Saridaki, Z., Weidhaas, J. B., Lenz, H. J., Laurent-Puig, P., Jacobs, B., De Schutter, J., et al. (2014). A let-7 microRNA-binding site polymorphism in KRAS predicts improved outcome in patients with metastatic colorectal cancer treated with salvage cetuximab/panitumumab monotherapy. *Clinical Cancer Research*, *20*, 4499–4510.

Saunders, M. A., Liang, H., & Li, W. H. (2007). Human polymorphism at microRNAs and microRNA target sites. *Proceedings of the National Academy of Sciences of the United States of America*, *104*, 3300–3305.

Schembri, F., Sridhar, S., Perdomo, C., Gustafson, A. M., Zhang, X., Ergun, A., et al. (2009). MicroRNAs as modulators of smoking-induced gene expression changes in human airway epithelium. *Proceedings of the National Academy of Sciences of the United States of America*, *106*, 2319–2324.

Schliesser, M. G., Claus, R., Hielscher, T., Grimm, C., Weichenhan, D., Blaes, J., et al. (2016). Prognostic relevance of miRNA-155 methylation in anaplastic glioma. *Oncotarget*, *7*, 82028–82045.

Sclafani, F., Chau, I., Cunningham, D., Peckitt, C., Lampis, A., Hahne, J. C., et al. (2015). Prognostic role of the LCS6 KRAS variant in locally advanced rectal cancer: Results of the EXPERT-C trial. *Annals of Oncology*, *26*, 1936–1941.

Sebio, A., Paré, L., Páez, D., Salazar, J., González, A., Sala, N., et al. (2013). The LCS6 polymorphism in the binding site of let-7 microRNA to the KRAS 3′-untranslated region: Its role in the efficacy of anti-EGFR-based therapy in metastatic colorectal cancer patients. *Pharmacogenetics and Genomics*, *23*, 142–147.

Sevignani, C., Calin, G. A., Nnadi, S. C., Shimizu, M., Davuluri, R. V., Hyslop, T., et al. (2007). MicroRNA genes are frequently located near mouse cancer susceptibility loci. *Proceedings of the National Academy of Sciences of the United States of America*, *104*, 8017–8022.

Sha, D., Lee, A. M., Shi, Q., Alberts, S. R., Sargent, D. J., Sinicrope, F. A., et al. (2014). Association study of the let-7 miRNA-complementary site variant in the 3′ untranslated region of the KRAS gene in stage III colon cancer (NCCTG N0147 Clinical Trial). *Clinical Cancer Research*, *20*, 3319–3327.

Shastry, B. S. (2002). SNP alleles in human disease and evolution. *Journal of Human Genetics*, *47*, 561–566.

Shen, J., Ambrosone, C. B., DiCioccio, R. A., Odunsi, K., Lele, S. B., & Zhao, H. (2008). A functional polymorphism in the miR-146a gene and age of familial breast/ovarian cancer diagnosis. *Carcinogenesis*, *29*, 1963–1966.

Shi, M., Du, L., Liu, D., Qian, L., Hu, M., Yu, M., et al. (2012). Glucocorticoid regulation of a novel HPV-E6-p53-miR-145 pathway modulates invasion and therapy resistance of cervical cancer cells. *The Journal of Pathology*, *228*, 148–157.

Sijen, T. (2015). Molecular approaches for forensic cell type identification: On mRNA, miRNA, DNA methylation and microbial markers. *Forensic Science International. Genetics*, *18*, 21–32.

Smits, K. M., Paranjape, T., Nallur, S., Wouters, K. A., Weijenberg, M. P., Schouten, L. J., et al. (2011). A let-7 microRNA SNP in the KRAS 3'UTR is prognostic in early-stage colorectal cancer. *Clinical Cancer Research*, *17*, 7723–7731.

Sun, G., Yan, J., Noltner, K., Feng, J., Li, H., Sarkis, D. A., et al. (2009). SNPs in human miRNA genes affect biogenesis and function. *RNA*, *15*, 1640–1651.

Tan, G. C., Chan, E., Molnar, A., Sarkar, R., Alexieva, D., Isa, I. M., et al. (2014). 5′ isomiR variation is of functional and evolutionary importance. *Nucleic Acids Research*, *42*, 9424–9435.

Tay, Y., Zhang, J., Thomson, A. M., Lim, B., & Rigoutsos, I. (2008). MicroRNAs to Nanog, Oct4 and Sox2 coding regions modulate embryonic stem cell differentiation. *Nature*, *455*, 1124–1128.

Tellez, C. S., Juri, D. E., Do, K., Picchi, M. A., Wang, T., Liu, G., et al. (2016). miR-196b is epigenetically silenced during the premalignant stage of lung Carcinogenesis. *Cancer Research*, *76*, 4741–4751.

Telonis, A. G., Loher, P., Jing, Y., Londin, E., & Rigoutsos, I. (2015). Beyond the one-locus-one-miRNA paradigm: MicroRNA isoforms enable deeper insights into breast cancer heterogeneity. *Nucleic Acids Research*, *43*, 9158–9175.

Tian, B., & Manley, J. L. (2017). Alternative polyadenylation of mRNA precursors. *Nature Reviews. Molecular Cell Biology*, *18*, 18–30.

Tian, T., Shu, Y., Chen, J., Hu, Z., Xu, L., Jin, G., et al. (2009). A functional genetic variant in microRNA-196a2 is associated with increased susceptibility of lung cancer in Chinese. *Cancer Epidemiology, Biomarkers & Prevention*, *18*, 1183–1187.

Tong, N., Chu, H., Wang, M., Xue, Y., Du, M., Lu, L., et al. (2016). Pri-miR-34b/c rs4938723 polymorphism contributes to acute lymphoblastic leukemia susceptibility in Chinese children. *Leukemia & Lymphoma*, *57*, 1436–1441.

Volinia, S., Calin, G. A., Liu, C. G., Ambs, S., Cimmino, A., Petrocca, F., et al. (2006). A microRNA expression signature of human solid tumors defines cancer gene targets. *Proceedings of the National Academy of Sciences of the United States of America*, *103*, 2257–2261.

Vrba, L., Munoz-Rodriguez, J. L., Stampfer, M. R., & Futscher, B. W. (2013). miRNA gene promoters are frequent targets of aberrant DNA methylation in human breast cancer. *PLoS One*, *8*, e54398.

Wang, L. Q., & Chim, C. S. (2015). DNA methylation of tumor-suppressor miRNA genes in chronic lymphocytic leukemia. *Epigenomics*, *7*, 461–473.

Wang, G., Wang, W., Gao, W., Lv, J., & Fang, J. (2014). Two functional polymorphisms in microRNAs and lung cancer risk: A meta-analysis. *Tumour Biology*, *35*, 2693–2699.

Wong, K. Y., Yu, L., & Chim, C. S. (2011). DNA methylation of tumor suppressor miRNA genes: A lesson from the miR-34 family. *Epigenomics*, *3*, 83–92.

Wu, H., Neilson, J. R., Kumar, P., Manocha, M., Shankar, P., Sharp, P. A., et al. (2007). miRNA profiling of naïve, effector and memory CD8 T cells. *PLoS One*, *2*, e1020.

Wu, H., Ye, C., Ramirez, D., & Manjunath, N. (2009). Alternative processing of primary microRNA transcripts by Drosha generates 5′ end variation of mature microRNA. *PLoS One*, *4*, e7566.

Xie, K., Liu, J., Zhu, L., Liu, Y., Pan, Y., Wen, J., et al. (2013). A potentially functional polymorphism in the promoter region of let-7 family is associated with survival of hepatocellular carcinoma. *Cancer Epidemiology*, *37*, 998–1002.

Xu, X., Ling, Q., Wang, J., Xie, H., Wei, X., Lu, D., et al. (2016). Donor miR-196a-2 polymorphism is associated with hepatocellular carcinoma recurrence after liver transplantation in a Han Chinese population. *International Journal of Cancer*, *138*, 620–629.

Xu, Y., Liu, L., Liu, J., Zhang, Y., Zhu, J., Chen, J., et al. (2011). A potentially functional polymorphism in the promoter region of miR-34b/c is associated with an increased risk for primary hepatocellular carcinoma. *International Journal of Cancer*, *128*, 412–417.

Xu, T., Zhu, Y., Wei, Q. K., Yuan, Y., Zhou, F., Ge, Y. Y., et al. (2008). A functional polymorphism in the miR-146a gene is associated with the risk for hepatocellular carcinoma. *Carcinogenesis*, *29*, 2126–2131.

Yanaihara, N., Caplen, N., Bowman, E., Seike, M., Kumamoto, K., Yi, M., et al. (2006). Unique microRNA molecular profiles in lung cancer diagnosis and prognosis. *Cancer Cell, 9*, 189–198.

Yang, H., Dinney, C. P., Ye, Y., Zhu, Y., Grossman, H. B., & Wu, X. (2008). Evaluation of genetic variants in microRNA-related genes and risk of bladder cancer. *Cancer Research, 68*, 2530–2537.

Yang, P. W., Huang, Y. C., Hsieh, C. Y., Hua, K. T., Huang, Y. T., Chiang, T. H., et al. (2014). Association of miRNA-related genetic polymorphisms and prognosis in patients with esophageal squamous cell carcinoma. *Annals of Surgical Oncology, 21*(Suppl. 4), S601–S609.

Ye, Y., Wang, K. K., Gu, J., Yang, H., Lin, J., Ajani, J. A., et al. (2008). Genetic variations in microRNA-related genes are novel susceptibility loci for esophageal cancer risk. *Cancer Prevention Research (Philadelphia, Pa.), 1*, 460–469.

Yin, Z., Cui, Z., Guan, P., Li, X., Wu, W., Ren, Y., et al. (2015). Interaction between polymorphisms in pre-MiRNA genes and cooking oil fume exposure on the risk of lung cancer in Chinese non-smoking female population. *PLoS One, 10*, e0128572.

Yin, Z., Cui, Z., Ren, Y., Xia, L., Li, H., & Zhou, B. (2017). MiR-146a polymorphism correlates with lung cancer risk in Chinese nonsmoking females. *Oncotarget, 8*, 2275–2283.

Yoon, K. A., Yoon, H., Park, S., Jang, H. J., Zo, J. I., Lee, H. S., et al. (2012). The prognostic impact of microRNA sequence polymorphisms on the recurrence of patients with completely resected non-small cell lung cancer. *The Journal of Thoracic and Cardiovascular Surgery, 144*, 794–807.

Zanetti, K. A., Haznadar, M., Welsh, J. A., Robles, A. I., Ryan, B. M., McClary, A. C., et al. (2012). 3′-UTR and functional secretor haplotypes in mannose-binding lectin 2 are associated with increased colon cancer risk in African Americans. *Cancer Research, 72*, 1467–1477.

Zeng, Y., Yi, R., & Cullen, B. R. (2005). Recognition and cleavage of primary microRNA precursors by the nuclear processing enzyme Drosha. *The EMBO Journal, 24*, 138–148.

Zhang, M. W., Jin, M. J., Yu, Y. X., Zhang, S. C., Liu, B., Jiang, X., et al. (2012b). Associations of lifestyle-related factors, hsa-miR-149 and hsa-miR-605 gene polymorphisms with gastrointestinal cancer risk. *Molecular Carcinogenesis, 51*(Suppl. 1). E21–31.

Zhang, W., Labonte, M. J., & Lenz, H. J. (2011). KRAS let-7 LCS6 SNP predicts cetuximab efficacy in KRASwt metastatic colorectal cancer patients: Does treatment combination partner matter? *Annals of Oncology, 22*, 484–485.

Zhang, S., Qian, J., Cao, Q., Li, P., Wang, M., Wang, J., et al. (2014). A potentially functional polymorphism in the promoter region of miR-34b/c is associated with renal cell cancer risk in a Chinese population. *Mutagenesis, 29*, 149–154.

Zhang, H., Su, Y. L., Yu, H., & Qian, B. Y. (2012a). Meta-analysis of the association between Mir-196a-2 polymorphism and cancer susceptibility. *Cancer Biology & Medicine, 9*, 63–72.

Ziebarth, J. D., Bhattacharya, A., & Cui, Y. (2012). Integrative analysis of somatic mutations altering microRNA targeting in cancer genomes. *PLoS One*, 7, e47137.

Zorc, M., Skok, D. J., Godnic, I., Calin, G. A., Horvat, S., Jiang, Z., et al. (2012). Catalog of microRNA seed polymorphisms in vertebrates. *PLoS One*, 7, e30737.

CHAPTER SIX

Role of the tRNA-Derived Small RNAs in Cancer: New Potential Biomarkers and Target for Therapy

Veronica Balatti*, Yuri Pekarsky*, Carlo M. Croce[†,1]
*The Ohio State University, Columbus, OH, United States
[†]Wexner Medical Center, The Ohio State University, Columbus, OH, United States
[1]Corresponding author: e-mail address: carlo.croce@osumc.edu

Contents

1. Regulatory Noncoding RNAs: Long ncRNAs, Small ncRNAs, and tRNA-Derived ncRNAs 174
2. tRNA-Derived ncRNAs: tRFs, tsRNAs, and tiRNA and the Differences in Their Biogenesis, Structures, and Nomenclature 175
   2.1 tRFs 176
   2.2 tsRNAs 177
   2.3 tiRNAs 179
3. tsRNA in Cancer: Signatures, Functions, and Mutations 179
   3.1 tsRNA Signatures in Cancer 180
   3.2 tsRNA Functions in Cancer 182
   3.3 tsRNA Mutations in Cancer 184
4. Concluding Remarks 184
References 185

Abstract

Noncoding RNAs are untranslated RNA molecules that can be divided into two main types: infrastructural, including transfer RNAs (tRNAs) and ribosomal RNAs (rRNAs), and regulatory, including long ncRNAs (lncRNAs) and small ncRNAs (sRNA). Among small ncRNA, the role of microRNAs (miRNAs) and Piwi-interacting RNAs (piRNAs) in cancer is well documented. Recently, other small ncRNAs have been described. In particular, tRNA-derived small RNAs (tsRNA) have been found to be frequently dysregulated in cancer. Since tsRNAs can be considered unique sequences and are able to bind both Argonaute proteins (like miRNAs) and Piwi proteins (like piRNAs), their dysregulation could play a critical role in cancer by interfering with gene expression regulation at different levels. Like microRNAs, *ts-53* (previously known as *miR-3676*) interacts with the 3′UTR of *TCL1*, therefore supporting a role for tsRNAs on the posttranscriptional

Advances in Cancer Research, Volume 135
ISSN 0065-230X
http://dx.doi.org/10.1016/bs.acr.2017.06.007

regulation of gene expression. Like piRNAs, tsRNAs are produced as single-stranded molecules and can interact with DNA and histone methylation machinery, suggesting a role in the pretranscriptional regulation of gene expression. Herein, we describe the most recent findings about the role of tsRNAs in cancer.

1. REGULATORY NONCODING RNAs: LONG ncRNAs, SMALL ncRNAs, AND tRNA-DERIVED ncRNAs

Noncoding RNAs are RNA molecules that are not translated into a protein (Birney et al., 2007). Recent studies revealed that about 90% of human genome is transcribed, but only 1%–3% of those transcripts encode for proteins while the majority of the transcripts are ncRNA, which can have either infrastructural or regulatory functions (Kaikkonen, Lam, & Glass, 2011; Ling, Girnita, Buda, & Calin, 2017). Infrastructural ncRNAs comprise transfer RNAs (tRNAs) and ribosomal RNAs (rRNAs). Regulatory RNAs include long ncRNAs (lncRNAs), which are longer than 200 nucleotides, and small ncRNAs (sRNA) which are shorter than 200 nucleotides (Kaikkonen et al., 2011). lncRNAs are not strongly conserved and little is known about their contribution to disease etiology (Chen et al., 2016). Indeed, even though it was revealed that in several types of cancer their expression level changes when comparing normal and tumor cells, their role in tumorigenesis is still under investigation (Gupta et al., 2010; Merry et al., 2015; Trimarchi et al., 2014; Wu et al., 2014; Xing et al., 2014; Yang et al., 2013). Small RNA includes microRNA (miRNA), Piwi-interacting RNA (piRNA), small interfering RNA (siRNA), small nucleolar RNA, small nuclear RNA (snRNAs, also referred to as U-RNA), and exosomal RNA (exRNAs) (Müller et al., 2015). Recently, a class of small ncRNAs was described to be derived from tRNA precursor or mature sequences. Since tRNA-derived ncRNA can originate either from a tRNA precursor sequence maturation process (originating small RNAs called tsRNAs) or from a tRNA mature sequence (originating either tRNA-derived fragment called tRF or tDRs, or tRNA halves called tiRNAs), their nomenclature is still debated and their function is still under investigation (Lee, Shibata, Malhotra, & Dutta, 2009; Pekarsky et al., 2016; Saikia & Hatzoglou, 2015; Selitsky & Sethupathy, 2015). In the next section of this chapter, we will describe all the three classes of tRNA-derived ncRNAs to establish a simple classification and a more accurate nomenclature and to clarify the differences in their biogenesis, structure, and biological role. In the last section, we will discuss the role of tsRNAs, which we recently found dysregulated in several types of cancers.

2. tRNA-DERIVED ncRNAs: tRFs, tsRNAs, AND tiRNA AND THE DIFFERENCES IN THEIR BIOGENESIS, STRUCTURES, AND NOMENCLATURE

Transfer RNA genes are interspersed in several copies throughout the entire genome. Since the genetic code contains multiple codons for each amino acid, tRNA genes can be grouped into 49 families according to the anticodon features. In humans, these 49 families are encoded by more than 500 genes that can be found on all chromosomes, except 22 and Y. Furthermore, human genome also harbors over 300 nonfunctional tRNA pseudogenes and 22 mitochondrial tRNA genes (Parisien, Wang, & Pan, 2013). In eukaryotic cells, tRNAs are transcribed by RNA polymerase III as pre-tRNAs in the nucleus. RNA polymerase III recognizes two highly conserved promoter sequences both located inside the tRNA mature sequence: the 5′ control region (A box) and the 3′ control region (B box) (Canella, Praz, Reina, Cousin, & Hernandez, 2010). The transcription typically terminates after a stretch of three or more Ts located 10–60 nucleotides downstream the 3′ end of the tRNA mature sequence, but other stop signals have been described (Arimbasseri, Rijal, & Maraia, 2013; Maraia & Lamichhane, 2011). Pre-tRNAs and mature tRNAs undergo extensive modifications before and after exportation to the cytoplasm, respectively (Ohira & Suzuki, 2016). These processing steps result in the production of three distinct classes of tRNA-derived ncRNAs: tRNA-derived small RNA (tsRNAs) (Pekarsky et al., 2016), tRNA-derived fragments (tRFs or tDRs) (Lee et al., 2009; Selitsky & Sethupathy, 2015), and tRNA halves (tiRNAs) (Ivanov, Emara, Villen, Gygi, & Anderson, 2011). tsRNAs are generated in the nucleus as a consequence of the pre-tRNA 3′-end cleavage, while tRFs and tiRNA are generated in the cytoplasm from mature tRNA (Fig. 1) (Kumar, Kuscu, & Dutta, 2016).

In the following sections, we will describe the main features of each of these tRNA-derived ncRNAs. Although, before describing the details of each class, it is important to clarify that in previous reports, tsRNAs were considered a type of tRFs (Lee et al., 2009). In those studies, tRFs were grouped in three categories: tRF-1, tRF-3, and tRF-5, but while tRF-3

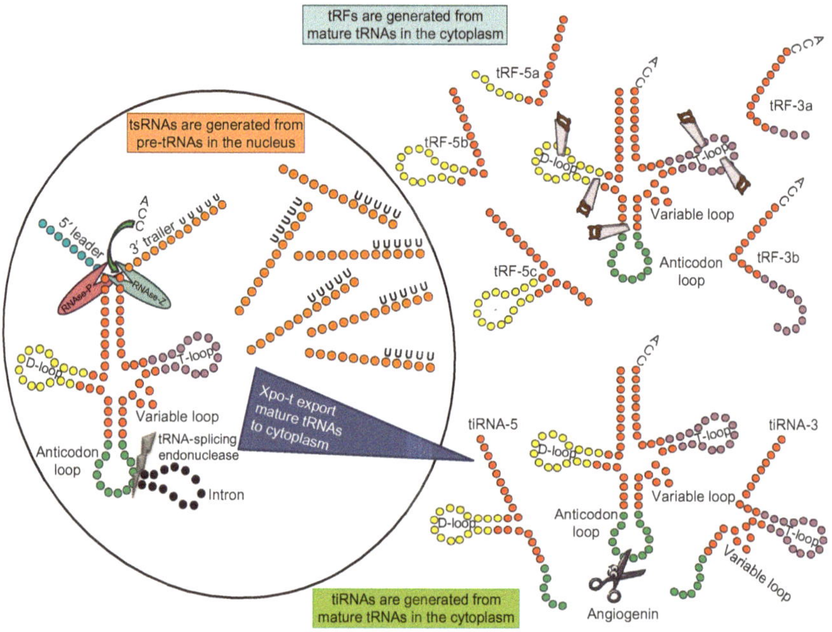

Fig. 1 tsRNA processing.

and tRF-5 are produced in the cytoplasm from mature tRNA sequences (thus can be considered as redundant sequences), tRF-1s are produced in the nucleus from the 3′ end of the tRNA precursor and are unique sequences. Because of these distinctive features, we proposed to distinguish these molecules from the other tRFs by calling them tsRNAs (Pekarsky et al., 2016).

2.1 tRFs

tRFs are divided into two subclass: *tRF-5s* derived from the 5′-end region of the tRNA and containing the A box of the RNA polymerase III promoter, and *tRF-3s* derived from the 3′-end region of the tRNA and containing the B box (Lee et al., 2009). In 2009, Cole et al. proposed a Dicer-dependent system to generate tRFs (Cole et al., 2009), but successive analysis in Dicer or Drosha incompetent cells did not show a significant decrease in tRF levels (Li et al., 2012). Thus, the enzymes responsible for the generation of the tRFs remain unknown. The size of tRF-5s and tRF-3s is well defined: tRF-5s can be generated by three cleavage sites located either in the D-loop or between the D-loop and anticodon loop. Thus, tRF-5s are divided into three subclasses: (i) tRF-5a: 14–16 bases, single stranded, with cleavage site right before the D-loop; (ii) tRF-5b: 22–24 bases, partially double stranded, with cleavage site right after the D-loop; and (iii) tRF-5c: 28–30 bases, double-stranded molecules, with cleavage site right before the anticodon loop (Kumar, Anaya, Mudunuri, & Dutta, 2014). tRF-3s are grouped accordingly to their size (iv) tRF-3a: usually 18 bases, with cleavage site right before the T-loop, or (v) tRF-3b: usually 22 bases, with cleavage sites within the T-loop. All tRF-3s are single-stranded molecules (Kumar et al., 2014). Since tRF-5 and tRF-3 are produced in the cytoplasm, it is expected that they would accumulate outside the nucleus. However tRF-5s were recently reported to be mostly located in the nucleus (Hopper & Shaheen, 2008). Since tRNAs can be transported back into the nucleus (Hopper & Shaheen, 2008), it has been speculated that tRF-5s may be imported into the nucleus by similar mechanisms and affect epigenetic regulation of genes (Haussecker et al., 2010; Zhang et al., 2016). Indeed, Zhang et al. revealed that in monocytes a tRF-5 (named *tRF-5030c*) associates with Piwi proteins and recruits histone methyltransferases to the CD1A promoter region to facilitate histone H3K9 methylation, repressing CD1A expression (Zhang et al., 2016). As for tRF-3, Maute et al. revealed that a particular tRF-3 (named *CU1276* or *tRF-3027*) is downregulated in human malignant B cells and lymphoma biopsies, while it is abundant in naïve, germinal, and memory B cells. This tRF-3 interacts with Ago proteins and its transient overexpression can repress in a miRNA-like manner the endogenous level of *RPA1*, a tumor suppressor that inhibits cell proliferation and modulates the DNA damage response (Maute et al., 2013). Although Couvillion et al. reported that in Tetrahymena many 18–22-base-long tRF-3s associate with Twi12, a Piwi family protein essential for growth (Couvillion, Sachidanandam, & Collins, 2010), lastly, a tRF-3 (*tRF-3006*) was shown to bind to the primer-binding site of HIV genomic RNA, serving as primer for reverse transcription during the viral life cycle (Yeung et al., 2009).

2.2 tsRNAs

The length of the tsRNAs, previously called tRF-1, is very variable when compared to that of the tRF-5 and tRF-3. Indeed, the Pol III stop signal is located 10–60 nucleotides downstream the end of the tRNA sequence (Arimbasseri et al., 2013; Nielsen, Yuzenkova, & Zenkin, 2013). In our studies, tsRNAs are defined as 16–48 nucleotides, single-stranded small RNA ending with a stretch of 3 or more Ts and produced in the nucleus as a consequence of the 3′-end cut of a tRNA precursor performed by the enzyme RNase Z (Balatti et al., 2017; Pekarsky et al., 2016). Since tsRNAs do not cover any of the redundant consensus sequences present within the tRNA itself, they can be considered unique sequences (Pekarsky et al., 2016). When tsRNAs accumulate in the nucleus, they can be exported to the cytoplasm suggesting that tsRNAs could regulate gene expression on different levels (Liao et al., 2010). Indeed, tsRNAs can interact with both Ago and Piwi proteins, and thus, they could affect the epigenetic regulation of gene expression while in the nucleus (similarly to piRNAs) and by 3′UTR targeting when they are exported to the cytoplasm (similarly to miRNAs) (Balatti et al., 2015; Haussecker et al., 2010; Pekarsky et al., 2016).

In 2009, Lee et al. showed that the expression of *ts-36*, which the author called *tRF-1001*, correlates with cell proliferation in prostate cancer cell lines and promotes the transition of these cells from the G2 phase of the cell cycle to the M phase (Lee et al., 2009). In this study, there was no evidence of a physical connection with Ago proteins; however, in 2010 Haussecker et al. reported that the same tsRNA (this author calls it *cand45*) associates with Ago proteins, but miRNA-like activity was not proven (Haussecker et al., 2010). In 2015, our group demonstrated that *ts-53* and *ts-101* (previously indicated as *miR-3676* and *miR-4521*, respectively) can interact with both Ago and Piwi protein (Pekarsky et al., 2016). The genes codifying for these tsRNAs are located in tandem within the region of chromosome 17 deleted in chronic lymphocytic leukemia (CLL) patients. We proved that *ts-53* targets the 3′UTR of *TCL1*, a key gene in development of aggressive CLL, and its downregulation in patients is inversely correlated to *TCL1* expression (Balatti et al., 2015). Furthermore, we described two mutations found in CLL patients at the *ts-53* locus: one of these is a loss-of-function mutation that failed to display the *TCL1* 3′UTR inhibition, and the second mutation creates a stretch of 4 Ts that prompts to a downregulation of *ts-53* expression (Balatti et al., 2015). *Ts-53* and *ts-101* expression pattern positively correlates with *ZAP-70* methylation, and *ts-101* sequence shows a complementarity of 13 out of 14 bases to *ZAP-70* promoter region, giving supplementary support to the hypothesis that tsRNAs could interfere with epigenetic regulation of genes by providing sequence specificity for the Piwi-ribonucleoprotein complexes to interact the promoter of the target gene. Lastly, both *ts-53* and *ts-101* were found downregulated in lung cancer tissue when compared to lung normal tissue from the same patient, and sequencing of their genomic region in a vast collection of lung cancer samples revealed a considerable number of mutations, mainly located in *ts-101* sequence (Balatti et al., 2017; Pekarsky et al., 2016). In addition, by developing a tsRNA-specific microarray chip, we verified that other tsRNAs are dysregulated in several cancer samples and we found *ts-46* and *ts-47* particularly downregulated in CLL and lung cancer. In very recent clonogenic assay experiments, we induced exogenous overexpression of *ts-46*, *ts-47*, and *ts-53* in lung cancer cell lines and observed a significantly reduction in their colony formation ability, further supporting their potential tumor suppressor activity (Pekarsky et al., 2016).

2.3 tiRNAs

tiRNAs are 30–40 bases long double-stranded fragment generated and located in the cytoplasm as a consequence of a specific cleavage in the anticodon loops of mature tRNAs (Ivanov et al., 2011). The cleavage, performed by the cytoplasmic ribonuclease angiogenin mainly in response to heat, UV exposure, starvation, and hypoxia, generates two subclasses of tiRNAs: 5-tiRNAs starting from the 5′ end of mature tRNA and ending in the anticodon loop, and 3-tiRNAs starting from the anticodon loop and ending at the 3′ end of mature tRNA (Emara et al., 2010; Lyons, Achorn, Kedersha, Anderson, & Ivanov, 2016). TiRNAs induced by stress were shown to repress translation and play an important role in the cellular intrinsic apoptotic pathway (Ivanov et al., 2011). In addition to these stress response pathways, tiRNAs are also speculated to participate in the cellular RNAi pathway by associating with Ago and other proteins involved in the silencing pathway (Ivanov et al., 2011). Lastly, several tiRNAs are expressed in a sex hormone-dependent manner in estrogen receptor-positive breast and androgen receptor-positive prostate cancer cell lines (Honda et al., 2015), although the mechanism by which these sex hormone-specific tiRNAs promote the proliferation of breast and prostate cancer remains to be identified (Honda & Kirino, 2016).

3. tsRNA IN CANCER: SIGNATURES, FUNCTIONS, AND MUTATIONS

As anticipated in the previous section, we studied the *miR-4521/3676* cluster (now renamed *ts-101/ts-53* cluster) in samples from CLL and its correlation with *TCL1* regulation and *ZAP-70* methylation (Balatti et al., 2015). We found that these molecules are associated with tRNAs, and then confirmed that they are not miRNAs but tsRNAs (*ts-101* and *ts-53*, respectively) (Pekarsky et al., 2016). To verify that these molecules were not miRNA, we performed northern blot analysis. Both the mature sequence of *miR-3676* and *miR-4521* were predicted to start exactly at the end of the tRNA to which they are associated, although while for *miR-4521* the predicted end was located before the transcription termination stop for RNA polymerase III, for *miR-3676* the end was exactly the same. Therefore, *ts-101* is longer than the predicted *miR-4521*, while *ts-53* and *miR-3676* have exactly the same size. To prove that these molecules are tsRNAs and not miRNAs, we designed probes to either target specifically the predicted miRNA precursors or the precursor of a tsRNA. Since the probe specific for the miRNA precursor failed to show a signal, these molecules were confirmed to be tsRNAs (Pekarsky et al., 2016). Both *ts-53* and *ts-101* are downregulated in CLL and *ts-53* targets three consecutive 28 base pair repeats within the 3′ UTR of *TCL1*, an key oncogene in CLL, in a miRNA-like fashion (Balatti et al., 2015). Since we established that these tsRNAs are dysregulated in CLLs, we studied their expression levels in lung cancer and determined that both of them are drastically downregulated in cancer samples vs matched normal tissues (Pekarsky et al., 2016). These results prompted us to investigate whether other tsRNAs are dysregulated in CLL, lung cancer, and possibly other tumors (Balatti et al., 2017).

3.1 tsRNA Signatures in Cancer

Since *ts-101* and *ts-53* are inactivated in CLL and lung cancer, we decided to examine the expression of other tsRNAs in different cancers (Balatti et al., 2015, 2017; Pekarsky et al., 2016). Thus, we retrieved all tRNA sequences (including tRNAs themselves and following unique sequences ending with a 4 Ts termination signal) and isolated unique sequences starting from the 3′ end of tRNAs and ending with the 4 Ts termination signal. We selected sequences of 16 base pair or longer since shorter sequences cannot be detected by microarray or real-time PCR and identified 120 potential tsRNAs sequences. We then designed a microarray chip to study tsRNAs expression in normal and tumor samples (Pekarsky et al., 2016). We first hybridized CLLs and lung cancer samples to confirm our previous results. We profiled 23 RNA samples from 11 indolent and 12 aggressive CLL patients and 8 normal controls extracted from CD19+ B cells of healthy donors and found that 9 tsRNAs are differentially expressed in aggressive CLL vs normal CD19+ B cells, 15 tsRNAs significantly differentially expressed in indolent CLL vs aggressive CLL, and that 10 tsRNAs are differentially expressed in indolent CLL vs normal CD19+ B cells. Then, we hybridized seven lung cancer samples and five normal lung tissue samples and identified six tsRNAs differentially expressed in lung cancer vs normal lung tissue. With these analysis, we discovered that *ts-46* and *ts-47* were strongly downregulated both in CLL and in lung cancer (Balatti et al., 2015; Pekarsky et al., 2016). Then, we proceeded to profile RNA extracted from paired samples of cancer and normal tissues from colon adenoma, colon adenocarcinoma, and breast cancer patients plus we hybridized ovarian cancer and normal ovary RNAs from tissues. We identified signatures of tsRNAs in each of these cancers (Balatti et al., 2017). In particular, we showed a signature of 8 tsRNAs characterizing adenomas and a signature of 7 tsRNAs for adenocarcinomas (Balatti et al., 2017). *ts-53* and *ts-101* were downregulated in adenomas but not in adenocarcinomas, which suggests a role for these two molecules in the initial phases of transformation (Balatti et al., 2017). Instead, *ts-40* was upregulated in both comparisons, indicating that this tsRNA could have an oncogenic role in colon cancer onset and staging (Balatti et al., 2017). Interestingly, *ts-36,* previously reported to be correlated with cell proliferation in prostate cell lines (Lee et al., 2009), was found over twofolds upregulated in colon carcinomas, indicating a role in the final step of malignant transformation (Balatti et al., 2017). We found only two tsRNAs significantly dysregulated in breast cancer samples: *ts-66* was upregulated when comparing cancer to normal tissue, while *ts-86* was downregulated in such comparison (Balatti et al., 2017). In ovarian cancer, we identified a signature of 10 tsRNAs where *ts-29* was overexpressed more than twofolds in cancer samples (Balatti et al., 2017). Lastly, we performed a wide analysis by investigating the signatures from all the types of tissues that we profiled: CLL and normal B cells, paired samples of normal and lung cancer tissues, paired samples of normal and colon adenomas plus colon carcinomas tissues, paired samples of normal and breast cancer tissues, ovary cancer samples and normal ovary tissues. Our analysis shown that in CLL and lung cancer *ts-46*, *ts-47*, *ts-49*, *ts-53*, *ts-101* are downregulated, while *ts-4* is upregulated (Balatti et al., 2017). *ts-53* and *ts-101* are downregulated also in colon adenomas, and *ts-3* is upregulated in CLL and ovarian cancer (Balatti et al., 2017). This analysis provided a final important result: a signature of 31 tsRNAs able to discriminate among different cancers and tissues types (Balatti et al., 2017). All these results confirm the important role of tsRNAs in cancer and indicate that the dysregulation of these molecules during carcinogenesis could represent a new marker for diagnosis (Balatti et al., 2015, 2017; Pekarsky et al., 2016).

3.2 tsRNA Functions in Cancer

To better understand the role of *ts-46* and *ts-101* in gene expression regulation, we generated *ts-101* and a *ts-46* knocked out stable clones from HEK-293 cells by using CRISPR technology (Balatti et al., 2017). Affymetrix gene expression profile was performed on RNA from two ts-46-KO clones, two ts-101-KO clones, and two controls transfected with the empty pCAS9-vector. We found 270 genes differentially expressed in the ts-46-KO clones and 216 genes differentially expressed in the ts-101-KO clones when compared to the WT. Some examples of upregulated genes in ts-46-KO and ts-101-KO clones are *miR-222*, *IRS4*, *MAP3K19*, *PLK4*, *MED28*, *EPS8L2*, *CDH3*, *GINS1*, *SMARCA5*, and *PKP3*. As expected *miR-4521* is strongly downregulated in ts-101-KO clones (Balatti et al., 2017). Among downregulated genes we found *TAF9B*, *METTL23*, *FILIP1*, *OCLN*, *BMP2*, *TAX1BP1*, *RASGRF2*, *MiR-613*, *OGN*, *SMC1A,* and all the components of the H3A1 histone subfamily *HIST1H3A*, *HIST1H3B*, *HIST1H3C*, *HIST1H3D*, *HIST1H3E*, *HIST1H3F*, *HIST1H3G*, *HIST1H3H*, *HIST1H3I*, and *HIST1H3J*. The downregulation of all the components of the H3A1 histone subfamily supports the hypothesis that tsRNAs can interfere with chromatin structure and thus with the epigenetic control of gene expression (Balatti et al., 2017). Interestingly, a very recent study reports that *tRF-5030c* interacts with Piwi proteins and DNA- and histone methyltransferases and affects methylation of genes and chromatin condensation (Zhang et al., 2016), although, as previously explained, tRFs are not unique sequences. Therefore, we investigated whether tsRNAs, which are unique sequences, can act like piRNAs. Our results showed that indeed this is the case (Pekarsky et al., 2016). To prove our theory, we performed an RNA immunoprecipitation experiment in cells cotransfected with tagged Ago or Piwi proteins and tsRNAs (*ts-53* and *ts-101*) or control miRNAs (*miR-92* and *miR-93*). We found an enrichment of *ts-53* and *ts-101* in complexes containing Piwi-like protein 2 (Piwi-L2), compared with the load of control miRNAs. Indeed, *ts-53* and *ts-101* were present in complexes containing Ago1 and Ago2 and in complexes containing Piwi-L2, while miRNAs are only loaded on protein complexes containing Ago1 and Ago2. This finding indicates that *ts-53* and *ts-101* can act as miRNAs and as piRNAs by interacting with both Ago and Piwi proteins (Pekarsky et al., 2016). Furthermore, *ts-53* targets *TCL1* in a miRNA-fashion manner and overexpression of *TCL1* inhibits de novo DNA methylation (Palamarchuk et al., 2012). Thus, downregulation of *ts-53* can interfere with epigenetic control of gene expression in two independent ways: it can result in a deficiency of methylation activity induced by *TCL1* upregulation, as *TCL1* functions as *DNMT3A* and *DNMT3B* inhibitor (Palamarchuk et al., 2012), or it can result in a deficiency of methylation activity through direct interaction with the epigenetic machinery (Haussecker et al., 2010; Pekarsky et al., 2016; Zhang et al., 2016).

Finally, we performed a functional enrichment study in *ts-46* and *ts-101* KO cells to evaluate changes in cancer-related pathways, and we found a significant activation of networks associated with cell proliferation and inhibition of networks associated with apoptosis (Balatti et al., 2017). Thus, tsRNAs can interfere and significantly alter the profile of gene expression in cells. However, we also demonstrated that tsRNA expression can be modulated by the activation of specific oncogenes (Balatti et al., 2017). Indeed, we investigated tsRNA expression patterns in human lymphocytes with and without activation of the *MYC* oncogene and found a signature of 15 tsRNAs among which *ts-47* was the most downregulated by *MYC* activation, indicating that *MYC* may be involved in cancer by turning off an epigenetic silencer belonging to the tsRNA gene family (Balatti et al., 2017). Given these results, we examined the effect of other oncogenes on tsRNA expression (Balatti et al., 2017). We studied the tsRNA expression profile of cell lines derived from normal breast epithelial cells carrying activating mutations of HRAS, *KRAS*, and *PIK3CA* genes and two cell lines from breast cancer: MCF7 cells, from an early-stage breast cancer, and MDA-MB-231 cells, from a late-stage breast cancer (Balatti et al., 2017). When performing an unsupervised analysis to compare all these cell lines together, we found a signature of 50 tsRNAs able to cluster in a hierarchical order and determined that *KRAS* activation significantly affects tsRNA expression profiles in breast cells (Balatti et al., 2017). Surprisingly, we found *ts-47* upregulated in all cell lines with a mutation of *KRAS*, and *ts-46* upregulated in all cell lines with a mutation of *PIK3CA*. When comparing normal with cancer cells, we found a signature of 34 tsRNAs able to distinguish the two groups (Balatti et al., 2017). Finally, we studied the expression of tsRNAs in cell lines derived from prostate cancer at different stages: LNCaP cells, derived from an early-stage p53 wild-type lymph node metastasis, and PC-3 cells, derived from a late-stage p53 defective bone metastasis (Balatti et al., 2017). When normal prostate epithelium cell line RWPE-1 was compared with LNCaP and with PC-3, a signature of 20 dysregulated tsRNAs was identified and *ts-18*, *ts-104*, and *ts-66* are upregulated in the p53-impaired cell line (Balatti et al., 2017). These results suggest that tsRNAs could be key effectors in pathways regulated by these oncogenes (Balatti et al., 2017) and that an oncogene-driven dysregulation of tsRNA could be involved in feedback loops as previously shown for miRNAs (Fabbri et al., 2011).

3.3 tsRNA Mutations in Cancer

We found that *ts-53* and *ts-101* are often mutated both in CLL and lung cancer, suggesting a key role of these molecules and tsRNAs in general in tumorigenesis. To determine if *ts-53* is mutated in CLL, we sequenced 549 CLL samples and also 146 normal DNAs that were extracted from leukocytes released from Red Cross leukocyte reduction filters. We found two heterozygous germline mutations within the mature sequence of *ts-53*. We demonstrated that these are loss-of-function mutations. Indeed, while WT *ts-53* inhibits *TCL1* expression, the mutants failed to display such function (Balatti et al., 2015). Also, one of these mutations creates a stretch of 4 Ts, a transcription termination stop for RNA polymerase III (Arimbasseri et al., 2013), and we verified that when this termination signal is used, *ts-53* expression is inhibited (Balatti et al., 2015). Moreover, we found a mutation in the sequence of *ts-101* within the region complementary to *ZAP-70* promoter in a CLL patient, which could inhibit the load of *ts-101* onto Piwi-L2 or impair its function as template for the Piwi–ribonucleoprotein complexes to target the promoter of *ZAP-70*. Lastly, we decided to determine if *ts-53* and *ts-101* are mutated in lung cancer, and thus we sequenced 300 lung cancer samples (Pekarsky et al., 2016). We found six different mutations in six lung cancer samples (~2%) within in the *ts-53*/*ts-101* locus. Three mutations found were germline, two mutations were somatic, and in one case we found a mutation that was identical to a known rare SNP (Rs 368503695) but was not present in PBMCs of the same patient. The presence of mutations at *ts-101*/*ts-53* locus in cancer samples supports the idea that these tsRNAs are tumor suppressor and that their functions in cancer need to be more deeply evaluated.

4. CONCLUDING REMARKS

tsRNAs are unique sequences derived from tRNA precursors and generated in the nucleus. These molecules can be transported to the cytoplasm where they associate with Ago proteins and act like miRNAs by inhibiting the expression of target genes via 3′UTR complementarity. In the nucleus tsRNAs can interact with Piwi proteins and potentially affect methylation of genes and chromatin structure. We found that *ts-53*, *ts-46*, and *ts-47* have tumor suppressor characteristics. These molecules are all downregulated in CLL and lung cancer and inhibit colony formation in lung cancer cell lines. Furthermore, *ts-53* and *ts-101* are frequently mutated CLL and lung cancer samples from patients. Also, *ts-53* targets *TCL1* in CLL and *ts-101* shows a high complementarity with *ZAP-70* promoter. Taken together, these evidences suggest that tsRNAs can represent a new diagnostic tool in cancer management.

REFERENCES

Arimbasseri, A. G., Rijal, K., & Maraia, R. J. (2013). Transcription termination by the eukaryotic RNA polymerase III. *Biochimica et Biophysica Acta*, *1829*, 318–330. http://dx.doi.org/10.1016/j.bbagrm.2012.10.006.

Balatti, V., et al. (2015). TCL1 targeting miR-3676 is codeleted with tumor protein p53 in chronic lymphocytic leukemia. *Proceedings of the National Academy of Sciences of the United States of America*, *112*, 2169–2174. http://dx.doi.org/10.1073/pnas.1500010112.

Balatti, V., et al. (2017). TsRNA signatures in cancer. *Proceedings of the National Academy of Sciences of the United States of America*. http://dx.doi.org/10.1073/pnas.1706908114. in press (ahead of print).

Birney, E., et al. (2007). Identification and analysis of functional elements in 1% of the human genome by the ENCODE pilot project. *Nature*, *447*, 799–816. http://dx.doi.org/10.1038/nature05874.

Canella, D., Praz, V., Reina, J. H., Cousin, P., & Hernandez, N. (2010). Defining the RNA polymerase III transcriptome: Genome-wide localization of the RNA polymerase III transcription machinery in human cells. *Genome Research*, *20*, 710–721. http://dx.doi.org/10.1101/gr.101337.109.

Chen, J., et al. (2016). Evolutionary analysis across mammals reveals distinct classes of long non-coding RNAs. *Genome Biology*, *17*, 19. http://dx.doi.org/10.1186/s13059-016-0880-9.

Cole, C., et al. (2009). Filtering of deep sequencing data reveals the existence of abundant Dicer-dependent small RNAs derived from tRNAs. *RNA*, *15*, 2147–2160. http://dx.doi.org/10.1261/rna.1738409.

Couvillion, M. T., Sachidanandam, R., & Collins, K. (2010). A growth-essential Tetrahymena Piwi protein carries tRNA fragment cargo. *Genes & Development*, *24*, 2742–2747. http://dx.doi.org/10.1101/gad.1996210.

Emara, M. M., et al. (2010). Angiogenin-induced tRNA-derived stress-induced RNAs promote stress-induced stress granule assembly. *The Journal of Biological Chemistry*, *285*, 10959–10968. http://dx.doi.org/10.1074/jbc.M109.077560.

Fabbri, M., et al. (2011). Association of a microRNA/TP53 feedback circuitry with pathogenesis and outcome of B-cell chronic lymphocytic leukemia. *JAMA*, *305*, 59–67. http://dx.doi.org/10.1001/jama.2010.1919.

Gupta, R. A., et al. (2010). Long non-coding RNA HOTAIR reprograms chromatin state to promote cancer metastasis. *Nature*, *464*, 1071–1076. http://dx.doi.org/10.1038/nature08975.

Haussecker, D., et al. (2010). Human tRNA-derived small RNAs in the global regulation of RNA silencing. *RNA*, *16*, 673–695. http://dx.doi.org/10.1261/rna.2000810.

Honda, S., & Kirino, Y. (2016). SHOT-RNAs: A novel class of tRNA-derived functional RNAs expressed in hormone-dependent cancers. *Molecular & Cellular Oncology, 3,* e1079672. http://dx.doi.org/10.1080/23723556.2015.1079672.

Honda, S., et al. (2015). Sex hormone-dependent tRNA halves enhance cell proliferation in breast and prostate cancers. *Proceedings of the National Academy of Sciences of the United States of America, 112,* E3816–E3825. http://dx.doi.org/10.1073/pnas.1510077112.

Hopper, A. K., & Shaheen, H. H. (2008). A decade of surprises for tRNA nuclear-cytoplasmic dynamics. *Trends in Cell Biology, 18,* 98–104. http://dx.doi.org/10.1016/j.tcb.2008.01.001.

Ivanov, P., Emara, M. M., Villen, J., Gygi, S. P., & Anderson, P. (2011). Angiogenin-induced tRNA fragments inhibit translation initiation. *Molecular Cell, 43,* 613–623. http://dx.doi.org/10.1016/j.molcel.2011.06.022.

Kaikkonen, M. U., Lam, M. T., & Glass, C. K. (2011). Non-coding RNAs as regulators of gene expression and epigenetics. *Cardiovascular Research, 90,* 430–440. http://dx.doi.org/10.1093/cvr/cvr097.

Kumar, P., Anaya, J., Mudunuri, S. B., & Dutta, A. (2014). Meta-analysis of tRNA derived RNA fragments reveals that they are evolutionarily conserved and associate with AGO proteins to recognize specific RNA targets. *BMC Biology, 12,* 78. http://dx.doi.org/10.1186/s12915-014-0078-0.

Kumar, P., Kuscu, C., & Dutta, A. (2016). Biogenesis and function of transfer RNA-related fragments (tRFs). *Trends in Biochemical Sciences, 41,* 679–689. http://dx.doi.org/10.1016/j.tibs.2016.05.004.

Lee, Y. S., Shibata, Y., Malhotra, A., & Dutta, A. (2009). A novel class of small RNAs: tRNA-derived RNA fragments (tRFs). *Genes & Development, 23,* 2639–2649. http://dx.doi.org/10.1101/gad.1837609.

Li, Z., et al. (2012). Extensive terminal and asymmetric processing of small RNAs from rRNAs, snoRNAs, snRNAs, and tRNAs. *Nucleic Acids Research, 40,* 6787–6799. http://dx.doi.org/10.1093/nar/gks307.

Liao, J. Y., et al. (2010). Deep sequencing of human nuclear and cytoplasmic small RNAs reveals an unexpectedly complex subcellular distribution of miRNAs and tRNA 3′ trailers. *PLoS One, 5,* e10563. http://dx.doi.org/10.1371/journal.pone.0010563.

Ling, H., Girnita, L., Buda, O., & Calin, G. A. (2017). Non-coding RNAs: The cancer genome dark matter that matters!. *Clinical Chemistry and Laboratory Medicine, 55,* 705–714. http://dx.doi.org/10.1515/cclm-2016-0740.

Lyons, S. M., Achorn, C., Kedersha, N. L., Anderson, P. J., & Ivanov, P. (2016). YB-1 regulates tiRNA-induced Stress Granule formation but not translational repression. *Nucleic Acids Research, 44,* 6949–6960. http://dx.doi.org/10.1093/nar/gkw418.

Maraia, R. J., & Lamichhane, T. N. (2011). 3′ processing of eukaryotic precursor tRNAs. *Wiley Interdisciplinary Reviews RNA, 2,* 362–375. http://dx.doi.org/10.1002/wrna.64.

Maute, R. L., et al. (2013). tRNA-derived microRNA modulates proliferation and the DNA damage response and is down-regulated in B cell lymphoma. *Proceedings of the National Academy of Sciences of the United States of America, 110,* 1404–1409. http://dx.doi.org/10.1073/pnas.1206761110.

Merry, C. R., et al. (2015). DNMT1-associated long non-coding RNAs regulate global gene expression and DNA methylation in colon cancer. *Human Molecular Genetics, 24,* 6240–6253. http://dx.doi.org/10.1093/hmg/ddv343.

Müller, S., et al. (2015). Next-generation sequencing reveals novel differentially regulated mRNAs, lncRNAs, miRNAs, sdRNAs and a piRNA in pancreatic cancer. *Molecular Cancer, 14,* 94. http://dx.doi.org/10.1186/s12943-015-0358-5.

Nielsen, S., Yuzenkova, Y., & Zenkin, N. (2013). Mechanism of eukaryotic RNA polymerase III transcription termination. *Science, 340,* 1577–1580. http://dx.doi.org/10.1126/science.1237934.

Ohira, T., & Suzuki, T. (2016). Precursors of tRNAs are stabilized by methylguanosine cap structures. *Nature Chemical Biology, 12*, 648–655. http://dx.doi.org/10.1038/nchembio.2117.

Palamarchuk, A., et al. (2012). Tcl1 protein functions as an inhibitor of de novo DNA methylation in B-cell chronic lymphocytic leukemia (CLL). *Proceedings of the National Academy of Sciences of the United States of America, 109*, 2555–2560. http://dx.doi.org/10.1073/pnas.1200003109.

Parisien, M., Wang, X., & Pan, T. (2013). Diversity of human tRNA genes from the 1000-genomes project. *RNA Biology, 10*, 1853–1867. http://dx.doi.org/10.4161/rna.27361.

Pekarsky, Y., et al. (2016). Dysregulation of a family of short noncoding RNAs, tsRNAs, in human cancer. *Proceedings of the National Academy of Sciences of the United States of America, 113*, 5071–5076. http://dx.doi.org/10.1073/pnas.1604266113.

Saikia, M., & Hatzoglou, M. (2015). The many virtues of tRNA-derived stress-induced RNAs (tiRNAs): Discovering novel mechanisms of stress response and effect on human health. *The Journal of Biological Chemistry, 290*, 29761–29768. http://dx.doi.org/10.1074/jbc.R115.694661.

Selitsky, S. R., & Sethupathy, P. (2015). tDRmapper: Challenges and solutions to mapping, naming, and quantifying tRNA-derived RNAs from human small RNA-sequencing data. *BMC Bioinformatics, 16*, 354. http://dx.doi.org/10.1186/s12859-015-0800-0.

Trimarchi, T., et al. (2014). Genome-wide mapping and characterization of Notch-regulated long noncoding RNAs in acute leukemia. *Cell, 158*, 593–606. http://dx.doi.org/10.1016/j.cell.2014.05.049.

Wu, G., et al. (2014). LincRNA-p21 regulates neointima formation, vascular smooth muscle cell proliferation, apoptosis, and atherosclerosis by enhancing p53 activity. *Circulation, 130*, 1452–1465. http://dx.doi.org/10.1161/circulationaha.114.011675.

Xing, Z., et al. (2014). lncRNA directs cooperative epigenetic regulation downstream of chemokine signals. *Cell, 159*, 1110–1125. http://dx.doi.org/10.1016/j.cell.2014.10.013.

Yang, L., et al. (2013). lncRNA-dependent mechanisms of androgen-receptor-regulated gene activation programs. *Nature, 500*, 598–602. http://dx.doi.org/10.1038/nature12451.

Yeung, M. L., et al. (2009). Pyrosequencing of small non-coding RNAs in HIV-1 infected cells: Evidence for the processing of a viral-cellular double-stranded RNA hybrid. *Nucleic Acids Research, 37*, 6575–6586. http://dx.doi.org/10.1093/nar/gkp707.

Zhang, X., et al. (2016). IL-4 inhibits the biogenesis of an epigenetically suppressive PIWI-interacting RNA to upregulate CD1a molecules on monocytes/dendritic cells. *Journal of Immunology, 196*, 1591–1603. http://dx.doi.org/10.4049/jimmunol.1500805.

CHAPTER SEVEN

MicroRNAs and Epigenetics

Catia Moutinho*, Manel Esteller*,†,‡,1
*Cancer Epigenetics and Biology Program (PEBC), Bellvitge Biomedical Research Institute (IDIBELL), Barcelona, Catalonia, Spain
†Institució Catalana de Recerca i Estudis Avançats (ICREA), Barcelona, Catalonia, Spain
‡School of Medicine and Health Sciences, University of Barcelona (UB), Catalonia, Spain
[1]Corresponding author: e-mail address: mesteller@idibell.cat

Contents

1. Introduction 190
2. Epigenetics and Its Mechanisms 191
   2.1 DNA Methylation 191
   2.2 Histones Modifications 192
3. Epigenetic Regulation of miRNAs in Cancer 193
   3.1 DNA Methylation of miRNAs and Functional Consequences 193
   3.2 Control of miRNAs Expression by Histone Modifiers 202
   3.3 miRNAs Epigenetic Regulation Through Transcription Factors 203
4. miRNAs Control of the Epigenetic Machinery 204
   4.1 Epi-miRNAs 204
   4.2 Transcriptional Epigenetic Modulation by miRNAs 208
5. Exploiting miRNAs Epigenetics to Fight Cancer 208
6. Conclusion 210
References 210

Abstract

MicroRNAs (miRNAs) are small noncoding RNAs that regulate gene expression mainly at the posttranscriptional level. Similar to protein-coding genes, their expression is also controlled by genetic and epigenetic mechanisms. Disruption of these control processes leads to abnormal expression of miRNAs in cancer. In this chapter, we discuss the supportive links between miRNAs and epigenetics in the context of carcinogenesis. miRNAs can be epigenetically regulated by DNA methylation and/or specific histone modifications. However, they can themselves (epi-miRNAs) repress key enzymes that drive epigenetic remodeling and also bind to complementary sequences in gene promoters, recruiting specific protein complexes that modulate chromatin structure and gene expression. All these issues affect the transcriptional landscape of cells. Most important, in the cancer clinical scenario, knowledge about miRNAs epigenetic dysregulation can not only be beneficial as a prognostic biomarker, but can also help in the design of new therapeutic approaches.

Advances in Cancer Research, Volume 135
ISSN 0065-230X
http://dx.doi.org/10.1016/bs.acr.2017.06.003

1. INTRODUCTION

MicroRNAs (miRNAs) are small single-strand noncoding RNA molecules, 18–25 nucleotides, which regulate gene expression mostly at a posttranscriptional level. They are involved in a diversity of biological processes including development, apoptosis, cell differentiation, and proliferation (Sampath et al., 2012). With the exception of miRNAs within the Alu repeats transcribed by polymerase III (Borchert, Lanier, & Davidson, 2006), most miRNAs are derived from primary miRNA transcripts (pri-miRNAs) produced by polymerase II, containing a 5′ cap and a poly(A) tail (Cai, Hagedorn, & Cullen, 2004; Lee et al., 2004). Pri-miRNAs are cleaved within the nucleus by a multiprotein complex called Microprocessor, which is composed of DROSHA and double-stranded RNA-binding domain protein DGCR8/PASHA (Denli, Tops, Plasterk, Ketting, & Hannon, 2004; Landthaler, Yalcin, & Tuschl, 2004; Lee et al., 2003), into ≍70 nucleotides hairpin precursors known as pre-miRNAs. Next, pre-miRNAs are exported into the cytoplasm by XPO5 via a RAN-GTP-dependent mechanism (Bohnsack, Czaplinski, & Gorlich, 2004; Lund, Guttinger, Calado, Dahlberg, & Kutay, 2004; Yi, Qin, Macara, & Cullen, 2003). Pre-miRNAs are further cleaved into mature ≍22 nucleotides miRNA: miRNA* duplexes by DICER, in association with its partners, TRBP/Loquacious, and PACT in human cells (Hutvagner et al., 2001; Ketting et al., 2001). Afterward, an RNA-induced silencing complex called RISC is assembled with Argonaute 2 (AGO2) (Gregory, Chendrimada, Cooch, & Shiekhattar, 2005; Maniataki & Mourelatos, 2005). The miRNA strand is selectively incorporated into RISC (Du & Zamore, 2005) and guides the complex specifically to its messenger RNA (mRNA) targets (Liu et al., 2004).

In the human genome, more than 1000 miRNAs are predicted to modulate the expression of approximately 60% of the protein-coding genes (Friedman, Farh, Burge, & Bartel, 2009). In addition, most transcripts can be modulated by multiple miRNAs (Lee & Dutta, 2009). Gene expression downregulation by miRNAs can occur in two ways, which is mainly dependent on their complementarity. miRNAs with perfect or near-perfect complementarity (Watsonian match) to the target sequence induce cleavage and degradation by deadenylation and decapping, respectively (Pillai, Bhattacharyya, & Filipowicz, 2007). Nevertheless, most miRNAs bind imperfectly to their target sequence and preferentially repress the translational process (Standart & Jackson, 2007; Wang, Love, Call, Doench, &

Novina, 2006). Regarding gene upregulation, a smaller group of miRNAs has been reported to perform this function. One example is miR-10, a miRNA that interacts with the 5′ untranslated region of mRNAs, encoding ribosomal proteins to enhance their targets' genes translation (Orom, Nielsen, & Lund, 2008; Vasudevan, Tong, & Steitz, 2007).

Not surprisingly, these small noncoding RNAs are also implicated in tumor development by regulating the cellular levels of specific oncogenes or tumor suppressor genes (Iorio & Croce, 2012; Lee & Dutta, 2009). Atypical expression of miRNAs in tumor cells occurs due to genetic or epigenetic mechanisms or even by aberrant expression of transcription factors. In addition, the activity of some miRNAs can also modulate the epigenetic regulation of multiple genes by controlling the levels of DNA methyltransferases (DNMTs) or histone deacetylases (HDACs), among multiple other epigenetic regulators (Iorio & Croce, 2012; Iorio, Piovan, & Croce, 2010; Kunej et al., 2011; Sato, Tsuchiya, Meltzer, & Shimizu, 2011). Different evidences suggest also that miRNAs are able to bind complementary sequences in gene promoters, recruiting specific protein complexes that modulate chromatin structure and gene expression (Garzon et al., 2009).

2. EPIGENETICS AND ITS MECHANISMS

Epigenetics is defined as mitotically and/or meiotically heritable changes in gene expression that are not accompanied by changes in DNA sequence. Epigenetic inheritance is important in many physiological processes including differentiation, silencing of chromosomal domains such as the X chromosome of female mammals (Xi), stem cell plasticity, aging, and genomic imprinting (Jones & Baylin, 2007). Pathologic epigenetic changes are considered as alternatives to mutations and chromosomal alterations in disrupting gene function. In cancer, these include global DNA hypomethylation in tumors, hypermethylation of the CpG islands in the promoter regions of tumor suppressor genes, chromatin alterations, and loss of imprinting (Hatziapostolou & Iliopoulos, 2011). Epigenetic regulation generally resides in two categories: DNA methylation and histone modifications. It is important to highlight that both modes of epigenetic regulation can cooperate tightly in their function (Malumbres, 2013).

2.1 DNA Methylation

Reversible DNA methylation occurs mainly at cytosine proceeded by guanine (CpG) dinucleotides where DNMTs add a methyl group to the 5 position of the cytosine. CpG sites are not randomly distributed in the genome; instead, there are CpG-rich regions known as CpG islands, which span the 5′ end of the regulatory region of many genes (Esteller, 2008). There are three major forms of DNMTs: DNMT1 that maintains the methylation patterns throughout each cell division and DNMT3A and 3B that catalyze de novo methylation during early development or in germ cells (Bestor, 2000). As indicated, DNA methylation is a reversible mechanism and few years ago ten–eleven translocation (TET) enzymes were described to perform DNA demethylation, modifying methylated cytosine into unmethylated ones, during DNA replication. However, these mechanisms are still poorly understood (Tu et al., 2015).

2.2 Histones Modifications

Histones comprise the protein backbone of chromatin. They can undergo numerous modifications, including lysine acetylation, lysine and arginine methylation, serine threonine and tyrosine phosphorylation, glutamic acid ADP-ribosylation, lysine ubiquitination, and sumoylation (Ting, McGarvey, & Baylin, 2006). Such covalent modifications can directly alter the physical properties of the chromatin fiber determining whether chromatin is in the accessible form (euchromatin), or in the inaccessible one (heterochromatin) (Goldberg, Allis, & Bernstein, 2007). These histone modifications that are responsible for the chromatin status constitute the so-called "histone code." For example, while acetylation of histones H3 and H4 and methylation of lysine 4 at histone H3 (H3K4me) are found in euchromatin leading to transcriptional activation, di- or trimethylation of hitone H3 lysine 9 (H3K9me) is a characteristic mark of heterochromatin, which mediates transcriptional repression (Bernstein, Meissner, & Lander, 2007). These histone modifications are catalyzed by several HDACs and histone methyltransferases (HMTs). Finally, the polycomb repressive complex 2 (PRC2) can mediate epigenetic gene silencing by trimethylating histone H3 lysine 27 (H3K27me3), a characteristic mark of heterochromatin (Wang, Kwong, et al., 2013; Wang, Yao, et al., 2013).

It is relevant to highlight the relationship between DNA methylation and histone modifications that cooperate in controlling gene expression. For example, methylation of histone H3 lysine 9 can be triggered by DNA methylation (Johnson, Cao, & Jacobsen, 2002). DNMTs have also been shown to interact in a complex network with HDAC, HMT, and methyl-cytosine-binding proteins (Fuks, Burgers, Brehm, Hughes-Davies, & Kouzarides, 2000; Fuks et al., 2003).

3. EPIGENETIC REGULATION OF miRNAs IN CANCER

3.1 DNA Methylation of miRNAs and Functional Consequences

Almost 50% of the miRNAs are surrounded by a CpG island; therefore, miRNAs in multiple cancers are targets of CpG hyper- or hypomethylation (Tuna, Machado, & Calin, 2016). miRNAs can be located within introns or exons of genes, or in intergenic regions (Kim, Han, & Siomi, 2009; Rodriguez, Griffiths-Jones, Ashurst, & Bradley, 2004). Intronic miRNAs may have their own promoters or share it with the host gene (Fig. 1). In the latter case, the expression of miRNAs is usually driven by the same promoter as is the corresponding host gene. Currently, an enormous number of miRNAs have been described to be regulated through epigenetic mechanisms, mainly by DNA methylation. Emphasis will be now placed on some of the most important miRNAs epigenetically regulated in cancer. In Table 1, a more extensive list of miRNAs can be found.

The first evidence of miRNAs epigenetic regulation in cancer was obtained by using epigenetic drugs, 5-AZA-2′-deoxycytidine (5-AZA) and 4-phenylbutyric acid (PBA) that reactivated miRNAs expression in bladder cancer cells. 5-AZA serves as an epigenetic regulator of gene expression due to DNA demethylation, since it is a chemical analogue of the cytosine that after incorporation at the DNA level blocks/inhibits the DNMTs. In contrast, PBA is a HDAC inhibitor. For example, miR-127 was upregulated about 49-fold in 5-AZA and PBA-treated cells. At closer inspection, this miRNA is localized in a CpG island region and is epigenetically silenced by both promoter hypermethylation and histone modifications in cancer cells, since its upregulation was only detected following combination treatment. A candidate target of miR-127, the proto-oncogene *BCL6*, was translationally inhibited after miR-127 upregulation. BCL6 is a transcriptional repressor that has been shown to be involved in p53 regulation (Saito et al., 2006).

A similar approach was used by our research group. Comparing HCT-116 colon cancer cells and HCT-116 *DNMT1* and *DNMT3B* double-knockout cells (DKO) expression miRNAs profiles, we uncovered 18 (out of 320) miRNAs that were upregulated in DKO cells. miR-124a was among them and was located in a CpG island, which is densely methylated in the tumor cells, but not in normal tissue. Upregulation of miR-124a reduced the levels of CDK6 kinase and impacted on the phosphorylation status of the CDK6 downstream effector Rb protein (Agirre et al.,

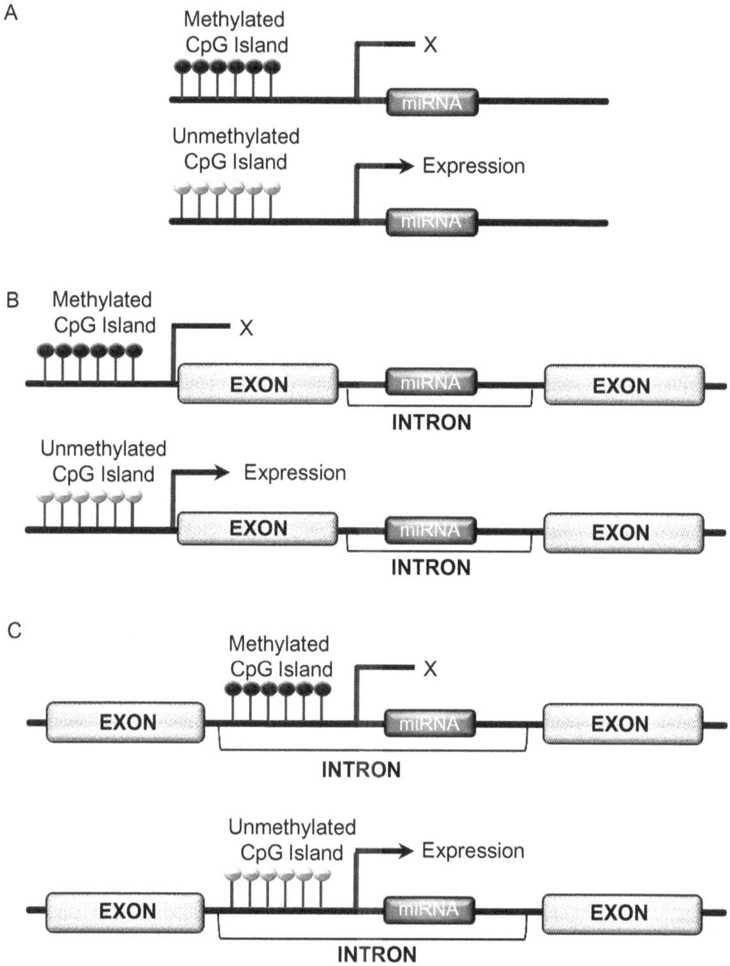

Fig. 1 MicroRNAs epigenetic regulation mechanisms. (A) Some microRNAs have their own promoter that can be altered by DNA methylation changes. Methylation of promoter CpGs leads to miRNA repression. (B) miRNAs localized in host genes can be under the regulation of the host gene promoter which makes their expression synchronized. (C) miRNAs in host genes can also have an independent promoter, being its expression autonomous.

2009; Lujambio et al., 2007). Hypermethylation of this miRNA is also involved in the formation of a gastric cancer epigenetic field defect, a predisposing condition characterized by the accumulation of abnormal DNA methylation in normal-appearing gastric mucosa, mostly induced by *Helicobacter pylori* infection (Ando et al., 2009). These observations suggest

Table 1 MicroRNAs Epigenetically Regulated

Cancer Type	Hypermethylated MicroRNAs	Hypomethylated MicroRNAS	References
Breast Cancer	miR-1258, miR-1247, miR-432, miR-140, miR-129-2	miR-15a/b, miR-16-1/2, miR-21, miR-29b-1, miR-3613, miR-1292, miR-106a, miR-181 b-1/2, miR-590, miR-455, miR-147b, miR-330, miR-643, miR-769, miR-3194, miR-2114, miR-509-3, miR-98	Li, Zhang, Li, Lu, et al. (2015)
	miR-203, miR-195, miR-497		Li et al. (2011)
	miR-34a		Lodygin et al. (2008)
	miR-91		Lehmann et al. (2008)
	miR-195, miR-497		Li et al. (2011)
	miR-148		Lujambio et al. (2008)
	miR-34a		Vogt et al. (2011)
	miR-34b/c		
	miR-335		
Basal-like and Estrogen Receptor-Negative Breast Cancer	Let7e-3p		Png et al. (2011)
		miR-93-5p, miR-30c-23p	Aure et al. (2013)
Triple Negative Breast Cancer	miR-31		Augoff, McCue, Plow, and Sossey-Alaoui (2012)
Endometrial Cancer	miR-129-2		Huang et al. (2009)
	miR-203		Huang et al. (2014)
	miR-152		Tsuruta et al. (2011)
Ovarian Cancer	miR-34a, miR-34b/c		Corney et al. (2010) and Vogt et al. (2011)
	Let7a-3		Lu, Katsaros, de la Longrais, Sochirca, and Yu (2007)
	miR-130b		Yang et al. (2012)
		miR-21	Iorio et al. (2007)

Continued

Table 1 MicroRNAs Epigenetically Regulated—cont'd

Cancer Type	Hypermethylated MicroRNAs	Hypomethylated MicroRNAS	References
Cervical Cancer	miR-124		Tanaka et al. (2012) and Wilting et al. (2010)
Lung Cancer	miR-34s		Tanaka et al. (2012)
	miR-34b/c		Wang, Chen, et al. (2011)
	miR-200c		Ceppi et al. (2010)
	miR-212		Incoronato et al. (2011)
	miR-9-3, miR-193a		Heller, Zielinski, and Zochbauer-Muller (2010)
	miR-503		Li, Zhang, Li, and Zhou (2014)
	miR-196a		Liu et al. (2012)
	miR-373a		Seol et al. (2014)
	miR-9s		Muraoka et al. (2013)
	miR-148		Lujambio et al. (2008)
	miR-124a-1, miR-124a-2, miR-124a-3		Lujambio et al. (2007)
	miR-141, miR-429		Lopez-Serra and Esteller (2012)
	miR-34a		Lodygin et al. (2008)
		Let7a-3	Brueckner et al. (2007)
Malignant Pleural Mesothelioma	miR-34b/c		Kubo, Koh, et al. (2011)
Oral Cancer	miR-34		Kozaki, Imoto, Mogi, Omura, and Inazawa (2008)
HNSCC	miR-148		Lujambio et al. (2008)
	miR-137		Langevin et al. (2010)
Esophageal Cancer	miR-34a		Cui et al. (2014)
	miR-129-2		

Table 1 MicroRNAs Epigenetically Regulated—cont'd

Cancer Type	Hypermethylated MicroRNAs	Hypomethylated MicroRNAS	References
Gastric Cancer	miR-10-b		Kim et al. (2011)
		miR-196b	Tsai et al. (2010)
	miR-34		Suzuki, Nishio, Katada, Sastre, and Niwa (2011)
	miR-34b/c, miR-129		Tsai, Wu, et al. (2011)
	miR-9		Tsai, Liao, et al. (2011)
	miR-335		Li, Li, et al. (2014)
	miR-137		Steponaitiene et al. (2016)
	miR-495		Li, Zhang, Li, Jie, et al. (2015)
	miR-148a		Zhu et al. (2012)
	miR-129-2		Shen, Pan, Qi, Lin, and Cheng (2010)
	miR-124		Ando et al. (2009)
	miR-181c		Hashimoto, Akiyama, Otsubo, Shimada, and Yuasa (2010)
Colorectal Cancer	miR-124		Lujambio et al. (2007)
	miR-34b/c		Toyota et al. (2008)
	miR-34a		Lodygin et al. (2008)
	miR-9-1, miR-192-2, miR-137		Bandres et al. (2009)
	miR-137		Balaguer et al. (2010)
	miR-200 family		Davalos et al. (2012)
	miR-34a, miR-34b/c		Vogt et al. (2011)
	miR-342		Grady et al. (2008)
	miR-484		Mei et al. (2015)

Continued

Table 1 MicroRNAs Epigenetically Regulated—cont'd

Cancer Type	Hypermethylated MicroRNAs	Hypomethylated MicroRNAS	References
Hepatocellular Carcinoma	miR-1		Datta et al. (2008)
		miR-519d	Fornari et al. (2012)
	miR-125b, miR-199a		Shen et al. (2015)
	miR-124, miR-203		Furuta et al. (2010)
	miR-335		Dohi et al. (2013)
	miR-205, miR-132		Wei et al. (2013) and Zhang, Zhang, et al. (2013)
	miR-34b		Xie et al. (2014)
Pancreatic Cancer	miR-124		Wang et al. (2014)
		miR-200a, miR-200b	Li et al. (2010)
	miR-34a, miR-34b, miR-34c		Vogt et al. (2011)
	miR-34a		Lodygin et al. (2008)
	miR-9-1		Omura et al. (2008)
	miR-132		Zhang et al. (2011)
	miR-107		Lee et al. (2009)
	miR-148a		Hanoun et al. (2010)
Clear Cell Renal Cancer	miR-9-3		Hildebrandt et al. (2010)
	miR-124		Gebauer et al. (2013)
	miR-34a		Vogt et al. (2011)
	miR-34b/c		Lodygin et al. (2008)
Bladder Cancer	miR-137, miR-124-2, miR-124-3, miR-9-3		Shimizu et al. (2013)
	miR-34a		Lodygin et al. (2008)
	miR-127		Saito et al. (2006)
	miR-200 family, miR-205		Wiklund et al. (2011)

Table 1 MicroRNAs Epigenetically Regulated—cont'd

Cancer Type	Hypermethylated MicroRNAs	Hypomethylated MicroRNAS	References
Urothelial Cancer	miR-34a		Vogt et al. (2011)
	miR-34b/c		
Prostate Cancer	miR-127		Saito et al. (2006)
	miR-34a		Lodygin et al. (2008)
	miR-132		Formosa et al. (2013)
	miR-145		Suh et al. (2011)
Melanoma	miR-148		Lujambio et al. (2008)
	miR-34a		Lodygin et al. (2008)
Soft Tissue Sarcoma	miR-34a		Vogt et al. (2011)
	miR-34b/c		
Rhabdomyosarcoma	miR-203		Diao et al. (2014)
Gliomas	miR-148a		Li, Huang, et al. (2014)
Testicular Tumor	miR-199a		Cheung et al. (2011)
Multiple Myeloma	miR-34b/c		Wong, Yim, et al. (2011)
	miR-34a		Chim et al. (2010)
	miR-203		Wong, Liang, et al. (2011)
ALL	miR-124a		Agirre et al. (2009)
	miR-9		Roman-Gomez et al. (2009)
	miR-203		Bueno et al. (2008)
	Let7b		Nishi et al. (2013)
AML	miR-223		Fazi et al. (2007)
	miR-663		Yan-Fang et al. (2013)
APL	miR-34a, miR-34b/c		Asmar et al. (2014)
Diffuse Large B-cell lymphoma	miR-34a		Deneberg et al. (2014)

Continued

Table 1 MicroRNAs Epigenetically Regulated—cont'd

Cancer Type	Hypermethylated MicroRNAs	Hypomethylated MicroRNAS	References
Chronic Lymphoblastic leukemia	miR-34b/c		Wang et al. (2014)
	miR-34a		Chim et al. (2010)
	miR-34b/c	miR-21, miR-34a, miR-155	Baer et al. (2012)
	miR-34a		Wang, Kwong, et al. (2013)
	miR-9-2, miR-124-2, miR-129-2		Bueno et al. (2008)
	miR-9		Chim et al. (2011)
CML	miR-203		Erdogan et al. (2013)
pH-negative Myeloproliferative Neoplasms	miR-34a, miR-34b/c, miR-124-1, miR-203 miR-140, miR-378, miR-632		Wang, Chen, et al. (2011)
Non-Hodgkin's Lymphoma	miR-34a		Chim et al. (2010)

SLCL: small lung cell carcinoma; NSCLC: non-small lung cell carcinoma; HNSC: head and neck squamous cell carcinoma; ALL: acute lymphoblastic leukemia; AML: acute myeloid leukemia; APL: acute promyelocytic leukemia.

that miR-124a promoter hypermethylation is an early event in gastric carcinogenesis. Methylation of miR-124a was also found in many other tumor types (Tuna et al., 2016). miR-107, another epigenetically controlled miRNA, also targets CDK6 and its deregulation affects this oncogenic pathway in pancreatic and gastric cancers (Feng, Xie, Zhang, & Wu, 2012; Lee et al., 2009).

miR-34 family members are other examples of DNA methylation regulation of miRNAs. These miRNAs appear hypermethylated in a diverse set of neoplasms (Corney et al., 2010; Gallardo et al., 2009; Lujambio et al., 2008; Tsai, Liao, et al., 2011; Tsai, Wu, et al., 2011; Tuna et al., 2016). It is not surprising since the miR-34 family members are direct transcriptional targets of p53 and constitute a part of p53 tumor suppressor network regulating cell-cycle arrest, apoptosis, and senescence. Indeed, miR-34 members are downregulated in TP53-null or TP53-mutant cells.

Accordingly, the miR-34 family is a key regulator of tumor suppression through p53 regulation (Kubo, Koh, et al., 2011; Kubo, Toyooka, et al., 2011). p53, the "guardian of the genome" causes G1 cell-cycle arrest, allowing damage to be repaired before replication or triggering apoptosis, if the damage cannot be repaired. Another target of miR-34 is *SIRT1*, an oncogene that decreases the ability of p53 to promote cell-cycle arrest by deacetylation. In this way, cells create a positive feedback loop between TP53 and miR-34 for tumor prevention, which is mediated by *SIRT1* silencing (Yamakuchi, Ferlito, & Lowenstein, 2008). Expression of miR-34a also induces senescence and cell-cycle arrest by targeting CDK6 in prostate and pancreatic cancer cells (Lodygin et al., 2008).

miR-200 family members, mainly associated with epithelial–mesenchymal transition (EMT), are also reported to be hypermethylated in several cancer types. Since they are master regulators of epithelial phenotypes by suppressing CDH1 through directly targeting ZEB1 and SIP, its' methylation may contribute to initiation of lung, breast, bladder, and colorectal cancers through EMT (Davalos et al., 2012; Gregory et al., 2008; Tellez et al., 2011; Wiklund et al., 2011).

As discussed previously, some miRNAs are encoded in host genes. Therefore, they are under the influence of the same regulatory elements and epigenetic controls as their host gene. Examples include miR-342 and its' host gene *EVL*, usually methylated in colorectal adenomas and tumors (Grady et al., 2008). miR-126 and its host gene *EGFL7* are hypermethylated in bladder and in prostate tumors, as are miR-548 and its host tumor suppressor gene *FHIT* (Grady et al., 2008; Hu et al., 2014; Saito et al., 2009). In all cases, miRNAs are located in introns and are reexpressed in synchronization with their host gene after cancer cells are treated with epigenetic drugs. In these cases, alterations in host genes and their regulatory machineries also affect the expression of miRNAs.

Aberrant promoter hypomethylation is also an important mechanism underlying the regulation of miRNAs with oncogenic features. Let-7a-3 was the first hypomethylated miRNA to be described in human lung adenocarcinoma. This miRNA is a master regulator of cell proliferation being normally silenced by promoter methylation in normal cells. In lung adenocarcinoma let-7a-3 promoter hypomethylation induced enhanced tumor phenotypes and oncogenic changes in transcription profiles (Brueckner et al., 2007). Hypermethylation of let-7a-3 promoter was, however, described in epithelial ovarian cancer, but was still associated with a good prognosis (Lu et al., 2007). Therefore, in this situation, DNA methylation acts as a protective mechanism by silencing miRNAs with oncogenic function (Brueckner et al., 2007). This also suggests that the methylation pattern (epigenetic regulation) of miRNAs is likely to be tissue specific (Tuna et al., 2016).

miR-21 is overexpressed in breast cancer due to hypomethylation (Lee et al., 2009). miR-29b is also controlled by hypomethylation and its elevated expression, also in breast cancer, accelerates cell migration and invasion (Wang, Bian, Wei, & Zhang, 2011; Wang, Chen, et al., 2011). The miR-196 family members are encoded in three paralogous loci in the mammalian HOX clusters and negatively regulate several of the HOX genes. The methylation status of a CpG island upstream of miR196b is correlated with transcriptional activity around the miRNA locus in cell lines, and miR-196b was demonstrated to be overexpressed in association with CpG hypomethylation in primary gastric tumors as compared to adjacent normal tissue (Tsai et al., 2010).

Finally, several studies have demonstrated that miRNAs affect the metastatic process by targeting metastasis-related genes (Huang et al., 2008; Ma, Teruya-Feldstein, & Weinberg, 2007; Tavazoie et al., 2008). Our group treated three lymph-node metastatic cell lines with 5-AZA and identified three miRNAs which showed cancer-specific CpG island hypermethylation: miR-148a, miR-34b/c, and miR-9. The reintroduction of miR-148a and miR-34b/c in cancer cells with epigenetic inactivation inhibited cell motility, their metastatic potential in xenograft models and it was associated with downregulation of miRNA oncogenic target genes, such as *C-MYC*, *E2F3*, *CDK6*, and *TGIF2*. Finally, these three miRNAs were also significantly associated with metastasis formation in human malignancies (Lujambio et al., 2008). miR-335 was also depicted as having a suppressive role on metastasis due to targeting SOX4 and to the extracellular matrix protein tenascin-C. Its downregulation through hypermethylation in breast cancer is related with the malignant progression of these tumors (Suh et al., 2011).

3.2 Control of miRNAs Expression by Histone Modifiers

Tumor cells frequently display not only DNA methylation dysregulation but also altered levels of histone modifiers. Thus, miRNAs can also be dysregulated as a consequence of aberrant expression of specific epigenetic regulators, such as HDACs, or polycomb repressor complexes (PRC1 or

PRC2) that will alter the tissue-specific histone code. Inhibition of HDACs resulted in transcriptional changes in about 40% of the miRNAs expressed in a particular cell line and was the first evidence that histone modifiers can alter miRNA expression (Scott, Mattie, Berger, Benz, & Benz, 2006). HDACs are overexpressed in specific classes of leukemias and may mediate the epigenetic silencing of several miRNAs such as miR-15a, miR-16, and miR-29b. Inhibition of these proteins induced the reexpression of the referred miRNAs and the subsequent decline in some of their antiapoptotic targets such as *MCL1*, mitochondrial dysfunction, and induction of cell death (Sampath et al., 2012). More recently, large-scale studies using high-resolution ChIP-sequencing for specific histone marks in colorectal cancer cell lines untreated or treated with -5-AZA and PBA, found over 20 miRNAs subjected to regulation by epigenetic marks without any evidences of dysregulated methylation (Suzuki, Nishio, et al., 2011; Suzuki, Takatsuka, et al., 2011).

In leukemia cells, overexpression of Polycomb proteins results in the epigenetic repression of miR-31, leading to the activation of *NF-KB* and apoptosis resistance (Yamagishi et al., 2012). EZH2, a component of PRC2, is able to suppress the expression of additional well-characterized tumor suppressor miRNAs, such as miR-139-5p, miR-125b, miR-101, let-7c, and miR-200b, in liver tumor cells. These EZH2-silenced miRNAs display a common effect in modulating cell motility and metastasis, suggesting that EZH2 can epigenetically silence multiple miRNAs that regulate metastasis from liver tumors (Au et al., 2012). The histone demethylase JUMONJI/ARID1 B (JARID1B) also contributes to tumor cell proliferation through the epigenetic repression of let-7 tumor suppressor miRNAs. Binding of JARID1B to the let-7e promoter results in the removal of H3K4me3 histone marks associated with active gene expression. Inhibition of JARID1B results in Let-7e reexpression and cell-cycle arrest (Mitra, Das, Huynh, & Jones, 2011).

3.3 miRNAs Epigenetic Regulation Through Transcription Factors

Sometimes in tumor cells the epigenetic regulators are not altered by themselves. Instead, their activity is modified by transcription factors leading to the epigenetic dysregulation of miRNAs. For instance, MYC can induce epigenetic repression of miR-29, a tumor suppressor that inhibits cell proliferation and promotes cancer cell apoptosis, through histone deacetylation and trimethylation in B-cell lymphomas, since it can recruit HDAC3 and EZH2 to the miR-29 promoter, forming a MYC/HDAC3/EZH2 corepressor complex. Without MYC, however, the lack of binding of HDAC3 and EZH2 to the miR-29 promoter results in increased miR-29 expression. Therefore, MYC plays an indispensable role in the epigenetic repression of miR-29 by inducing histone deacetylation and histone trimethylation. Meanwhile, EZH2 can also repress miR-494 that normally targets MYC, in order to create a positive feedback loop which in turn increases MYC abundance and then sustains miR-29 repression in B-cell lymphomas (Zhang et al., 2012).

BRCA1, a well-known tumor suppressor that acts at the DNA repair pathway, can mediate the epigenetic repression of miRNAs, modulating the activity of HDACs. When BRCA1 protein is associated with HDAC2, which in turn acetylates histones H2A and H3 in the miR-155 promoter, it leads to the repression of this miRNA. It is noteworthy that miR-155 is an oncogenic miRNA (Chang et al., 2011).

4. miRNAs CONTROL OF THE EPIGENETIC MACHINERY

In the previous section, we described how miRNAs expression is regulated by epigenetic mechanisms. Now we will highlight their epigenetic functions and how by themselves they can control the epigenetic machinery (Fig. 2). First, a subset of miRNAs that controls important epigenetic regulators expression will be discussed. The expression of these miRNAs, known as epi-miRNAs, have important consequences in the epigenetic regulation of multiple cellular pathways and processes. Second, recent evidences suggest that miRNAs may also have direct epigenetic functions by recruiting specific protein complexes to genomic DNA, and specifically to some promoters, modulating the expression of the corresponding genes.

4.1 Epi-miRNAs

Epi-miRNAs are defined as those miRNAs which target, directly or indirectly, effectors of the epigenetic machinery, including DNMTs, HDACs, histone methyltransferases (HMTs), or TETs (Table 2). miR-29b was the first epi-miRNA reported. It induces global DNA hypomethylation by decreasing DNMTs expression, through direct target of DNMT3A and DNMT3B 3′-UTR regions (Fig. 2A) or indirectly in the case of DNMT1 by targeting Sp1, a transactivator of the *DNMT1* gene (Fabbri et al., 2007; Garzon et al., 2009). Interestingly, tumor suppressor genes that are

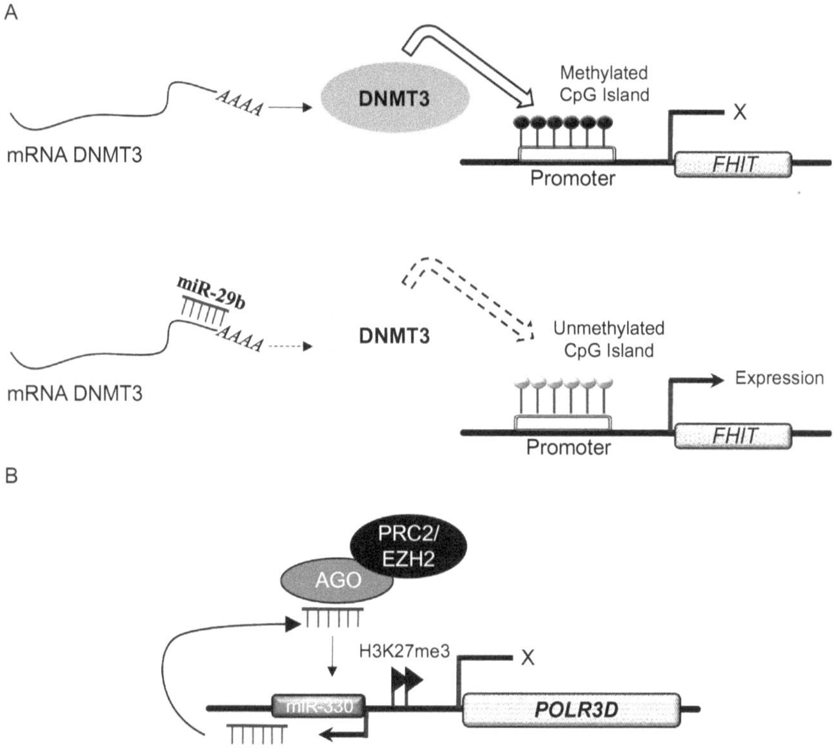

Fig. 2 Direct effects of miRNAs as epigenetic regulators. (A) Epi-miRNAs can directly regulate DNMTs, HDACs, HMTs, or TETs. Epi-miR29 directly target DNMT3 leading to hypomethylation of downstream gene targets and consequently reexpression, which occurs in *FHIT*. (B) Some miRNAs were recently described to transcriptionally regulate gene expression. For example, miR-330 is encoded within the promoter region of the *POLR3D* gene in the antisense orientation. After transcribed, the mature miR-330 binds to a complementary sequence in the *POLR3D* DNA promoter region and recruits AGO1 proteins and the PRC2 component EZH2. This results in H3K27 trimethylation and transcriptional silencing.

epigenetically silenced by promoter hypermethylation like FHIT and WWOX in lung cancer, were reexpressed as a consequence of their promoter CpG island demethylation occurring after miR-29b reintroduction. This also led to cancer cell apoptosis and growth inhibition (Fabbri et al., 2007; Garzon et al., 2009). miR-29 was also found to target TET 3, increasing global DNA methylation levels in some tumors and decreasing it in others (Zhang, Huang, Xu, & Sessa, 2013; Zhang, Zhang, et al., 2013). This suggests that miR-29 modulates tumorigenesis by protecting against changes in the existing DNA methylation status rather than by preventing de novo

Table 2 Epi-MicroRNAs and Respective Epigenetic Targets

MicroRNA	Target	Cancer	References
miR-1	HDAC4	Skeletal muscle tissue	Chen et al. (2006)
miR-22	HDAC4	Hepatocellular carcinoma	Zhang et al. (2010)
miR-26a	EZH2	Lymphoma	Hatziapostolou and Iliopoulos (2011)
miR-29a/b/c	DNMT3A, DNMT3B, DNMT1	Lung cancer, AML	Fabbri et al. (2007) and Garzon et al. (2009)
miR-101	EZH2	Prostatic, Bladder, Breast cancers	Fabbri et al. (2007), Friedman, Liang, et al. (2009), and Varambally et al. (2008)
miR-128	BMI1	Glioma, Pancreatic, and Colorectal cancers	Hatziapostolou and Iliopoulos (2011)
miR-132	MECP2	Myofibroblast	Huang et al. (2014)
miR-140	HDAC4	Osteosarcoma and Colon cancer	Tuddenham et al. (2006)
miR-143	DNMT3A	Colorectal cancer	Ng et al. (2009) and Suh et al. (2011)
miR-144	EZH2	Bladder cancer	Hatziapostolou and Iliopoulos (2011)
miR-148a/b	DNMT3	Testicular germ cell tumors, cervical cancers	Duursma, Kedde, Schrier, le Sage, and Agami (2008)
miR-185	DNMT3B, DNMT1	Cervical cancer, Cholangiocarcinoma	Braconi, Huang, and Patel (2010) and Duursma et al. (2008)
miR-152	DNMT1	Endometrial cancer, liver, cholangiocarcinoma	Braconi et al. (2010), Hatziapostolou and Iliopoulos (2011), and Tsuruta et al. (2011)
miR-148a/b	DNMT1, DNMT3B	Glioma; cervical cancer	Braconi et al. (2010) and Duursma et al. (2008)
miR-200a,b	SUZ12, SIRT1, HDAC4	Liver tumors	Hatziapostolou and Iliopoulos (2011)
miR-301	DNMT1	Cholangiocarcinoma	Braconi et al. (2010)
miR-449a	HDAC1	Prostatic cancer	Li, Huang, et al. (2014) and Noonan et al. (2009)
MiR-22	TET1, TET2	Breast and Hematopoietic cancers	Feng et al. (2012) and Song, Ito, et al. (2013)

AML: acute myeloid leukemia.

methylation of DNA. Thus, like in other scenarios, the role of the miR-29 cluster is tissue/tumor dependent (Tu et al., 2015).

DNMT3A is also a target of miR-143, a miRNA frequently downregulated in colorectal cancer. More significantly, enforced expression of miR-143 in colon cancer cells decreases tumor cell growth and soft-agar colony formation (Ng et al., 2009). Other miRNAs that directly control the activity of DNMTs include miR-152 or miR-301. miR-148a, which is regulated by DNA methylation is also able to modulate the levels of DNA methyltransferases suggesting the presence of regulatory loops that tightly link miRNAs and DNA methylation (Braconi et al., 2010; Duursma et al., 2008).

HDAC protein levels are also modulated by several miRNAs. HDAC4 is a direct target of both miR-1 and miR-140, whereas miR-449a binds to the 3′-UTR region of HDAC1 (Chen et al., 2006; Noonan et al., 2009; Tuddenham et al., 2006). HDAC1 is upregulated in several kinds of cancer, and miR-449a reexpression in prostate tumors induces cancer cell-cycle arrest, apoptosis, and a senescent-like phenotype, by reducing the levels of HDAC1 (Noonan et al., 2009).

Different studies have also revealed that EZH2, the catalytic domain PRC2 (HMT), is responsible for heterochromatin formation by trimethylating histone H3 lysine 27 (H3K27me3), leading to the silencing of several tumor suppressor genes. EZH2 is a target of miR-101, miR-26a, and miR-214. In prostate cancer, the expression of miR-101 decreases during cancer progression, inversely correlating with an increase in EZH2 expression (Varambally et al., 2008). The same phenomenon was subsequently demonstrated in bladder cancer models (Friedman, Liang, et al., 2009). These miRNAs that control adjustments in the epigenetic machinery could in turn affect the expression not only of protein-coding genes but also of epigenetically regulated miRNAs. As example, decreased expression of EZH2 by miR-101 and miR-26a repression leads to decreased miR-181a, miR-181b, miR200b, miR-200c, and miR-203, attenuating cancer invasion (Friedman, Liang, et al., 2009). Furthermore, this group of miRNAs target different components of the PRC1, such as BMI1, linking the activity of PRC1 and PRC2 through the trimethylation of H3K27 (PRC2) or the ubiquitination of histone H2 with the transcriptional repression of multiple genes (Cao et al., 2011).

miR-22 was recently suggested to direct target 3′UTR-region of TET1 and TET2 inhibiting their functions. This fact induces EMT and expression of oncogenes in mammary and hematopoietic stem cells, leading to neoplasms formation (Song, Ito, et al., 2013; Song, Poliseno, et al., 2013). miR-26a was also found to directly target TETs, decreasing the levels of methylated cytosines in pancreatic cells and promoting stem cell features (Fu et al., 2013).

4.2 Transcriptional Epigenetic Modulation by miRNAs

Besides mRNA posttranscriptional regulation in the cytoplasm, different studies demonstrated that endogenous miRNAs can directly affect gene expression in the nucleus at the transcriptional level. For example, as shown in Fig. 2B miR-320, is encoded within the promoter region of the *POLR3D* gene in the antisense orientation. The mature miRNA can direct the recruitment of AGO1 proteins and the PRC2 component EZH2 at the *POLR3D* promoter, resulting in H3K27 trimethylation and transcriptional silencing of *POLR3D* (Kim, Saetrom, Snove, & Rossi, 2008). Another example is miR-423-5p that can recognize target sites at the *PR* promoter and participates in the recruitment of AGO2 to a noncoding transcript that overlaps the promoter region. This suggests that the recognition of nascent noncoding RNA transcripts serves as a mechanism that miRNAs use for transcriptional gene silencing (Younger & Corey, 2011).

Finally, an additional example corroborates that miRNAs can also induce promoter activation as well as repression. The cyclin B1 (*CCNB1*) promoter contains highly complementary sites for 21 miRNAs. Among them, miR-744, miR-1186, and miR-466d-3p can induce *CCNB1* expression. Conversely, knockdown of endogenous miR-744 leads to decreased CCNB1 levels. miR-744 favors the recruitment of AGO1 and RNA POL II to the *CCNB1* promoter, and results in H3K4 trimethylation activating chromatin mark at the *CCNB1* transcription start site (Huang et al., 2012). Thus, endogenous miRNAs can either induce or repress the expression of specific genes by recognizing specific target sites at the promoter regions.

5. EXPLOITING miRNAs EPIGENETICS TO FIGHT CANCER

In the clinical setting, hypermethylation of miRNAs can be used as a prognostic marker, predicting treatment effectiveness and/or disease outcome. For instance, hypermethylation of multiple miRNAs are associated with lymph node metastasis, as is the case of miR-9-1and miR-34 in colorectal cancer and miR-148 in melanoma, breast, lung, and head and neck cancers; while hypermethylation of miR-9-3 is associated with an increased risk of recurrence in clear cell renal cell carcinoma (Bandres et al., 2009; Hildebrandt et al., 2010; Lujambio et al., 2008; Siemens et al., 2013). Similarly, hypermethylation of miR-124 is associated with poor prognosis in ALL and with disease recurrence in clear cell renal carcinoma (Agirre et al., 2009; Gebauer et al., 2013). In head and neck squamous cell carcinoma, the hypermethylation of miR-137 is associated with shorter overall survival (Langevin et al., 2011). Recently, miR-129-2 was found to be a potential marker for hepatocellular early diagnosis and also an outcome biomarker in lymphoid malignancies (Wong et al., 2013). DNA hypermethylation-associated transcriptional silencing of miR-22, known to target HDAC4, was correlated with decreased disease-free survival and worse prognosis in hepatocellular cancer (Zhang et al., 2010).

miR-21 DNA hypomethylation associated with transcriptional activation in breast cancer is associated with mastectomy, larger tumor size, higher tumor stage and grade, and estrogen receptor status (Lee et al., 2009). With respect to histone modifications, silencing of miR-212 due to alterations in the methylation status of histone tails in lung cancer is correlated with later stages of the disease (Incoronato et al., 2011).

The modulation of miRNA expression is a promising novel strategy in cancer therapy (Garofalo & Croce, 2011; Garzon, Marcucci, & Croce, 2010; Trang, Weidhaas, & Slack, 2008). An intriguing feature relating to epigenetic changes is that, unlike genetic changes, they are potentially reversible by treatment with certain drugs. For miRNAs that function as tumor suppressors, the aim is to restore their "normal" expression levels in cancer cells. This can be achieved in tumors by using DNA methylation inhibitors such as 5-AZA derivatives that are already approved for the treatment of specific hematopoietic malignancies in the United States (Mack, 2006). HDAC inhibitors also have antitumor potential, and they have begun to constitute another promising group of agents for the epigenetic therapy of cancer. One of the main therapeutic mechanisms lies in their transcriptional reactivation of "dormant" tumor suppressor genes, such as p21WAF1 (Esteller, 2007). Indeed, the Food and Drug Administration has approved the first drug of this type, suberoylanilide hydroxamic acid, for the treatment of cutaneous T cell lymphoma (Mack, 2006; Thompson, 2006). The development of new HDAC inhibitor drugs is now being actively pursued by pharmaceutical companies (Bolden, Peart, & Johnstone, 2006). Nevertheless, epigenetic drugs such as DNMT or HDAC inhibitors display a profound global effect on the epigenome and their mechanisms of action are difficult to be attributed to a specific coding- or noncoding gene.

6. CONCLUSION

The breakthrough confirmation that miRNAs can undergo epigenetic regulation similar to protein-coding genes, and that miRNAs themselves can control epigenetic effectors reveals new layers of complexity to (epi-)genome regulation. An enhanced comprehension of these mechanisms and a solid understanding of the link between epigenetics–miRNAs are mandatory. This will allow us to appreciate how the human genome is regulated and how gene expression aberrations that contribute to carcinogenesis can be therapeutically corrected.

REFERENCES

Agirre, X., Vilas-Zornoza, A., Jimenez-Velasco, A., Martin-Subero, J. I., Cordeu, L., Garate, L., et al. (2009). Epigenetic silencing of the tumor suppressor microRNA Hsa-miR124a regulates CDK6 expression and confers a poor prognosis in acute lymphoblastic leukemia. *Cancer Research*, *69*, 4443–4453.

Ando, T., Yoshida, T., Enomoto, S., Asada, K., Tatematsu, M., Ichinose, M., et al. (2009). DNA methylation of microRNA genes in gastric mucosae of gastric cancer patients: Its possible involvement in the formation of epigenetic field defect. *International Journal of Cancer*, *124*, 2367–2374.

Asmar, F., Hother, C., Kulosman, G., Treppendahl, M. B., Nielsen, H. M., Ralfkiaer, U., et al. (2014). Diffuse large B-cell lymphoma with combined TP53 mutation and MIR34A methylation: Another "double hit" lymphoma with very poor outcome? *Oncotarget*, *5*, 1912–1925.

Au, S. L., Wong, C. C., Lee, J. M., Fan, D. N., Tsang, F. H., Ng, I. O., et al. (2012). Enhancer of zeste homolog 2 epigenetically silences multiple tumor suppressor microRNAs to promote liver cancer metastasis. *Hepatology*, *56*, 622–631.

Augoff, K., McCue, B., Plow, E. F., & Sossey-Alaoui, K. (2012). miR-31 and its host gene lncRNA LOC554202 are regulated by promoter hypermethylation in triple-negative breast cancer. *Molecular Cancer*, *11*, 5.

Aure, M. R., Leivonen, S. K., Fleischer, T., Zhu, Q., Overgaard, J., Alsner, J., et al. (2013). Individual and combined effects of DNA methylation and copy number alterations on miRNA expression in breast tumors. *Genome Biology*, *14*, R126.

Baer, C., Claus, R., Frenzel, L. P., Zucknick, M., Park, Y. J., Gu, L., et al. (2012). Extensive promoter DNA hypermethylation and hypomethylation is associated with aberrant microRNA expression in chronic lymphocytic leukemia. *Cancer Research*, *72*, 3775–3785.

Balaguer, F., Link, A., Lozano, J. J., Cuatrecasas, M., Nagasaka, T., Boland, C. R., et al. (2010). Epigenetic silencing of miR-137 is an early event in colorectal carcinogenesis. *Cancer Research*, *70*, 6609–6618.

Bandres, E., Agirre, X., Bitarte, N., RamiRez, N., Zarate, R., Roman-Gomez, J., et al. (2009). Epigenetic regulation of microRNA expression in colorectal cancer. *International Journal of Cancer*, *125*, 2737–2743.

Bernstein, B. E., Meissner, A., & Lander, E. S. (2007). The mammalian epigenome. *Cell*, *128*, 669–681.

Bestor, T. H. (2000). The DNA methyltransferases of mammals. *Human Molecular Genetics*, *9*, 2395–2402.

Bohnsack, M. T., Czaplinski, K., & Gorlich, D. (2004). Exportin 5 is a RanGTP-dependent dsRNA-binding protein that mediates nuclear export of pre-miRNAs. *RNA*, *10*, 185–191.

Bolden, J. E., Peart, M. J., & Johnstone, R. W. (2006). Anticancer activities of histone deacetylase inhibitors. *Nature Reviews. Drug Discovery*, *5*, 769–784.

Borchert, G. M., Lanier, W., & Davidson, B. L. (2006). RNA polymerase III transcribes human microRNAs. *Nature Structural & Molecular Biology*, *13*, 1097–1101.

Braconi, C., Huang, N., & Patel, T. (2010). MicroRNA-dependent regulation of DNA methyltransferase-1 and tumor suppressor gene expression by interleukin-6 in human malignant cholangiocytes. *Hepatology*, *51*, 881–890.

Brueckner, B., Stresemann, C., Kuner, R., Mund, C., Musch, T., Meister, M., et al. (2007). The human let-7a-3 locus contains an epigenetically regulated microRNA gene with oncogenic function. *Cancer Research*, *67*, 1419–1423.

Bueno, M. J., Perez de Castro, I., Gomez de Cedron, M., Santos, J., Calin, G. A., Cigudosa, J. C., et al. (2008). Genetic and epigenetic silencing of microRNA-203 enhances ABL1 and BCR-ABL1 oncogene expression. *Cancer Cell*, *13*, 496–506.

Cai, X., Hagedorn, C. H., & Cullen, B. R. (2004). Human microRNAs are processed from capped, polyadenylated transcripts that can also function as mRNAs. *RNA*, *10*, 1957–1966.

Cao, Q., Mani, R. S., Ateeq, B., Dhanasekaran, S. M., Asangani, I. A., Prensner, J. R., et al. (2011). Coordinated regulation of polycomb group complexes through microRNAs in cancer. *Cancer Cell*, *20*, 187–199.

Ceppi, P., Mudduluru, G., Kumarswamy, R., Rapa, I., Scagliotti, G. V., Papotti, M., et al. (2010). Loss of miR-200c expression induces an aggressive, invasive, and chemoresistant phenotype in non-small cell lung cancer. *Molecular Cancer Research*, *8*, 1207–1216.

Chang, S., Wang, R. H., Akagi, K., Kim, K. A., Martin, B. K., Cavallone, L., et al. (2011). Tumor suppressor BRCA1 epigenetically controls oncogenic microRNA-155. *Nature Medicine*, *17*, 1275–1282.

Chen, J. F., Mandel, E. M., Thomson, J. M., Wu, Q., Callis, T. E., Hammond, S. M., et al. (2006). The role of microRNA-1 and microRNA-133 in skeletal muscle proliferation and differentiation. *Nature Genetics*, *38*, 228–233.

Cheung, H. H., Davis, A. J., Lee, T. L., Pang, A. L., Nagrani, S., Rennert, O. M., et al. (2011). Methylation of an intronic region regulates miR-199a in testicular tumor malignancy. *Oncogene*, *30*, 3404–3415.

Chim, C. S., Wan, T. S., Wong, K. Y., Fung, T. K., Drexler, H. G., & Wong, K. F. (2011). Methylation of miR-34a, miR-34b/c, miR-124-1 and miR-203 in Ph-negative myeloproliferative neoplasms. *Journal of Translational Medicine*, *9*, 197.

Chim, C. S., Wong, K. Y., Qi, Y., Loong, F., Lam, W. L., Wong, L. G., et al. (2010). Epigenetic inactivation of the miR-34a in hematological malignancies. *Carcinogenesis*, *31*, 745–750.

Corney, D. C., Hwang, C. I., Matoso, A., Vogt, M., Flesken-Nikitin, A., Godwin, A. K., et al. (2010). Frequent downregulation of miR-34 family in human ovarian cancers. *Clinical Cancer Research*, *16*, 1119–1128.

Cui, X., Zhao, Z., Liu, D., Guo, T., Li, S., Hu, J., et al. (2014). Inactivation of miR-34a by aberrant CpG methylation in Kazakh patients with esophageal carcinoma. *Journal of Experimental & Clinical Cancer Research*, *33*, 20.

Datta, J., Kutay, H., Nasser, M. W., Nuovo, G. J., Wang, B., Majumder, S., et al. (2008). Methylation mediated silencing of MicroRNA-1 gene and its role in hepatocellular carcinogenesis. *Cancer Research*, *68*, 5049–5058.

Davalos, V., Moutinho, C., Villanueva, A., Boque, R., Silva, P., Carneiro, F., et al. (2012). Dynamic epigenetic regulation of the microRNA-200 family mediates epithelial and mesenchymal transitions in human tumorigenesis. *Oncogene*, *31*, 2062–2074.

Deneberg, S., Kanduri, M., Ali, D., Bengtzen, S., Karimi, M., Qu, Y., et al. (2014). microRNA-34b/c on chromosome 11q23 is aberrantly methylated in chronic lymphocytic leukemia. *Epigenetics*, *9*, 910–917.

Denli, A. M., Tops, B. B., Plasterk, R. H., Ketting, R. F., & Hannon, G. J. (2004). Processing of primary microRNAs by the microprocessor complex. *Nature*, *432*, 231–235.

Diao, Y., Guo, X., Jiang, L., Wang, G., Zhang, C., Wan, J., et al. (2014). miR-203, a tumor suppressor frequently down-regulated by promoter hypermethylation in rhabdomyosarcoma. *The Journal of Biological Chemistry*, *289*, 529–539.

Dohi, O., Yasui, K., Gen, Y., Takada, H., Endo, M., Tsuji, K., et al. (2013). Epigenetic silencing of miR-335 and its host gene MEST in hepatocellular carcinoma. *International Journal of Oncology*, *42*, 411–418.

Du, T., & Zamore, P. D. (2005). microPrimer: The biogenesis and function of microRNA. *Development*, *132*, 4645–4652.

Duursma, A. M., Kedde, M., Schrier, M., le Sage, C., & Agami, R. (2008). miR-148 targets human DNMT3b protein coding region. *RNA*, *14*, 872–877.

Erdogan, B., Bosompem, A., Peng, D., Han, L., Smith, E., Kennedy, M. E., et al. (2013). Methylation of promoters of microRNAs and their host genes in myelodysplastic syndromes. *Leukemia & Lymphoma*, *54*, 2720–2727.

Esteller, M. (2007). Cancer epigenomics: DNA methylomes and histone-modification maps. *Nature Reviews. Genetics*, *8*, 286–298.

Esteller, M. (2008). Epigenetics in cancer. *The New England Journal of Medicine*, *358*, 1148–1159.

Fabbri, M., Garzon, R., Cimmino, A., Liu, Z., Zanesi, N., Callegari, E., et al. (2007). MicroRNA-29 family reverts aberrant methylation in lung cancer by targeting DNA methyltransferases 3A and 3B. *Proceedings of the National Academy of Sciences of the United States of America*, *104*, 15805–15810.

Fazi, F., Zardo, G., Gelmetti, V., Travaglini, L., Ciolfi, A., Di Croce, L., et al. (2007). Heterochromatic gene repression of the retinoic acid pathway in acute myeloid leukemia. *Blood*, *109*, 4432–4440.

Feng, L., Xie, Y., Zhang, H., & Wu, Y. (2012). miR-107 targets cyclin-dependent kinase 6 expression, induces cell cycle G1 arrest and inhibits invasion in gastric cancer cells. *Medical Oncology*, *29*, 856–863.

Formosa, A., Lena, A. M., Markert, E. K., Cortelli, S., Miano, R., Mauriello, A., et al. (2013). DNA methylation silences miR-132 in prostate cancer. *Oncogene*, *32*, 127–134.

Fornari, F., Milazzo, M., Chieco, P., Negrini, M., Marasco, E., Capranico, G., et al. (2012). In hepatocellular carcinoma miR-519d is up-regulated by p53 and DNA hypomethylation and targets CDKN1A/p21, PTEN, AKT3 and TIMP2. *The Journal of Pathology*, *227*, 275–285.

Friedman, R. C., Farh, K. K., Burge, C. B., & Bartel, D. P. (2009). Most mammalian mRNAs are conserved targets of microRNAs. *Genome Research*, *19*, 92–105.

Friedman, J. M., Liang, G., Liu, C. C., Wolff, E. M., Tsai, Y. C., Ye, W., et al. (2009). The putative tumor suppressor microRNA-101 modulates the cancer epigenome by repressing the polycomb group protein EZH2. *Cancer Research*, *69*, 2623–2629.

Fu, X., Jin, L., Wang, X., Luo, A., Hu, J., Zheng, X., et al. (2013). MicroRNA-26a targets ten eleven translocation enzymes and is regulated during pancreatic cell differentiation. *Proceedings of the National Academy of Sciences of the United States of America*, *110*, 17892–17897.

Fuks, F., Burgers, W. A., Brehm, A., Hughes-Davies, L., & Kouzarides, T. (2000). DNA methyltransferase Dnmt1 associates with histone deacetylase activity. *Nature Genetics*, *24*, 88–91.

Fuks, F., Hurd, P. J., Wolf, D., Nan, X., Bird, A. P., & Kouzarides, T. (2003). The methyl-CpG-binding protein MeCP2 links DNA methylation to histone methylation. *The Journal of Biological Chemistry*, *278*, 4035–4040.

Furuta, M., Kozaki, K. I., Tanaka, S., Arii, S., Imoto, I., & Inazawa, J. (2010). miR-124 and miR-203 are epigenetically silenced tumor-suppressive microRNAs in hepatocellular carcinoma. *Carcinogenesis*, *31*, 766–776.

Gallardo, E., Navarro, A., Vinolas, N., Marrades, R. M., Diaz, T., Gel, B., et al. (2009). miR-34a as a prognostic marker of relapse in surgically resected non-small-cell lung cancer. *Carcinogenesis*, *30*, 1903–1909.

Garofalo, M., & Croce, C. M. (2011). microRNAs: Master regulators as potential therapeutics in cancer. *Annual Review of Pharmacology and Toxicology*, *51*, 25–43.

Garzon, R., Liu, S., Fabbri, M., Liu, Z., Heaphy, C. E., Callegari, E., et al. (2009). MicroRNA-29b induces global DNA hypomethylation and tumor suppressor gene reexpression in acute myeloid leukemia by targeting directly DNMT3A and 3B and indirectly DNMT1. *Blood*, *113*, 6411–6418.

Garzon, R., Marcucci, G., & Croce, C. M. (2010). Targeting microRNAs in cancer: Rationale, strategies and challenges. *Nature Reviews. Drug Discovery*, *9*, 775–789.

Gebauer, K., Peters, I., Dubrowinskaja, N., Hennenlotter, J., Abbas, M., Scherer, R., et al. (2013). Hsa-miR-124-3 CpG island methylation is associated with advanced tumours and disease recurrence of patients with clear cell renal cell carcinoma. *British Journal of Cancer*, *108*, 131–138.

Goldberg, A. D., Allis, C. D., & Bernstein, E. (2007). Epigenetics: A landscape takes shape. *Cell*, *128*, 635–638.

Grady, W. M., Parkin, R. K., Mitchell, P. S., Lee, J. H., Kim, Y. H., Tsuchiya, K. D., et al. (2008). Epigenetic silencing of the intronic microRNA hsa-miR-342 and its host gene EVL in colorectal cancer. *Oncogene*, *27*, 3880–3888.

Gregory, P. A., Bert, A. G., Paterson, E. L., Barry, S. C., Tsykin, A., Farshid, G., et al. (2008). The miR-200 family and miR-205 regulate epithelial to mesenchymal transition by targeting ZEB1 and SIP1. *Nature Cell Biology*, *10*, 593–601.

Gregory, R. I., Chendrimada, T. P., Cooch, N., & Shiekhattar, R. (2005). Human RISC couples microRNA biogenesis and posttranscriptional gene silencing. *Cell*, *123*, 631–640.

Hanoun, N., Delpu, Y., Suriawinata, A. A., Bournet, B., Bureau, C., Selves, J., et al. (2010). The silencing of microRNA 148a production by DNA hypermethylation is an early event in pancreatic carcinogenesis. *Clinical Chemistry*, *56*, 1107–1118.

Hashimoto, Y., Akiyama, Y., Otsubo, T., Shimada, S., & Yuasa, Y. (2010). Involvement of epigenetically silenced microRNA-181c in gastric carcinogenesis. *Carcinogenesis*, *31*, 777–784.

Hatziapostolou, M., & Iliopoulos, D. (2011). Epigenetic aberrations during oncogenesis. *Cellular and Molecular Life Sciences*, *68*, 1681–1702.

Heller, G., Zielinski, C. C., & Zochbauer-Muller, S. (2010). Lung cancer: From single-gene methylation to methylome profiling. *Cancer Metastasis Reviews*, *29*, 95–107.

Hildebrandt, M. A., Gu, J., Lin, J., Ye, Y., Tan, W., Tamboli, P., et al. (2010). Hsa-miR-9 methylation status is associated with cancer development and metastatic recurrence in patients with clear cell renal cell carcinoma. *Oncogene*, *29*, 5724–5728.

Hu, B., Ying, X., Wang, J., Piriyapongsa, J., Jordan, I. K., Sheng, J., et al. (2014). Identification of a tumor-suppressive human-specific microRNA within the FHIT tumor-suppressor gene. *Cancer Research*, *74*, 2283–2294.

Huang, Q., GumiReddy, K., Schrier, M., le Sage, C., Nagel, R., Nair, S., et al. (2008). The microRNAs miR-373 and miR-520c promote tumour invasion and metastasis. *Nature Cell Biology*, *10*, 202–210.

Huang, Y. W., Kuo, C. T., Chen, J. H., Goodfellow, P. J., Huang, T. H., Rader, J. S., et al. (2014). Hypermethylation of miR-203 in endometrial carcinomas. *Gynecologic Oncology*, *133*, 340–345.

Huang, Y. W., Liu, J. C., Deatherage, D. E., Luo, J., Mutch, D. G., Goodfellow, P. J., et al. (2009). Epigenetic repression of microRNA-129-2 leads to overexpression of SOX4 oncogene in endometrial cancer. *Cancer Research*, *69*, 9038–9046.

Huang, V., Place, R. F., Portnoy, V., Wang, J., Qi, Z., Jia, Z., et al. (2012). Upregulation of cyclin B1 by miRNA and its implications in cancer. *Nucleic Acids Research*, *40*, 1695–1707.

Hutvagner, G., McLachlan, J., Pasquinelli, A. E., Balint, E., Tuschl, T., & Zamore, P. D. (2001). A cellular function for the RNA-interference enzyme Dicer in the maturation of the let-7 small temporal RNA. *Science*, *293*, 834–838.

Incoronato, M., Urso, L., Portela, A., Laukkanen, M. O., Soini, Y., Quintavalle, C., et al. (2011). Epigenetic regulation of miR-212 expression in lung cancer. *PLoS One*, *6*, e27722.

Iorio, M. V., & Croce, C. M. (2012). MicroRNA dysregulation in cancer: Diagnostics, monitoring and therapeutics. A comprehensive review. *EMBO Molecular Medicine*, *4*, 143–159.

Iorio, M. V., Piovan, C., & Croce, C. M. (2010). Interplay between microRNAs and the epigenetic machinery: An intricate network. *Biochimica et Biophysica Acta*, *1799*, 694–701.

Iorio, M. V., Visone, R., Di Leva, G., Donati, V., Petrocca, F., Casalini, P., et al. (2007). MicroRNA signatures in human ovarian cancer. *Cancer Research*, *67*, 8699–8707.

Johnson, L., Cao, X., & Jacobsen, S. (2002). Interplay between two epigenetic marks. DNA methylation and histone H3 lysine 9 methylation. *Current Biology*, *12*, 1360–1367.

Jones, P. A., & Baylin, S. B. (2007). The epigenomics of cancer. *Cell*, *128*, 683–692.

Ketting, R. F., Fischer, S. E., Bernstein, E., Sijen, T., Hannon, G. J., & Plasterk, R. H. (2001). Dicer functions in RNA interference and in synthesis of small RNA involved in developmental timing in *C. elegans*. *Genes & Development*, *15*, 2654–2659.

Kim, V. N., Han, J., & Siomi, M. C. (2009). Biogenesis of small RNAs in animals. *Nature Reviews. Molecular Cell Biology*, *10*, 126–139.

Kim, K., Lee, H. C., Park, J. L., Kim, M., Kim, S. Y., Noh, S. M., et al. (2011). Epigenetic regulation of microRNA-10b and targeting of oncogenic MAPRE1 in gastric cancer. *Epigenetics*, *6*, 740–751.

Kim, D. H., Saetrom, P., Snove, O., Jr., and Rossi, J. J. (2008). MicroRNA-directed transcriptional gene silencing in mammalian cells. Proceedings of the National Academy of Sciences of the United States of America, 105, 16230–16235.

Kozaki, K., Imoto, I., Mogi, S., Omura, K., & Inazawa, J. (2008). Exploration of tumor-suppressive microRNAs silenced by DNA hypermethylation in oral cancer. *Cancer Research*, *68*, 2094–2105.

Kubo, A., Koh, Y., Kawaguchi, T., Isa, S., Okamoto, I., Fukuoka, J., et al. (2011). Malignant pleural effusion from lung adenocarcinoma treated by gefitinib. *Internal Medicine*, *50*, 745–748.

Kubo, T., Toyooka, S., Tsukuda, K., Sakaguchi, M., Fukazawa, T., Soh, J., et al. (2011). Epigenetic silencing of microRNA-34b/c plays an important role in the pathogenesis of malignant pleural mesothelioma. *Clinical Cancer Research*, *17*, 4965–4974.

Kunej, T., Godnic, I., Ferdin, J., Horvat, S., Dovc, P., & Calin, G. A. (2011). Epigenetic regulation of microRNAs in cancer: An integrated review of literature. *Mutation Research*, *717*, 77–84.

Landthaler, M., Yalcin, A., & Tuschl, T. (2004). The human DiGeorge syndrome critical region gene 8 and its *D. melanogaster* homolog are required for miRNA biogenesis. *Current Biology*, *14*, 2162–2167.

Langevin, S. M., Stone, R. A., Bunker, C. H., Grandis, J. R., Sobol, R. W., & Taioli, E. (2010). MicroRNA-137 promoter methylation in oral rinses from patients with squamous cell carcinoma of the head and neck is associated with gender and body mass index. *Carcinogenesis*, *31*, 864–870.

Langevin, S. M., Stone, R. A., Bunker, C. H., Lyons-Weiler, M. A., LaFramboise, W. A., Kelly, L., et al. (2011). MicroRNA-137 promoter methylation is associated with poorer overall survival in patients with squamous cell carcinoma of the head and neck. *Cancer, 117*, 1454–1462.

Lee, Y., Ahn, C., Han, J., Choi, H., Kim, J., Yim, J., et al. (2003). The nuclear RNase III Drosha initiates microRNA processing. *Nature, 425*, 415–419.

Lee, Y. S., & Dutta, A. (2009). MicroRNAs in cancer. *Annual Review of Pathology, 4*, 199–227.

Lee, Y., Kim, M., Han, J., Yeom, K. H., Lee, S., Baek, S. H., et al. (2004). MicroRNA genes are transcribed by RNA polymerase II. *The EMBO Journal, 23*, 4051–4060.

Lee, K. H., Lotterman, C., Karikari, C., Omura, N., Feldmann, G., Habbe, N., et al. (2009). Epigenetic silencing of MicroRNA miR-107 regulates cyclin-dependent kinase 6 expression in pancreatic cancer. *Pancreatology, 9*, 293–301.

Lehmann, U., Hasemeier, B., Christgen, M., Muller, M., Romermann, D., Langer, F., et al. (2008). Epigenetic inactivation of microRNA gene hsa-miR-9-1 in human breast cancer. *The Journal of Pathology, 214*, 17–24.

Li, H. P., Huang, H. Y., Lai, Y. R., Huang, J. X., Chang, K. P., Hsueh, C., et al. (2014). Silencing of miRNA-148a by hypermethylation activates the integrin-mediated signaling pathway in nasopharyngeal carcinoma. *Oncotarget, 5*, 7610–7624.

Li, Z., Li, D., Zhang, G., Xiong, J., Jie, Z., Cheng, H., et al. (2014). Methylation-associated silencing of MicroRNA-335 contributes tumor cell invasion and migration by interacting with RASA1 in gastric cancer. *American Journal of Cancer Research, 4*, 648–662.

Li, A., Omura, N., Hong, S. M., Vincent, A., Walter, K., Griffith, M., et al. (2010). Pancreatic cancers epigenetically silence SIP1 and hypomethylate and overexpress miR-200a/200b in association with elevated circulating miR-200a and miR-200b levels. *Cancer Research, 70*, 5226–5237.

Li, Z., Zhang, G., Li, D., Jie, Z., Chen, H., Xiong, J., et al. (2015). Methylation-associated silencing of miR-495 inhibit the migration and invasion of human gastric cancer cells by directly targeting PRL-3. *Biochemical and Biophysical Research Communications, 456*, 344–350.

Li, Y., Zhang, Y., Li, S., Lu, J., Chen, J., Wang, Y., et al. (2015). Genome-wide DNA methylome analysis reveals epigenetically dysregulated non-coding RNAs in human breast cancer. *Scientific Reports, 5*, 8790.

Li, N., Zhang, F., Li, S., & Zhou, S. (2014). Epigenetic silencing of MicroRNA-503 regulates FANCA expression in non-small cell lung cancer cell. *Biochemical and Biophysical Research Communications, 444*, 611–616.

Li, D., Zhao, Y., Liu, C., Chen, X., Qi, Y., Jiang, Y., et al. (2011). Analysis of MiR-195 and MiR-497 expression, regulation and role in breast cancer. *Clinical Cancer Research, 17*, 1722–1730.

Liu, J., Carmell, M. A., Rivas, F. V., Marsden, C. G., Thomson, J. M., Song, J. J., et al. (2004). Argonaute2 is the catalytic engine of mammalian RNAi. *Science, 305*, 1437–1441.

Liu, X. H., Lu, K. H., Wang, K. M., Sun, M., Zhang, E. B., Yang, J. S., et al. (2012). MicroRNA-196a promotes non-small cell lung cancer cell proliferation and invasion through targeting HOXA5. *BMC Cancer, 12*, 348.

Lodygin, D., Tarasov, V., Epanchintsev, A., Berking, C., Knyazeva, T., Korner, H., et al. (2008). Inactivation of miR-34a by aberrant CpG methylation in multiple types of cancer. *Cell Cycle, 7*, 2591–2600.

Lopez-Serra, P., & Esteller, M. (2012). DNA methylation-associated silencing of tumor-suppressor microRNAs in cancer. *Oncogene, 31*, 1609–1622.

Lu, L., Katsaros, D., de la Longrais, I. A., Sochirca, O., & Yu, H. (2007). Hypermethylation of let-7a-3 in epithelial ovarian cancer is associated with low insulin-like growth factor-II expression and favorable prognosis. *Cancer Research, 67*, 10117–10122.

Lujambio, A., Calin, G. A., Villanueva, A., Ropero, S., Sanchez-Cespedes, M., Blanco, D., et al. (2008). A microRNA DNA methylation signature for human cancer metastasis. *Proceedings of the National Academy of Sciences of the United States of America, 105*, 13556–13561.

Lujambio, A., Ropero, S., Ballestar, E., Fraga, M. F., Cerrato, C., Setien, F., et al. (2007). Genetic unmasking of an epigenetically silenced microRNA in human cancer cells. *Cancer Research, 67*, 1424–1429.

Lund, E., Guttinger, S., Calado, A., Dahlberg, J. E., & Kutay, U. (2004). Nuclear export of microRNA precursors. *Science, 303*, 95–98.

Ma, L., Teruya-Feldstein, J., & Weinberg, R. A. (2007). Tumour invasion and metastasis initiated by microRNA-10b in breast cancer. *Nature, 449*, 682–688.

Mack, G. S. (2006). Epigenetic cancer therapy makes headway. *Journal of the National Cancer Institute, 98*, 1443–1444.

Malumbres, M. (2013). miRNAs and cancer: An epigenetics view. *Molecular Aspects of Medicine, 34*, 863–874.

Maniataki, E., & Mourelatos, Z. (2005). A human, ATP-independent, RISC assembly machine fueled by pre-miRNA. *Genes & Development, 19*, 2979–2990.

Mei, Q., Xue, G., Li, X., Wu, Z., Li, X., Yan, H., et al. (2015). Methylation-induced loss of miR-484 in microsatellite-unstable colorectal cancer promotes both viability and IL-8 production via CD137L. *The Journal of Pathology, 236*, 165–174.

Mitra, D., Das, P. M., Huynh, F. C., & Jones, F. E. (2011). Jumonji/ARID1 B (JARID1B) protein promotes breast tumor cell cycle progression through epigenetic repression of microRNA let-7e. *The Journal of Biological Chemistry, 286*, 40531–40535.

Muraoka, T., Soh, J., Toyooka, S., Maki, Y., Shien, K., Furukawa, M., et al. (2013). Impact of aberrant methylation of microRNA-9 family members on non-small cell lung cancers. *Molecular and Clinical Oncology, 1*, 185–189.

Ng, E. K., Tsang, W. P., Ng, S. S., Jin, H. C., Yu, J., Li, J. J., et al. (2009). MicroRNA-143 targets DNA methyltransferases 3A in colorectal cancer. *British Journal of Cancer, 101*, 699–706.

Nishi, M., Eguchi-Ishimae, M., Wu, Z., Gao, W., Iwabuki, H., Kawakami, S., et al. (2013). Suppression of the let-7b microRNA pathway by DNA hypermethylation in infant acute lymphoblastic leukemia with MLL gene rearrangements. *Leukemia, 27*, 389–397.

Noonan, E. J., Place, R. F., Pookot, D., Basak, S., Whitson, J. M., Hirata, H., et al. (2009). miR-449a targets HDAC-1 and induces growth arrest in prostate cancer. *Oncogene, 28*, 1714–1724.

Omura, N., Li, C. P., Li, A., Hong, S. M., Walter, K., Jimeno, A., et al. (2008). Genome-wide profiling of methylated promoters in pancreatic adenocarcinoma. *Cancer Biology & Therapy, 7*, 1146–1156.

Orom, U. A., Nielsen, F. C., & Lund, A. H. (2008). MicroRNA-10a binds the 5'UTR of ribosomal protein mRNAs and enhances their translation. *Molecular Cell, 30*, 460–471.

Pillai, R. S., Bhattacharyya, S. N., & Filipowicz, W. (2007). Repression of protein synthesis by miRNAs: How many mechanisms? *Trends in Cell Biology, 17*, 118–126.

Png, K. J., Yoshida, M., Zhang, X. H., Shu, W., Lee, H., Rimner, A., et al. (2011). MicroRNA-335 inhibits tumor reinitiation and is silenced through genetic and epigenetic mechanisms in human breast cancer. *Genes & Development, 25*, 226–231.

Rodriguez, A., Griffiths-Jones, S., Ashurst, J. L., & Bradley, A. (2004). Identification of mammalian microRNA host genes and transcription units. *Genome Research, 14*, 1902–1910.

Roman-Gomez, J., Agirre, X., Jimenez-Velasco, A., Arqueros, V., Vilas-Zornoza, A., Rodriguez-Otero, P., et al. (2009). Epigenetic regulation of microRNAs in acute lymphoblastic leukemia. *Journal of Clinical Oncology, 27*, 1316–1322.

Saito, Y., Friedman, J. M., Chihara, Y., Egger, G., Chuang, J. C., & Liang, G. (2009). Epigenetic therapy upregulates the tumor suppressor microRNA-126 and its host gene

EGFL7 in human cancer cells. *Biochemical and Biophysical Research Communications, 379*, 726–731.
Saito, Y., Liang, G., Egger, G., Friedman, J. M., Chuang, J. C., Coetzee, G. A., et al. (2006). Specific activation of microRNA-127 with downregulation of the proto-oncogene BCL6 by chromatin-modifying drugs in human cancer cells. *Cancer Cell, 9*, 435–443.
Sampath, D., Liu, C., Vasan, K., Sulda, M., Puduvalli, V. K., Wierda, W. G., et al. (2012). Histone deacetylases mediate the silencing of miR-15a, miR-16, and miR-29b in chronic lymphocytic leukemia. *Blood, 119*, 1162–1172.
Sato, F., Tsuchiya, S., Meltzer, S. J., & Shimizu, K. (2011). MicroRNAs and epigenetics. *The FEBS Journal, 278*, 1598–1609.
Scott, G. K., Mattie, M. D., Berger, C. E., Benz, S. C., & Benz, C. C. (2006). Rapid alteration of microRNA levels by histone deacetylase inhibition. *Cancer Research, 66*, 1277–1281.
Seol, H. S., Akiyama, Y., Shimada, S., Lee, H. J., Kim, T. I., Chun, S. M., et al. (2014). Epigenetic silencing of microRNA-373 to epithelial–mesenchymal transition in non-small cell lung cancer through IRAK2 and LAMP1 axes. *Cancer Letters, 353*, 232–241.
Shen, R., Pan, S., Qi, S., Lin, X., & Cheng, S. (2010). Epigenetic repression of microRNA-129-2 leads to overexpression of SOX4 in gastric cancer. *Biochemical and Biophysical Research Communications, 394*, 1047–1052.
Shen, J., Wang, S., Siegel, A. B., Remotti, H., Wang, Q., Sirosh, I., et al. (2015). Genome-wide expression of MicroRNAs is regulated by DNA methylation in Hepatocarcinogenesis. *Gastroenterology Research and Practice, 2015*, 230642.
Shimizu, T., Suzuki, H., Nojima, M., Kitamura, H., Yamamoto, E., Maruyama, R., et al. (2013). Methylation of a panel of microRNA genes is a novel biomarker for detection of bladder cancer. *European Urology, 63*, 1091–1100.
Siemens, H., Neumann, J., Jackstadt, R., Mansmann, U., Horst, D., Kirchner, T., et al. (2013). Detection of miR-34a promoter methylation in combination with elevated expression of c-Met and beta-catenin predicts distant metastasis of colon cancer. *Clinical Cancer Research, 19*, 710–720.
Song, S. J., Ito, K., Ala, U., Kats, L., Webster, K., Sun, S. M., et al. (2013). The oncogenic microRNA miR-22 targets the TET2 tumor suppressor to promote hematopoietic stem cell self-renewal and transformation. *Cell Stem Cell, 13*, 87–101.
Song, S. J., Poliseno, L., Song, M. S., Ala, U., Webster, K., Ng, C., et al. (2013). MicroRNA-antagonism regulates breast cancer stemness and metastasis via TET-family-dependent chromatin remodeling. *Cell, 154*, 311–324.
Standart, N., & Jackson, R. J. (2007). MicroRNAs repress translation of m7Gppp-capped target mRNAs in vitro by inhibiting initiation and promoting deadenylation. *Genes & Development, 21*, 1975–1982.
Steponaitiene, R., Kupcinskas, J., Langner, C., Balaguer, F., Venclauskas, L., Pauzas, H., et al. (2016). Epigenetic silencing of miR-137 is a frequent event in gastric carcinogenesis. *Molecular Carcinogenesis, 55*, 376–386.
Suh, S. O., Chen, Y., Zaman, M. S., Hirata, H., Yamamura, S., Shahryari, V., et al. (2011). MicroRNA-145 is regulated by DNA methylation and p53 gene mutation in prostate cancer. *Carcinogenesis, 32*, 772–778.
Suzuki, K., Nishio, T., Katada, N., Sastre, G., & Niwa, M. (2011). Ammonia IRMS-TPD measurements on Bronsted acidity of proton-formed SAPO-34. *Physical Chemistry Chemical Physics, 13*, 3311–3318.
Suzuki, H., Takatsuka, S., Akashi, H., Yamamoto, E., Nojima, M., Maruyama, R., et al. (2011). Genome-wide profiling of chromatin signatures reveals epigenetic regulation of MicroRNA genes in colorectal cancer. *Cancer Research, 71*, 5646–5658.

Tanaka, N., Toyooka, S., Soh, J., Kubo, T., Yamamoto, H., Maki, Y., et al. (2012). Frequent methylation and oncogenic role of microRNA-34b/c in small-cell lung cancer. *Lung Cancer, 76*, 32–38.

Tavazoie, S. F., Alarcon, C., Oskarsson, T., Padua, D., Wang, Q., Bos, P. D., et al. (2008). Endogenous human microRNAs that suppress breast cancer metastasis. *Nature, 451*, 147–152.

Tellez, C. S., Juri, D. E., Do, K., Bernauer, A. M., Thomas, C. L., Damiani, L. A., et al. (2011). EMT and stem cell-like properties associated with miR-205 and miR-200 epigenetic silencing are early manifestations during carcinogen-induced transformation of human lung epithelial cells. *Cancer Research, 71*, 3087–3097.

Thompson, C. A. (2006). Vorinostat approved for rare lymphoma. *American Journal of Health-System Pharmacy, 63*, 2168.

Ting, A. H., McGarvey, K. M., & Baylin, S. B. (2006). The cancer epigenome—Components and functional correlates. *Genes & Development, 20*, 3215–3231.

Toyota, M., Suzuki, H., Sasaki, Y., Maruyama, R., Imai, K., Shinomura, Y., et al. (2008). Epigenetic silencing of microRNA-34b/c and B-cell translocation gene 4 is associated with CpG island methylation in colorectal cancer. *Cancer Research, 68*, 4123–4132.

Trang, P., Weidhaas, J. B., & Slack, F. J. (2008). MicroRNAs as potential cancer therapeutics. *Oncogene, 27*(Suppl. 2), S52–S57.

Tsai, K. W., Hu, L. Y., Wu, C. W., Li, S. C., Lai, C. H., Kao, H. W., et al. (2010). Epigenetic regulation of miR-196b expression in gastric cancer. *Genes, Chromosomes & Cancer, 49*, 969–980.

Tsai, K. W., Liao, Y. L., Wu, C. W., Hu, L. Y., Li, S. C., Chan, W. C., et al. (2011). Aberrant hypermethylation of miR-9 genes in gastric cancer. *Epigenetics, 6*, 1189–1197.

Tsai, K. W., Wu, C. W., Hu, L. Y., Li, S. C., Liao, Y. L., Lai, C. H., et al. (2011). Epigenetic regulation of miR-34b and miR-129 expression in gastric cancer. *International Journal of Cancer, 129*, 2600–2610.

Tsuruta, T., Kozaki, K., Uesugi, A., Furuta, M., Hirasawa, A., Imoto, I., et al. (2011). miR-152 is a tumor suppressor microRNA that is silenced by DNA hypermethylation in endometrial cancer. *Cancer Research, 71*, 6450–6462.

Tu, J., Liao, J., Luk, A. C., Tang, N. L., Chan, W. Y., & Lee, T. L. (2015). MicroRNAs mediated targeting on the Yin-yang dynamics of DNA methylation in disease and development. *The International Journal of Biochemistry & Cell Biology, 67*, 115–120.

Tuddenham, L., Wheeler, G., Ntounia-Fousara, S., Waters, J., Hajihosseini, M. K., Clark, I., et al. (2006). The cartilage specific microRNA-140 targets histone deacetylase 4 in mouse cells. *FEBS Letters, 580*, 4214–4217.

Tuna, M., Machado, A. S., & Calin, G. A. (2016). Genetic and epigenetic alterations of microRNAs and implications for human cancers and other diseases. *Genes, Chromosomes & Cancer, 55*, 193–214.

Varambally, S., Cao, Q., Mani, R. S., Shankar, S., Wang, X., Ateeq, B., et al. (2008). Genomic loss of microRNA-101 leads to overexpression of histone methyltransferase EZH2 in cancer. *Science, 322*, 1695–1699.

Vasudevan, S., Tong, Y., & Steitz, J. A. (2007). Switching from repression to activation: MicroRNAs can up-regulate translation. *Science, 318*, 1931–1934.

Vogt, M., Munding, J., Gruner, M., Liffers, S. T., Verdoodt, B., Hauk, J., et al. (2011). Frequent concomitant inactivation of miR-34a and miR-34b/c by CpG methylation in colorectal, pancreatic, mammary, ovarian, urothelial, and renal cell carcinomas and soft tissue sarcomas. *Virchows Archiv, 458*, 313–322.

Wang, C., Bian, Z., Wei, D., & Zhang, J. G. (2011). miR-29b regulates migration of human breast cancer cells. *Molecular and Cellular Biochemistry, 352*, 197–207.

Wang, Z., Chen, Z., Gao, Y., Li, N., Li, B., Tan, F., et al. (2011). DNA hypermethylation of microRNA-34b/c has prognostic value for stage non-small cell lung cancer. *Cancer Biology & Therapy, 11*, 490–496.

Wang, P., Chen, L., Zhang, J., Chen, H., Fan, J., Wang, K., et al. (2014). Methylation-mediated silencing of the miR-124 genes facilitates pancreatic cancer progression and metastasis by targeting Rac1. *Oncogene, 33*, 514–524.

Wang, L. Q., Kwong, Y. L., Kho, C. S., Wong, K. F., Wong, K. Y., Ferracin, M., et al. (2013). Epigenetic inactivation of miR-9 family microRNAs in chronic lymphocytic leukemia—Implications on constitutive activation of NFkappaB pathway. *Molecular Cancer, 12*, 173.

Wang, B., Love, T. M., Call, M. E., Doench, J. G., & Novina, C. D. (2006). Recapitulation of short RNA-directed translational gene silencing in vitro. *Molecular Cell, 22*, 553–560.

Wang, Z., Yao, H., Lin, S., Zhu, X., Shen, Z., Lu, G., et al. (2013). Transcriptional and epigenetic regulation of human microRNAs. *Cancer Letters, 331*, 1–10.

Wei, X., Tan, C., Tang, C., Ren, G., Xiang, T., Qiu, Z., et al. (2013). Epigenetic repression of miR-132 expression by the hepatitis B virus x protein in hepatitis B virus-related hepatocellular carcinoma. *Cellular Signalling, 25*, 1037–1043.

Wiklund, E. D., Bramsen, J. B., Hulf, T., Dyrskjot, L., Ramanathan, R., Hansen, T. B., et al. (2011). Coordinated epigenetic repression of the miR-200 family and miR-205 in invasive bladder cancer. *International Journal of Cancer, 128*, 1327–1334.

Wilting, S. M., van Boerdonk, R. A., Henken, F. E., Meijer, C. J., Diosdado, B., Meijer, G. A., et al. (2010). Methylation-mediated silencing and tumour suppressive function of hsa-miR-124 in cervical cancer. *Molecular Cancer, 9*, 167.

Wong, K. Y., Liang, R., So, C. C., Jin, D. Y., Costello, J. F., & Chim, C. S. (2011). Epigenetic silencing of MIR203 in multiple myeloma. *British Journal of Haematology, 154*, 569–578.

Wong, K. Y., Yim, R. L., Kwong, Y. L., Leung, C. Y., Hui, P. K., Cheung, F., et al. (2013). Epigenetic inactivation of the MIR129-2 in hematological malignancies. *Journal of Hematology & Oncology, 6*, 16.

Wong, K. Y., Yim, R. L., So, C. C., Jin, D. Y., Liang, R., & Chim, C. S. (2011). Epigenetic inactivation of the MIR34B/C in multiple myeloma. *Blood, 118*, 5901–5904.

Xie, K., Liu, J., Chen, J., Dong, J., Ma, H., Liu, Y., et al. (2014). Methylation-associated silencing of microRNA-34b in hepatocellular carcinoma cancer. *Gene, 543*, 101–107.

Yamagishi, M., Nakano, K., Miyake, A., Yamochi, T., Kagami, Y., Tsutsumi, A., et al. (2012). Polycomb-mediated loss of miR-31 activates NIK-dependent NF-kappaB pathway in adult T cell leukemia and other cancers. *Cancer Cell, 21*, 121–135.

Yamakuchi, M., Ferlito, M., & Lowenstein, C. J. (2008). miR-34a repression of SIRT1 regulates apoptosis. *Proceedings of the National Academy of Sciences of the United States of America, 105*, 13421–13426.

Yan-Fang, T., Jian, N., Jun, L., Na, W., Pei-Fang, X., Wen-Li, Z., et al. (2013). The promoter of miR-663 is hypermethylated in Chinese pediatric acute myeloid leukemia (AML). *BMC Medical Genetics, 14*, 74.

Yang, C., Cai, J., Wang, Q., Tang, H., Cao, J., Wu, L., et al. (2012). Epigenetic silencing of miR-130b in ovarian cancer promotes the development of multidrug resistance by targeting colony-stimulating factor 1. *Gynecologic Oncology, 124*, 325–334.

Yi, R., Qin, Y., Macara, I. G., & Cullen, B. R. (2003). Exportin-5 mediates the nuclear export of pre-microRNAs and short hairpin RNAs. *Genes & Development, 17*, 3011–3016.

Younger, S. T., & Corey, D. R. (2011). Transcriptional gene silencing in mammalian cells by miRNA mimics that target gene promoters. *Nucleic Acids Research, 39*, 5682–5691.

Zhang, S., Hao, J., Xie, F., Hu, X., Liu, C., Tong, J., et al. (2011). Downregulation of miR-132 by promoter methylation contributes to pancreatic cancer development. *Carcinogenesis, 32*, 1183–1189.

Zhang, P., Huang, B., Xu, X., & Sessa, W. C. (2013). Ten–eleven translocation (Tet) and thymine DNA glycosylase (TDG), components of the demethylation pathway, are direct targets of miRNA-29a. *Biochemical and Biophysical Research Communications*, *437*, 368–373.

Zhang, J., Yang, Y., Yang, T., Liu, Y., Li, A., Fu, S., et al. (2010). microRNA-22, down-regulated in hepatocellular carcinoma and correlated with prognosis, suppresses cell proliferation and tumourigenicity. *British Journal of Cancer*, *103*, 1215–1220.

Zhang, T., Zhang, J., Cui, M., Liu, F., You, X., Du, Y., et al. (2013). Hepatitis B virus X protein inhibits tumor suppressor miR-205 through inducing hypermethylation of miR-205 promoter to enhance carcinogenesis. *Neoplasia*, *15*, 1282–1291.

Zhang, X., Zhao, X., Fiskus, W., Lin, J., Lwin, T., Rao, R., et al. (2012). Coordinated silencing of MYC-mediated miR-29 by HDAC3 and EZH2 as a therapeutic target of histone modification in aggressive B-cell lymphomas. *Cancer Cell*, *22*, 506–523.

Zhu, A., Xia, J., Zuo, J., Jin, S., Zhou, H., Yao, L., et al. (2012). MicroRNA-148a is silenced by hypermethylation and interacts with DNA methyltransferase 1 in gastric cancer. *Medical Oncology*, *29*, 2701–2709.

Kolibri 360 GmbH
Ferdinand-Jühlke-Straße 7
99095 Erfurt, Deutschland
produktsicherheit@kolibri360.de